modern transistor circuits

MODERN TRANSISTOR CIRCUITS

BY JOHN M. CARROLL

MANAGING EDITOR, *Electronics*

McGRAW-HILL BOOK COMPANY, INC.

NEW YORK **TORONTO** **LONDON**

1959

modern transistor circuits

VI

preface

TRANSISTORS ARE USED TODAY in nearly every kind of electronic circuit—from portable radios to satellite telemeters. Transistorized equipment is operational, too; not just experimental. Probably the only kind of electronic gear not using transistors in one form or another is high-power transmitters.

All signs point to increasing use of transistors: nearly 10 million were sold in 1956; more than 22 million in 1957. Sales in 1958 ran more than 3.5 million a month. Industry leaders now feel earlier estimates of 450 million transistors a year by 1965 are definitely on the conservative side. All this means that a thorough understanding of transistor circuit design is now an essential part of every electronic engineer's bag-of-tricks.

Reasons for the transistor's ever widening acceptance arise both from better ways of making transistors and better ways of using them. Transistor prices have fallen from $10 or more apiece to, in some cases, less than $.50. The handful of experimental types available in 1952 has grown to upward of 300 commercially available types for every conceivable application: Collector power dissipations range from a few milliwatts to a kilowatt. Alpha cutoff frequencies go from d-c to over 100 megacycles. Modern transistors are noted for their low noise figures as well as small size, low power requirements and absence of warm-up time.

Reliability is one of the most attractive features of present-day transistors. Indeed, their long life and ability to withstand rigorous environments contribute greatly to missile control systems and space satellite communications. Users of transistors now report lifetimes in excess of 10,000 hours. Transistors properly compensated against thermal effects have functioned in equipment subjected to the most extreme temperatures. When properly installed, transistors are practically immune to the effects of shock and vibration.

Several trends in transistor circuit design and device development soon become apparent in preparing a book of this nature. Transistor manufacturers are devoting increasing attention to developing high-frequency units and units that can withstand effectively both extremely high temperatures and high levels of nuclear radiation. Perhaps anticipating a role in control of thermonuclear power.

In circuit design, there are more high-frequency and very-high-frequency transistor circuits, taking advantages of new high-frequency transistors now available. Greater use is being made of complementary symmetry. This circuit configuration was once thought to be of academic interest only, because of difficulties in matching transistor pairs. The trend to complementary symmetry indicates how well transistor manufacturers have improved control over the quality of their products.

Also evident are more hybrid circuits—that is, circuits that use electron tubes and/or magnetic amplifiers as well as transistors. No longer are the supply-voltage-regulation problems and the need to hold chassis temperature down so urgent as to preclude hybrid design. Today the engineer can choose the electron device that best satisfies the needs of the circuit he is currently designing.

A glance at the contents of this book indicates the diverse applications of transistors today. New home-entertainment applications include portable and automobile radio receivers that use fewer parts and do a better job, and miniaturized phonograph preamplifiers. An all-transistor television receiver has been announced.

For the broadcaster there will be sound mixers, wireless microphones and television studio equipment. Communicators will find a large variety of portable transmitters, transceivers, telephone and telegraph terminal equipment. New transistorized instruments include transistor testers, portable frequency meters and extremely small microammeters. New military equipment includes telemetering and missile-guidance equipment, jet engine controls and advanced simulation equipment for training military specialists.

In the fast-growing industrial control field, transistors are doing important work in servo amplifiers, control relays, nuclear reactor controls, metal detectors and measuring instruments.

Transistors are playing an essential role in man's dramatic conquest of outer space. The longest single chapter in this book—28 pages—deals with telemetering circuits for high-speed aircraft, guided missiles and earth satellites. Many other communications, measuring and control circuits presented in other chapters have derived from guided missiles research and development.

Medical electronics is still in its infancy. But the transistor has already become an essential part of telemeters to probe the digestive tract and sensitive amplifiers for diagnostic recording. Transistorized instruments aid scientists in studying ocean currents and behavior of wild life.

The transistor has found a natural niche in the booming computer industry. Both analog and digital computers are using increasing numbers of transistors. Indeed, the transistor's small size alone makes possible computers of a scale thought impracticable only a few years ago. Transistors are also important in analog-to-digital converters and newly designed input/output equipment that may soon permit computers to work at or near their inherently high speeds—not limited to the pace of human or mechanical input and output.

The 101 articles reprinted in this book appeared in *Electronics* magazine during the years 1956, 1957 and 1958. They are published here largely in their entirety. In most cases, circuit schematic diagrams with all component parts values are supplied to short-cut design time. Special thanks are due Bill MacDonald and the staff of *Electronics* magazine whose skill and editorial judgment have made possible the timeliness, scope and completeness of this material.

John M. Carroll

contents

modern transistor circuits

HOW TRANSISTORS OPERATE UNDER ATOMIC RADIATION

By ROBERT L. RIDDLE
Senior Engineer, Haller, Raymond and Brown, Inc., State College, Pa.

Results of tests exposing transistor amplifier and single transistor to radiation from nuclear reactor show that degrading effects of irradiation can be controlled to some degree by use of negative feedback when applicable. Radiation effects on coaxial cables showed no noticeable change in r-f transmission characteristics

APPLICATION of nuclear energy to propulsion of ships, planes and other devices will require electronic equipment to operate under a wide range of nuclear radiation levels. For circuits to function properly, it will be necessary for designers to compensate for irradiation effects on active circuit elements.

Investigations have been made to find the effects of combined gamma and neutron flux on semiconductor devices. The facilities of the Pennsylvania State University pool-type research reactor were used for an experiment involving the effects of reactor radiation on an all-transistor amplifier and crystal video detector. Arrangement of the test is shown in block form in Fig. 1A.

Test Method

In general, the information desired was the overall performance of the combined crystal detector and transistor amplifier shown in Fig. 1B. Measurements included tangential sensitivity and transfer ratio of the system (ratio of video output voltage to r-f input voltage). Also determined were r-f attenuation in a coaxial cable extending from the center of the active pile area to the top of the pool and the h-parameters and I_{co} measurements on a single separate transistor in the active pile area. Noise in a properly terminated coaxial cable in the active area was also measured.

The tangential sensitivity was determined under the conditions of a 1,000-mc carrier pulsed by 10-

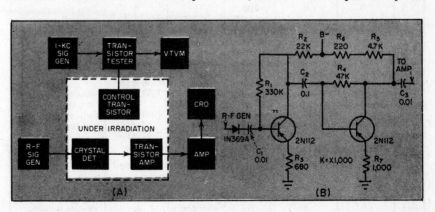

FIG. 1—Equipment setup used in irradiation study (A) with circuit of detector and transistor amplifier (B)

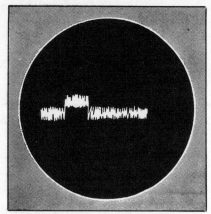

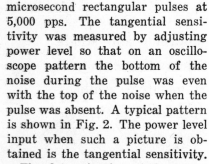

FIG. 2—Oscilloscope pattern used in determination of tangential sensitivity

Pool-type reactor. Amplifier was placed in aluminum tube for insertion in reactor

microsecond rectangular pulses at 5,000 pps. The tangential sensitivity was measured by adjusting power level so that on an oscilloscope pattern the bottom of the noise during the pulse was even with the top of the noise when the pulse was absent. A typical pattern is shown in Fig. 2. The power level input when such a picture is obtained is the tangential sensitivity.

The determination of the transfer ratio was made from the calibrated attenuator on the signal generator and a calibrated oscilloscope.

The control measurements were included so that the primary effects of radiation on the amplifier-detector system could be separated from possible extraneous effects.

The irradiation schedule is shown in Table I with irradiation times listed for each power level. Neutron flux ϕ, in neutrons per cm²-sec, is the combined thermal and resonance flux as determined by a cadmium ratio of 9.4 and $\phi_{thermal} = 2.07 \times 10^6 \times P$ as found by activation of foils. This gives $\phi = 2.32 \times 10^6 \times P$, where P is the reactor power in watts. Total flux ϕ includes approximately those neutrons between thermal energies and 2 ev.

The duration of each test in seconds and the integrated neutron flux is also given in the table. Gamma dose was determined from previous calibrations to be equivalent to a dose in roentgens per hour of $4.93 \times 10 \, P$. The gamma dose shown in the table is calculated from this equation. The integrated gamma dose was $3.2 \times 10^6 R$.

Test Results

The results of this experiment are shown in Fig. 3. The tangential sensitivity, the most important measurement on a crystal video system, is shown in Fig. 3A and Fig. 3B. Sensitivity decreased as the experiment progressed. There is a recovery in sensitivity as soon as the reactor is turned off as shown by tests 7 and 8. Test 9, which was performed several hours later, still shows about the same sensitivity.

Upon turning the reactor on again, the sensitivity again decreased with increasing flux. The recovery after removing the apparatus from the flux field, represented by test 15, is not as pronounced as that of test 8. Slight improvement is noticed, however, several hours later.

The results of this test show that the sensitivity is affected by flux density as well as integrated flux. The dashed line represents a guess at the probable effects of integrated flux, and the solid line the actual measurements which represent a combination of permanent and temporary effects. The distance between the solid line and dashed line approximates the effects of the temporary degradation as a result of the flux density. This is indicated by the amount of effective

Table I—Exposure Time for Transistors in Reactor

Test No.	Power in watts	ϕ in Neutrons per cm²-sec	γ flux in γ/hr	Test duration in sec	Integrated flux density in neutrons per cm²	Remarks
1	0	0	>100		0	Zero power, residual γ only
2	0.3	0.7×10^6	300	420	2.9×10^8	
3	10	2.3×10^7	793	840	1.95×10^{10}	
4	10^2	2.3×10^8	4.9×10^3	540	1.25×10^{11}	
5	10^3	2.3×10^8	4.9×10^4	420	9.75×10^{11}	
6	10^4	2.3×10^{10}	4.9×10^5	420	9.75×10^{12}	
7	5×10^4	1.16×10^{11}	2.5×10^6	480	5.56×10^{13}	
8	10^2	2.3×10^8	$\cong 5\times10^3$	660	1.53×10^{11}	Reactor off
9	Reactor off	....	$\cong 300$	58,440		Overnight decay
—	10^2	2.3×10^8	4.9×10^3	900	2.09×10^{11}	Calibrate
—	0 13	3.1×10^5	$\cong 300$	14,280	4.3×10^9	Calibrate & off
10	10^4	2.3×10^{10}	4.9×10^5	1,020	2.37×10^{13}	
11	10^5	2.3×10^{11}	4.9×10^6			
12	10^5	2.3×10^{11}	4.9×10^6			
13	10^5	2.3×10^{11}	4.9×10^6	1,920	4.46×10^{14}	
14	10^5	2.3×10^{11}	4.9×10^6			
15	0	0	0	840	0	Pulled sample out of aluminum tube
16	0	0	0	1,020	0	
17	0	0	0	63,360	0	

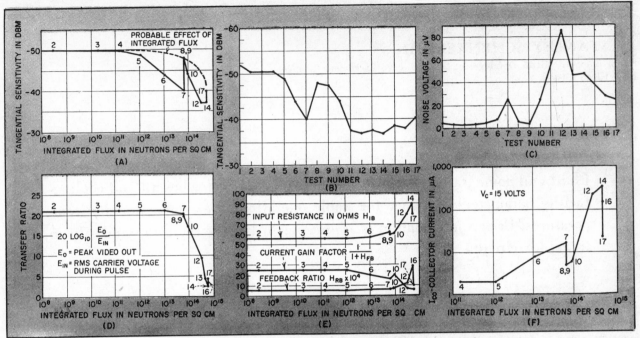

FIG. 3—Results of irradiation study shown graphically are descirbed in text. Numbers on chart indicate test

noise voltage present at the input to the crystal shown in Fig. 3C. This noise voltage is large in those tests where flux density is large. The presence of this noise at these high flux levels results in the decreased tangential sensitivity shown in Fig. 3A and Fig. 3B.

Transfer Ratio

The transfer ratio, measured as the peak-to-peak pulse output to the rms r-f input, is shown in Fig. 3D. No degradation in this ratio took place until an integrated neutron flux in excess of 10^{13} neutrons per sq cm was reached. From this point on, the transfer ratio dropped rapidly as the integrated flux increased. This same general trend is represented in Fig. 3E which shows the effect on the parameters of the control transistor. This transistor started to degrade at an integrated flux of slightly larger than 10^{12} neutrons per sq cm.

The crystal video circuit maintained its gain for an order of magnitude longer. This is probably a result of the degeneration present in the circuit. The transfer ratio recovered slightly after the device was taken from the flux field. This recovery was gradual and may possibly be due to thermal annealing of the defects produced by irradiation. This same effect is noticed in the control transistor parameters.

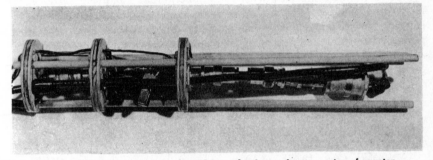

Transistor amplifier used in test is slipped into aluminum pipe to center of reactor

The I_{co} of the control transistor was also monitored throughout the test and variation is shown in Fig. 3F. The increase in I_{co} may be in part due to internal gamma heating in the transistor.

There was no noticeable change in the r-f transmission through the coaxial cable used as a control. This cable was not monitored for attenuation of video signals.

The noise power produced in a 12-mc bandwidth by the properly terminated cable was —84 dbm residual. The maximum reading on this cable was —82 dbm. This is well below the level of the r-f signal used in the test and also was measured in a much broader bandwidth than that of the crystal detector and transistor amplifier.

Several conclusions may be drawn from this experiment. Fission spectrum irradiation upon semiconductor devices has essentially three

effects. These are transient effects resulting from flux density and gamma heating, semipermanent effects resulting from integrated flux and permanent effects resulting from integrated flux after annealing.

The transient effects caused by flux density are effective mainly in producing noise and increasing I_{co}. The gamma heating, the second transient effect, appears to increase I_{co} and degrades the transistor in the same manner as any increase in temperature. These effects disappear soon after removal from the flux field.

The semipermanent effects caused by integrated flux result from lattice damage and transmutations. The semipermanent damage results in an overall change in the characteristics of the semiconductor devices which usually degrade their operation.

THERMAL STABILITY DESIGN NOMOGRAPHS

By STANLEY SCHENKERMAN

Senior Engineer, Missile Development Division, Ford Instrument Company,
Long Island City, N. Y.

Chart and nomographs simplify calculation of transistor circuit and cooling-facility parameters necessary for stable operation at elevated junction temperatures. Design information is applicable to both germanium and silicon transistor circuits

To ACHIEVE THERMAL STABILITY of transistorized equipment at elevated junction temperatures, the designer must provide a circuit compatible with its cooling facility. The accompanying graph and nomographs permit the rapid determination of suitable circuit and cooling facility parameters.

Theory

The thermal stability nomograph, Fig. 1, is based upon the criteria for thermal stability[1]

$$SV_cI_s\theta < 13 \qquad (1A)$$

for germanium and

$$SV_cI_s\theta < 23 \qquad (1B)$$

for silcon.

The stability factor[2] S is the change in quiescent collector current caused by a change in the temperature sensitive component, V_c is collector voltage, I_s is the temperature sensitive component of collector current and cooling-facility characteristic θ is the thermal resistance from collector junction to ambient in deg C per watt. The designer's objective is to obtain compatible values of S and θ.

Temperature sensitive current I_s increases exponentially with temperature, doubling every 9 C for germanium and every 16 C for silicon. The value of I_s at any temperature T may be found from Fig. 2 if the value of I_{so} at any temperature T_o is known; the latter values can be obtained from the manufacturer's data sheet.

Thermal resistance θ consists of the resistance from collector junction to mounting base, θ_{jm}, (specified by the manufacturer) and the resistance from mounting base to ambient, θ_{ma}. The latter may be determined experimentally from

$$\theta_{ma} = \Delta T/P_d \qquad (2)$$

where ΔT is the mounting base temperature rise above ambient in degrees centigrade and P_d is the transistor dissipation in watts. Then

$$\theta = \theta_{ma} + \theta_{jm} \qquad (3)$$

The thermal resistance nomo-

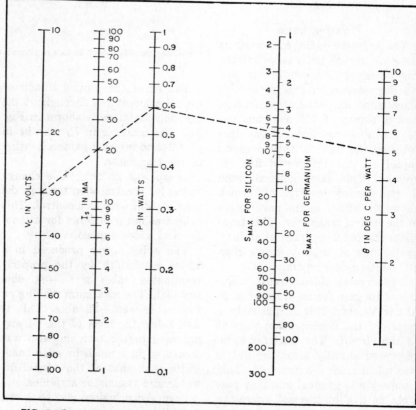

FIG. 1—Range of thermal stability nomograph is extended by powers of 10

graph, Fig. 3, facilitates the solution of Eq. 2 and 3.

From measurements, it is determined that at 4 w dissipation, the mounting base temperature of a transistor, when installed in the cooling facility, is 8 C. The resistance from junction to mounting base is specified as 3 C per watt.

Thermal Resistance

Using Fig. 3, connect the point $\Delta T = 8$ C with the point $P_d = 4$ w with a straight line that intersects $\theta_{ma} = 2$ C per watt. Next, connect $\theta_{ma} = 2$ C per watt and $\theta_{jm} = 3$ C per watt with a straight line. This line intersects $\theta = 5$ C per watt which is the total thermal resistance of the cooling facility.

The value of S_{max} may now be found from the thermal stability nomograph of Fig. 1. The circuit S must be less than S_{max} for stability. Conversely, for a given S the maximum permissible θ may be determined.

The transistor of the previous example is a germanium unit with $I_{so} = 0.2$ ma at $T_o = 25$ C. It is desired to operate this unit with $V_c = 30$ v at 85 C in the cooling facility for which $\theta = 5$ C per watt. Since $T - T_o$ is 60 C, I_s/I_{so} is 100 as determined from Fig. 2. Then I_s is 20 ma at 85 C.

On the stability nomograph, Fig. 1, connect $V_c = 30$ and $I_s = 20$ with a straight line that intersects the P scale at 0.6. Now connect this point by a straight line with $\theta = 5$. This line intersects S_{max} for germanium at 4.3. This is the maximum value of S for which the circuit is thermally stable.

REFERENCES

(1) Specification for H-5, H-6, and H-7 Power Transistors, Transistor Div., Minneapolis Honeywell Regulator Co., p 9, May 1956.

(2) R. F. Shea, "Principles of Transistor Circuits", John Wiley Sons, Inc., p 97, 1953.

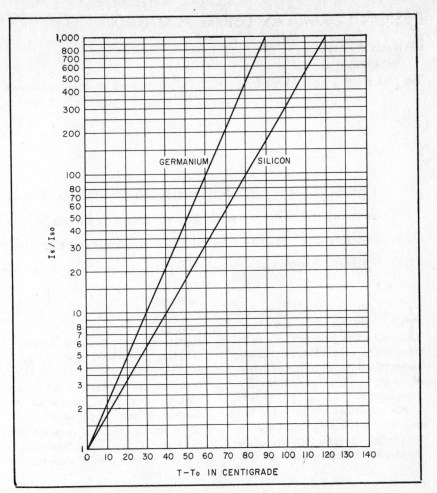

FIG. 2—Curves give normalized temperature-sensitive leakage current as function of temperature change

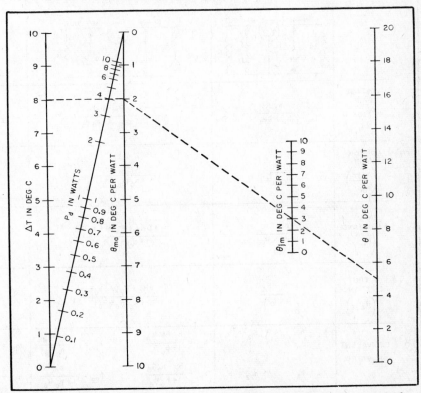

FIG. 3—Range of thermal resistance nomograph is extended by powers of 10

DESIGN FORMULAS USING H-MATRIX PARAMETERS

By ALBERT E. HAYES, JR. Project Manager, Mechanical Division, General Mills, Inc., Minneapolis, Minnesota

Charts list formulas for small-signal performance of transistor with external resistances in series with one or more of its leads in grounded-base, grounded-emitter and grounded-collector configurations. Formulas have been derived without approximations; simplifications may be made to fit specific situations

THE FORMULAS given here define the performance of a transistor, with external resistances in series with one or more of its leads, in terms of h-matrix parameters.

The equations are intended for use in the low-frequency region where the parameters are substantially independent of frequency (≈ 1 per cent of f_{aco}). Equations may be used at any frequency provided that the parameters are measured at the frequency of operation, the complex value of each parameter is measured and impedances and admittances are used in place of resistances and conductances.

Common-base small-signal, short-circuit input resistance is designated by h_i, open-circuit reverse-transfer voltage ratio by h_r, short-circuit forward-transfer current ratio by h_f and open-circuit output conductance by h_o. Quantity D is the determinant of the h matrix ($D = h_i h_o - h_f h_r$), and $M = D + h_f - h_r + 1$.

COMMON-BASE ARRANGEMENTS

VOLTAGE GAIN $\left(\dfrac{V_{OUT}}{V_{IN}}\right)$	$\dfrac{-h_f R_L}{h_i + D R_L}$	$\dfrac{R_L(h_o R_B - h_f)}{h_i + D R_L + R_B(M + h_o R_L)}$	$\dfrac{R_L(h_o R_B - h_f)}{h_i + D R_L + R_B(M + h_o R_L + h_o R_E) + R_E(1 + h_o R_L)}$
CURRENT GAIN $\left(\dfrac{i_{OUT}}{i_{IN}}\right)$	$\dfrac{-h_f}{h_o R_L + 1}$	$\dfrac{h_o R_B - h_f}{1 + h_o(R_B + R_L)}$	$\dfrac{h_o R_B - h_f}{1 + h_o(R_B + R_L)}$
INPUT RESISTANCE $\left(R_{IN}\right)$	$\dfrac{h_i + D R_L}{1 + h_o R_L}$	$\dfrac{h_i + D R_L + R_B(M + h_o R_L)}{1 + h_o(R_B + R_L)}$	$R_E + \dfrac{h_i + D R_L + R_B(M + h_o R_L)}{1 + h_o(R_B + R_L)}$
OUTPUT RESISTANCE $\left(R_{OUT}\right)$	$\dfrac{h_i + R_G}{D + h_o R_G}$	$\dfrac{h_i + R_G + R_B(M + h_o R_G)}{D + h_o(R_B + R_G)}$	$\dfrac{h_i + R_G + R_E + R_B[M + h_o(R_G + R_E)]}{D + h_o(R_B + R_E + R_G)}$

EQUIVALENT h MATRIX

$h_{11} = h_i$ $\quad h_{12} = h_r$	$h_{11} = \dfrac{h_i + M R_B}{1 + h_o R_B}$ $\quad h_{12} = \dfrac{h_r + M R_B}{1 + h_o R_B}$	$h_{11} = R_E + \dfrac{h_i + M R_B}{1 + h_o R_B}$ $\quad h_{12} = \dfrac{h_r + h_o R_B}{1 + h_o R_B}$
$h_{21} = h_f$ $\quad h_{22} = h_o$	$h_{21} = \dfrac{h_f - h_o R_B}{1 + h_o R_B}$ $\quad h_{22} = \dfrac{h_o}{1 + h_o R_B}$	$h_{21} = \dfrac{h_f - h_o R_B}{1 + h_o R_B}$ $\quad h_{22} = \dfrac{h_o}{1 + h_o R_B}$

COMMON-EMITTER ARRANGEMENTS

	Circuit 1	Circuit 2	Circuit 3
VOLTAGE GAIN $\left(\dfrac{V_{OUT}}{V_{IN}}\right)$	$\dfrac{(D+h_f)R_L}{h_i + DR_L}$	$\dfrac{(D+h_f+h_0 R_E)R_L}{h_i + DR_L + R_E(1+h_0 R_L)}$	$\dfrac{(D+h_f+h_0 R_E)R_L}{h_i + DR_L + R_E[1+h_0(R_B+R_L)] + R_B(M+h_0 R_L)}$
CURRENT GAIN $\left(\dfrac{i_{OUT}}{i_{IN}}\right)$	$\dfrac{D+h_f}{h_0 R_L + M}$	$\dfrac{D+h_f+h_0 R_E}{h_0(R_E+R_L)+M}$	$\dfrac{D+h_f+h_0 R_E}{h_0(R_E+R_L)+M}$
INPUT RESISTANCE $\left(R_{IN}\right)$	$\dfrac{DR_L + h_i}{h_0 R_L + M}$	$\dfrac{DR_L + h_i + R_E(1+h_0 R_L)}{h_0(R_E+R_L)+M}$	$R_B + \dfrac{DR_L + h_i + R_E(1+h_0 R_L)}{h_0(R_E+R_L)+M}$
OUTPUT RESISTANCE $\left(R_{OUT}\right)$	$\dfrac{h_i + MR_G}{D + h_0 R_G}$	$\dfrac{h_i + MR_G + R_E(1+h_0 R_G)}{D + h_0(R_E+R_G)}$	$\dfrac{h_i + M(R_B + R_G) + R_E[1+h_0(R_B+R_G)]}{D + h_0(R_B+R_E+R_G)}$
EQUIVALENT h MATRIX	$h_{11} = \dfrac{h_i}{M}$ $h_{12} = \dfrac{D-h_r}{M}$ $h_{21} = \dfrac{-(D+h_f)}{M}$ $h_{22} = \dfrac{h_0}{M}$	$h_{11} = \dfrac{h_i + R_E}{M+h_0 R_E}$ $h_{12} = \dfrac{D-h_r+h_0 R_E}{M+h_0 R_E}$ $h_{21} = \dfrac{-(D+h_f+h_0 R_E)}{M+h_0 R_E}$ $h_{22} = \dfrac{h_0}{M+h_0 R_E}$	$h_{11} = R_B + \dfrac{h_i + R_E}{M+h_0 R_E}$ $h_{12} = \dfrac{D-h_r+h_0 R_E}{M+h_0 R_E}$ $h_{21} = \dfrac{-(D+h_f+h_0 R_E)}{M+h_0 R_E}$ $h_{22} = \dfrac{h_0}{M+h_0 R_E}$

COMMON-COLLECTOR ARRANGEMENTS

	Circuit 1	Circuit 2	Circuit 3
VOLTAGE GAIN $\left(\dfrac{V_{OUT}}{V_{IN}}\right)$	$\dfrac{(1-h_r)R_L}{h_i + R_L}$	$\dfrac{(1-h_r+h_0 R_C)R_L}{h_i + R_L(1+h_0 R_C) + DR_C}$	$\dfrac{(1-h_r+h_0 R_C)R_L}{h_i + DR_C + R_L[1+h_0(R_B+R_C)] + R_B(M+h_0 R_C)}$
CURRENT GAIN $\left(\dfrac{i_{OUT}}{i_{IN}}\right)$	$\dfrac{1-h_r}{h_0 R_L + M}$	$\dfrac{1-h_r+h_0 R_C}{h_0(R_C+R_L)+M}$	$\dfrac{1-h_r+h_0 R_C}{h_0(R_C+R_L)+M}$
INPUT RESISTANCE $\left(R_{IN}\right)$	$\dfrac{h_i + R_L}{h_0 R_L + M}$	$\dfrac{h_i + R_L + R_C(D+h_0 R_L)}{h_0(R_C+R_L)+M}$	$R_B + \dfrac{h_i + R_L + R_C(D+h_0 R_L)}{h_0(R_C+R_L)+M}$
OUTPUT RESISTANCE $\left(R_{OUT}\right)$	$\dfrac{h_i + MR_G}{1 + h_0 R_G}$	$\dfrac{h_i + MR_G + R_C(D+h_0 R_G)}{1 + h_0(R_C+R_G)}$	$\dfrac{h_i + (M+h_0 R_C)(R_B+R_G) + DR_C}{1 + h_0(R_B+R_C+R_G)}$
EQUIVALENT h MATRIX	$h_{11} = \dfrac{h_i}{M}$ $h_{12} = \dfrac{1+h_f}{M}$ $h_{21} = \dfrac{h_r-1}{M}$ $h_{22} = \dfrac{h_0}{M}$	$h_{11} = \dfrac{h_i + DR_C}{M+h_0 R_C}$ $h_{12} = \dfrac{1+h_f+h_0 R_C}{M+h_0 R_C}$ $h_{21} = \dfrac{h_r-1-h_0 R_C}{M+h_0 R_C}$ $h_{22} = \dfrac{h_0}{M+h_0 R_C}$	$h_{11} = R_B + \dfrac{h_i + DR_C}{M+h_0 R_C}$ $h_{12} = \dfrac{1+h_f+h_0 R_C}{M+h_0 R_C}$ $h_{21} = \dfrac{h_r-1-h_0 R_C}{M+h_0 R_C}$ $h_{22} = \dfrac{h_0}{M+h_0 R_C}$

CONVERSION FORMULAS FOR HYBRID PARAMETERS

By T. P. SYLVAN

Transistor Application Engineer Advanced Electronics Center,
General Electric Co., Ithaca, New York

Charts summarize formulas required to convert transistor *H*-parameters to equivalent *T*-parameters in common base, emitter and collector configurations. Equivalence between new and old notations is given

HYBRID or *h*-parameters are coming into increasing favor as a basis for specifying the small-signal characteristics of junction transistors. They are compatible with the inherent characteristics of junction transistors and can be measured conveniently and accurately.[1]

Table I presents equations for the calculation of single-stage transistor amplifier characteristics directly from the corresponding *h*-parameters. Table II permits conversion between the familiar equivalent-T circuit and the *h*-parameters measured in the common-base, emitter or collector configuration. Typical numerical values for each parameter are given.

Recently adopted IRE standards on semiconductor symbols are used in these tables.[2] The first subscript designates the characteristic, i for input, o for output, f for forward transfer and r for reverse transfer, while the second subscript designates the circuit configuration, b for common base, e for common emitter and c for common collector. The corresponding old symbols are indicated on the left side of Table II.

REFERENCES

(1) A.W. Lo, "Transistor Electronics", p 55, Prentice Hall, 1955.
(2) IRE Standards on Letter Symbols for Semiconductor Devices, *Proc IRE*, p 934, July 1956.

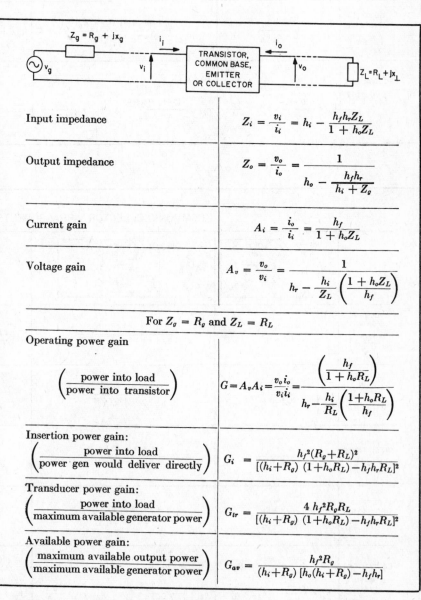

Table I—Single Stage Transistor Amplifier Characteristics in Terms of H-Parameters

Input impedance
$$Z_i = \frac{v_i}{i_i} = h_i - \frac{h_f h_r Z_L}{1 + h_o Z_L}$$

Output impedance
$$Z_o = \frac{v_o}{i_o} = \frac{1}{h_o - \dfrac{h_f h_r}{h_i + Z_g}}$$

Current gain
$$A_i = \frac{i_o}{i_i} = \frac{h_f}{1 + h_o Z_L}$$

Voltage gain
$$A_v = \frac{v_o}{v_i} = \frac{1}{h_r - \dfrac{h_i}{Z_L}\left(\dfrac{1 + h_o Z_L}{h_f}\right)}$$

For $Z_g = R_g$ and $Z_L = R_L$

Operating power gain
$$\left(\frac{\text{power into load}}{\text{power into transistor}}\right)$$
$$G = A_v A_i = \frac{v_o i_o}{v_i i_i} = \frac{\left(\dfrac{h_f}{1 + h_o R_L}\right)}{h_r - \dfrac{h_i}{R_L}\left(\dfrac{1 + h_o R_L}{h_f}\right)}$$

Insertion power gain:
$$\left(\frac{\text{power into load}}{\text{power gen would deliver directly}}\right)$$
$$G_i = \frac{h_f{}^2(R_g + R_L)^2}{[(h_i + R_g)(1 + h_o R_L) - h_f h_r R_L]^2}$$

Transducer power gain:
$$\left(\frac{\text{power into load}}{\text{maximum available generator power}}\right)$$
$$G_{tr} = \frac{4\, h_f{}^2 R_g R_L}{[(h_i + R_g)(1 + h_o R_L) - h_f h_r R_L]^2}$$

Available power gain:
$$\left(\frac{\text{maximum available output power}}{\text{maximum available generator power}}\right)$$
$$G_{av} = \frac{h_f{}^2 R_g}{(h_i + R_g)[h_o(h_i + R_g) - h_f h_r]}$$

$$v_i = h_{ij}i_i + h_{rj}v_o$$
$$i_o = h_{fj}i_i + h_{oj}v_o$$

(j = COMMON ELECTRODE, e, b or c)

Circuit: i_i, i_o, v_i, j, v_o — ai_e, i_e, r_e, r_b, r_c, $r_m = ar_c$

Symbols (Old / New)	New	Common Emitter	Common Base	Common Collector	Equivalent T Circuit
h_{11e} / $1/y_{11e}$	h_{ie}	2,000 ohms	$\dfrac{h_{ib}}{(1+h_{fb})(1-h_{rb})+h_{ob}h_{ib}}$	h_{ic}	$r_b + \dfrac{r_e r_c}{r_e + r_c - r_m}$
h_{12e} / μ_{bc}	h_{re}	9.3×10^{-4}	$\dfrac{-h_{rb}(1+h_{fb})+h_{ib}h_{ob}}{(1+h_{fb})(1-h_{rb})+h_{ob}h_{ib}}$	$1 - h_{rc}$	$\dfrac{r_e}{r_e + r_c - r_m}$
h_{21e} / α_{cb},β	h_{fe}	49	$\dfrac{-h_{fb}(1-h_{rb})-h_{ob}h_{ib}}{(1+h_{fb})(1-h_{rb})+h_{ob}h_{ib}}$	$-(1+h_{fc})$	$\dfrac{r_m - r_e}{r_e + r_c - r_m}$
h_{22e} / $1/z_{22e}$	h_{oe}	33×10^{-6} mhos	$\dfrac{h_{ob}}{(1+h_{fb})(1-h_{rb})+h_{ob}h_{ib}}$	h_{oc}	$\dfrac{1}{r_e + r_c - r_m}$
h_{11} / $1/y_{11}$	h_{ib}	$\dfrac{h_{ie}}{(1+h_{fe})(1-h_{re})+h_{ie}h_{oe}}$	40 ohms	$\dfrac{h_{ic}}{h_{ic}h_{oc}-h_{fc}h_{rc}}$	$r_e + (1-\alpha)r_b$
h_{12} / μ_{ec}	h_{rb}	$\dfrac{-h_{re}(1+h_{fe})+h_{ie}h_{oe}}{(1+h_{fe})(1-h_{re})+h_{ie}h_{oe}}$	4×10^{-4}	$\dfrac{h_{fc}(1-h_{rc})+h_{ic}h_{oc}}{h_{ic}h_{oc}-h_{fc}h_{rc}}$	$\dfrac{r_b}{r_c + r_b}$
h_{21} / α_{ce},α	h_{fb}	$\dfrac{-h_{fe}(1-h_{re})-h_{ie}h_{oe}}{(1+h_{fe})(1-h_{re})+h_{ie}h_{oe}}$	-0.98	$\dfrac{h_{rc}(1+h_{fc})-h_{ic}h_{oc}}{h_{ic}h_{oc}-h_{fc}h_{rc}}$	$-\alpha$
h_{22} / $1/z_{22}$	h_{ob}	$\dfrac{h_{oe}}{(1+h_{fe})(1-h_{re})+h_{ie}h_{oe}}$	0.67×10^{-6} mhos	$\dfrac{h_{oc}}{h_{ic}h_{oc}-h_{fc}h_{rc}}$	$\dfrac{1}{r_c + r_b}$
h_{11c} / $1/y_{11c}$	h_{ic}	h_{ie}	$\dfrac{h_{ib}}{(1+h_{fb})(1-h_{rb})+h_{ob}h_{ib}}$	2,000 ohms	$r_b + \dfrac{r_e r_c}{r_e + r_c - r_m}$
h_{12c} / μ_{be}	h_{rc}	$1 - h_{re}$	$\dfrac{1+h_{fb}}{(1+h_{fb})(1-h_{rb})+h_{ob}h_{ib}}$	1.0	$\dfrac{r_c - r_m}{r_e + r_c - r_m}$
h_{21c} / α_{eb}	h_{fc}	$-(1+h_{fe})$	$\dfrac{h_{rb}-1}{(1+h_{fb})(1-h_{rb})+h_{ob}h_{ib}}$	-50	$-\dfrac{r_c}{r_e + r_c - r_m}$
h_{22c} / $1/z_{22c}$	h_{oc}	h_{oe}	$\dfrac{h_{ob}}{(1+h_{fb})(1-h_{rb})+h_{ob}h_{ib}}$	33×10^{-6} mhos	$\dfrac{1}{r_e + r_c - r_m}$
$\alpha = a$		$\dfrac{h_{fe}(1-h_{re})+h_{ie}h_{oe}}{(1+h_{fe})(1-h_{re})+h_{ie}h_{oe}}$	$-h_{fb}$	$\dfrac{h_{ic}h_{oc}-h_{rc}(1+h_{fc})}{h_{ic}h_{oc}-h_{fc}h_{rc}}$	$(r_m+r_b)/(r_c+r_b)$; 0.98
$r_c = \dfrac{r_m}{a}$		$\dfrac{h_{fe}+1}{h_{oe}}$	$\dfrac{1-h_{rb}}{h_{ob}}$	$-\dfrac{h_{fc}}{h_{oc}}$	1.5 meg
r_e		$\dfrac{h_{re}}{h_{oe}}$	$h_{ib}-(1+h_{fb})\dfrac{h_{rb}}{h_{ob}}$	$\dfrac{1-h_{rc}}{h_{oc}}$	28 ohms
r_b		$h_{ie}-\dfrac{h_{re}(1+h_{fe})}{h_{oe}}$	$\dfrac{h_{rb}}{h_{ob}}$	$h_{ic}+\dfrac{h_{fc}(1-h_{rc})}{h_{oc}}$	600 ohms
a		$\dfrac{h_{fe}+h_{re}}{1+h_{fe}}$	$-\dfrac{h_{fb}+h_{rb}}{1-h_{rb}}$	$\dfrac{h_{fc}+h_{rc}}{h_{fc}}$	0.98

Table II—Exact Conversion Formulas for H Parameters and Transistor Equivalent Circuits

AUDIO TRANSFORMER DESIGN NOMOGRAPHS

By C. J. SAVANT and C. A. SAVANT

Servomechanisms, Inc., Western Division, Hawthorne, California

Critical parameters of volts per turn and magnetizing inductance for transistor-driven audio coupling transformers are conveniently given by individual nomographs. Final chart gives wire size

TRANSISTORS operate with lower input impedance and higher output impedance than tubes, hence transformer coupling often becomes more expedient than R-C coupling for audio circuits.

The transformers required are only a fraction of a cubic inch in volume and weigh only a fraction of an ounce. The accompanying nomographs simplify the design of these subminiature a-f transformers for use in transistor circuits.

Turns and Core Area

For sinusoidal flux changes the basic transformer equation is

$$e = N K A B f / 3{,}490 \qquad (1)$$

where e is rms primary volts, N is number of primary turns, K is stacking factor, A is core cross-sectional area in sq in., B is peak flux density in kilogauss and f is frequency in cps.

Having selected the core material and the maximum value of flux density allowable from physical considerations, the designer must determine the number of turns and the core area from the operating voltage and frequency. The stacking factor is usually specified by the core lamination manufacturer and is essentially constant.

The volts-per-turn design nomograph in Fig. 1 is readily constructed by rearranging Eq. 1 in the form

$$(e/N)(3{,}490) = (KAB)(f) \qquad (2)$$

The operating frequency is known from the circuit applica-

tion of the transformer. To get KAB, the maximum flux density is taken from the B-H characteristics for the particular iron, K is obtained from the iron specifications and the approximate area is taken from the known size requirement. From these values of f and KAB, the volts per turn can be read directly on the nomograph. Division of primary volts by the

value of e/N gives the number of primary turns.

Magnetizing Inductance

In a transformer-coupled amplifier, the loading effect of the transformer working through the high output impedance of a grounded-emitter transistor may be serious. For low-power transformers at audio frequencies the most significant parameter in the determination of this load-

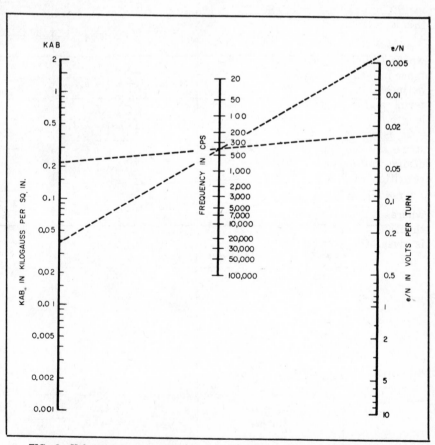

FIG. 1—Volts-per-turn nomograph for subminiature iron-core coupling transformers used in transistor circuits. Dashed lines correspond to design problem worked out as example

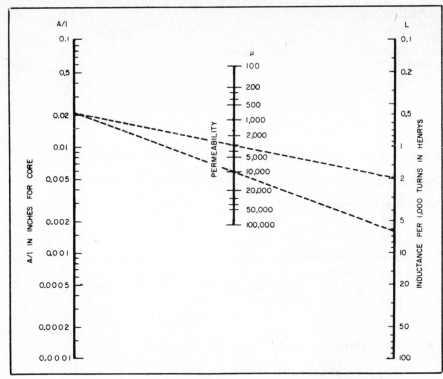

FIG. 2—Magnetizing inductance nomograph for transistor coupling transformers

ing is the shunt magnetizing inductance, L_m, across the primary. A practical equation for this is

$$L_m = K\ 3.2\ N^2 A\ 10^8\ \mu\ /\ l \quad (3)$$

where L_m is magnetizing inductance in henrys, μ is incremental permeability for the core material and l is magnetic path length.

It is again possible to construct a nomograph to facilitate design calculation, if Eq. 3 is rewritten as

$$L_m' = (A/l)\ (\mu)\ (K \times 320) \quad (4)$$

where L_m' is inductance per thousand turns. The nomograph is given in Fig. 2. The minimum

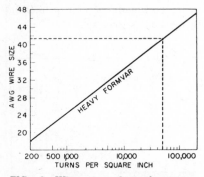

FIG. 3—Wire size chart for transformers using heavy Formvar insulation on windings. For single Formvar or plain enamel, increase turns per sq in. by six percent

allowable inductance per thousand turns is determined from the circuit application of the transformer. Transformer dimensions, the turns ratio and the initial permeability are then chosen to satisfy the nomograph, and wire size to fit the available window area is found from Fig. 3.

Example of Use

A transistor transformer is to be designed for matching a 5,000-ohm source to a 500,000-ohm load, using a 10-v, 400-cps primary and a 6.6-v secondary. Maximum phase shift is 10 deg.

When the transformer is operated from the specified source impedance, neglecting transformer losses, the magnetizing inductance required for 10-deg phase shift at 400 cps is

$$L = \frac{5,000}{(2\pi)\ 400\ \tan 10°} = 11.3\ \text{h}$$

The core type is selected next, largely from experience, as a laminated core, 187 E-I, in standard 14-mil audio A. For this material, A/l is 0.0216 in., KA is 0.0309 sq in. and permeability is 10,000 at 7 kilogauss. The value of KAB is then

0.0309 × 7 or 0.216. Using this and a frequency of 400 cps in Fig. 1 gives 0.023 volts per turn for e/N.

With a primary voltage of 10 v, primary turns are 435 and must produce an inductance of 11.3 h. The inductance per 1,000 turns is 11.3 $(1,000/435)^2$ or 59.7 h.

To check the design, use 0.0216 for A/l and 10,000 for permeability in Fig. 2. This gives 6.3 h for inductance per 1,000 turns instead of 59.7 h, hence more turns will be necesary. After one or two trials, 2,400 turns were decided upon for the primary. From Fig. 2, with 11.3 h per 2,400 turns or 1.96 h per 1,000 turns, at A/l of 0.0216 the permeability can be as low as 3,000. From Fig. 1, with e/N at 10/2,400 or 0.00417 and frequency at 400 cps, KAB is 0.037. Flux density B then is 0.037/ 0.0309 or 1.19 kilogauss. At this flux density, the material characteristics indicate that the permeability is actually 4,000, hence the magnetizing inductance of at least 11.3 h can easily be achieved.

The secondary turns will now be 0.66 × 2,400 or 1,580. Total turns are then 3,980. The window area for 187 E-I is 0.0822 sq in., hence the turns per square inch are 48,400. From Fig. 3, the allowable wire size is No. 42 heavy Formvar.

Thus, the final design is a primary of 2,400 turns and a secondary of 1,580 turns, both of No. 42 heavy Formvar, with core of 14-mil 187 E-I audio A.

The nomographs in themselves accomplish a simple multiplication. This fact alone would make them of little value. However, the designer of these transformers must manipulate between the circuit requirements, core type and size and wire size until all equations and requirements are satisfied. Actual experience indicates that this trial-and-error process is greatly speeded and simplified by using these nomographs.

MATCHING TRANSISTOR-DIODES

By ARTHUR GILL   Research Division, Raytheon Manufacturing Company,
Waltham, Massachusetts

Emitter-to-base circuit of transistor has characteristics comparable to ordinary diode. Variable resistor connected between collector and base alters characteristics to achieve matching

MISMATCH between a pair of diodes in a modulator or demodulator sets a lower limit to sensitivity and linearity.

Difficulties normally encountered with ordinary diodes are overcome by employing emitter diodes of transistors. Matching is accomplished by varying a resistor connected between the collector and base of the transistor. Since active elements are not involved, most rejected transistors can serve as transistorized diodes.

Control

The diode in the emitter-base circuit of a transistor has characteristics comparable to those of an ordinary diode of similar size and material. Forward and reverse resistances of a transistorized diode are lower when the collector is shorted to the base than when the collector is open circuited.

Experimental investigations with 2N131 a-f germanium transistors showed that a certain amount of control can be exercised over the diode characteristics with a variable resistor R_c connected between the collector and the base.

Varying R_c between zero and infinity increases the forward resistance r_f by a factor of 2.6 and the reverse resistance r_r by a factor of 1.5.

Resistance Change

Most of the change in r_f occurs when R_c is varied between 50 and 500 ohms and most of the change in r_r occurs when R_c is varied between 1 to 50,000 ohms. Consequently, adjusting r_f leaves r_r virtually intact and vice versa. Varying R_c rotates the entire diode characteristics about the origin of the V-I field.

A graph of r_f, at $I = 5$ ma and r_r at $V = -6$ v as functions of R_c for a 2N131 transistor is shown in Fig. 1A. The volt-ampere characteristics of a similar transistorized diode with R_c equal to zero and infinity are shown above in Fig. 1B.

Performance

The circut used to evaluate the matching conditions between transistorized diodes is shown in Fig. 2. Since the 10,000-ohm resistors were precision components, the null voltage V_n was taken as a measure of the mismatch between the two diodes. Source voltage and impedance were chosen for a peak forward current of about 5 ma through the diodes and a peak reverse voltage of about 12 v across them.

Components

A pair of 2N131 transistors, Q_1 and Q_2, were found to have the following characteristics: $40 < r_{f1} < 110$ ohms, $1.6 < r_{r1} < 2.2$ megohms, $44 < r_{f2} < 120$ ohms and $1.8 < r_{r2} < 3$ megohms. Matching by adjusting r_{r1} and r_{f2} was accomplished by using 50,000-ohm and 500-ohm potentiometers for R_{c1} and R_{c2} respectively.

With $R_{c1} = R_{c2} = \infty$, V_n was 1.1 v p-p. With $R_{c1} = R_{c2} = 0$, V_n was 1 v p-p. With $R_{c1} = 10,000$ ohms and $R_{c2} = 300$ ohms, the minimum null voltage of 0.1 v p-p was obtained. Balance condition was thus improved by a factor of ten.

Where the initial mismatch in characteristics is more pronounced, the improvement using this matching method is greater.

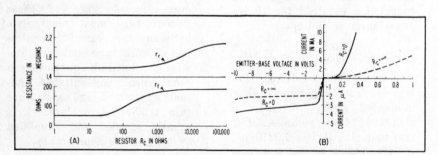

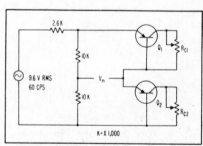

FIG. 1—Forward and reverse resistances of transistor-diode are plotted as functions of variable resistor R_c (A). Varying R_c rotates characteristics about origin (B)

FIG. 2—Matched transistor-diodes have R_c between collector and base

Chapter 2

TRANSISTOR AMPLIFIERS

THERMISTORS COMPENSATE PUSH-PULL AMPLIFIERS

By A. J. WHEELER
Semiconductor Division, Radio Corp. of America, Somerville, New Jersey

Temperature compensation of class-B push-pull transistor amplifiers is necessary to minimize distortion and prevent runaway. Typical compensating circuits using thermistors are described and equations for calculating component values and restrictions on use of two types of thermistor materials are given. Design technique gives an approximation to the desired linear decrease in bias with increase in ambient temperature

TRANSISTOR CLASS-B push-pull output stages usually use the common-emitter configuration because it permits high power gain and conserves battery power.

Design Method Development

A transistor circuit using thermistor compensation is shown in Fig. 1. Series and parallel resistances are used to shape the bias curve to a desired temperature function.

The thermistor bias circuit must provide an approximation to the desired change of 2.5 mv per degree C for alloy-junction transistors. Because the variation of thermistor resistance with temperature is approximately exponential, a shaping network must be used to obtain this variation.

Figure 2 shows a comparison between thermistor resistance as a function of temperature and the resistance of a thermistor circuit in which the parallel and series resistances were chosen to give an approximation to a linear change with temperature.

Tracking Conditions

In the circuit shown in Fig. 3A, the bias voltage is a constant ratio of the voltage between m and n, V_{mn}, which decreases with increasing temperature due to the drop in thermistor resistance R_T. Resistor

R_2 must equal a certain fraction of the thermistor resistance at room temperature (25 C) for tracking to occur at selected temperature points.

Three points are used: 0 C, 25 C and 50 C. The relative value of R_2 with respect to the thermistor resistance at 25 C is dependent on the type of thermistor material

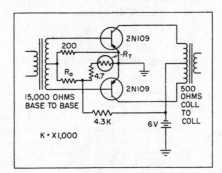

FIG. 1—Class-B push-pull amplifier

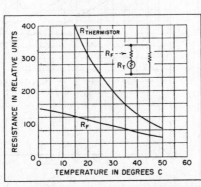

FIG. 2—Modified thermistor characteristic

used, and may be determined from calculations based on Fig. 4.

The nonlinear curves of Fig. 4 show the conditions for tracking at the selected temperature points for several values of room-temperature bias. The factors f_0 and f_{50} represent the relative change in resistance of the thermistor branch of the bias circuit shown in Fig. 3A. The factor f_{50} is equal to the ratio of the resistance of the branch at 50 C to its resistance at 25 C; f_0 equals the ratio of its resistance at 0 C to its resistance at 25 C. For a particular room-temperature bias, any point f_{50}, f_0 falling on the appropriate curve will allow tracking at the selected points.

The relation from which these curves are calculated my be derived in the form of

$$ f_0 = \frac{f_{50}(k_1 - 1)}{(k_2 - 1) + f_{50}(k_1 - k_2)} $$

where k_1 is the ratio of the bias voltage desired at 25 C to that desired at 50 C, and k_2 is the ratio of the bias voltage desired at 25 C to that desired at 0 C.

Restrictions on Material

The dashed lines in Fig. 4 show the variation obtainable in f_{50}, f_0 as the value of R_2 is increased from zero to some higher value for two types of thermistor material. The

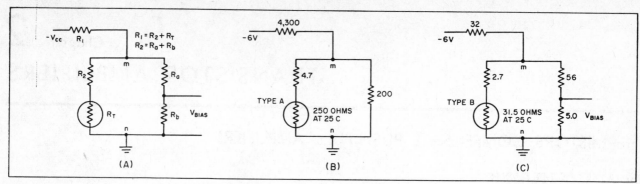

FIG. 3—Generalized bias circuit (A), bias circuit when R_a is zero (B) and bias circuit when R_a does not equal zero

point for R_2 equals zero, designated f'_{50}, f'_0, is dependent on the temperature coefficient of the particular type of thermistor material selected. The slope of the dashed line is determined by the relation

$$m = (f'_0 - 1)/(1 - f'_{50}) \qquad (1)$$

The ratio of R_2 to the thermistor resistance at 25 C as the resistance of R_2 is increased may be expressed as p. As an example, values of p are marked off on the dashed line for type A material. The intersection of the dashed line with the solid curves in Fig. 4 determines the proper value of R_2 expressed in terms of p. The value of p at the intersection may be calculated from the equation

$$p = (f'_{50} - f_{50})/(f_{50} - 1) \qquad (2)$$

The choice of thermistor material is subject to restrictions based on the curves shown in Fig. 4. If the point f'_0, f'_{50} lies to the right of a desired bias curve, negative values of R_2 are required. Consequently, the choice of a thermistor material which provides a point f'_0, f'_{50} to the left of or coincident with the desired bias curve is necessary to obtain a realizable value for R_2.

Restriction on R_a and R_o

When R_a in the circuit of Fig. 3A is equal to zero

$$R_b = \frac{V_{cc} \times R_3}{(V_{cc} \times K_1) - [V_{B25} \times (K_1 + 1)]} \qquad (3)$$

where V_{cc} = collector supply voltage, V_{B25} = bias voltage at 25 C, R_3 = resistance of $(R_2 + R_t)$ at 25 C and

$$K_1 = \frac{1 - (k_1 \times f_{50})}{f_{50} \times (k_1 - 1)} \qquad (4)$$

where k_1 = ratio of desired 25 C bias voltage to that desired at 50 C and f_{50} = ratio of R_3 at 50 C to R_3 at 25 C.

Because the equation for R_b is independent of R_1, R_b may be determined as soon as a thermistor has been chosen and R_2 has been calculated. As a first approximation, the value of thermistor resistance at 25 C should be about twice the desired bias-circuit resistance between base and emitter at 25 C.

Calculation

The value of R_1 may be calculated from

$$R_1 = \left(\frac{V_{cc}}{V_{B25}} - 1\right)\left(\frac{R_3 \times R_b}{R_3 + R_b}\right) \qquad (5)$$

This calculation is the final step in determining the bias network when R_a is equal to zero.

When R_a is not equal to zero, the following inequality must exist so R_a is a realizable value.

$$R_1 \leq \frac{1}{4} \times \frac{K_1^2}{K_1 + 1} \times \frac{V_{cc}}{V_{B25}} \times R_b \qquad (6)$$

Provided this inequality is met

by the choice of suitable values for thermistor resistance, R_2 and R_b, the value of R_a may be determined from the relation

$$R_2 = \frac{1}{2}\left[R_b (NK_1 - _2) + \sqrt{(NK_1 R_b)^2 - (4NR_b R_3)}\right] \qquad (7)$$

where $N = V_{cc}/V_{B25}(1 + K_1) \qquad (8)$

The value of R_1 is determined by

$$R_1 = \left(\frac{R_3 \times R_4}{R_3 + R_4}\right)\left[\left(\frac{V_{cc}}{V_{B25}} \times \frac{R_b}{R_4}\right) - 1\right] \qquad (9)$$

where $R_4 = R_a + R_b \qquad (10)$

Selection of Thermistors

When R_a is equal to zero, the thermistor resistance at 25 C is chosen so that the resistance of R_3 and R_b in parallel is low enough to avoid excessive loss of input power but high enough so that battery life is not appreciably shortened.

As a first approximation, the thermistor resistance at 25 C is

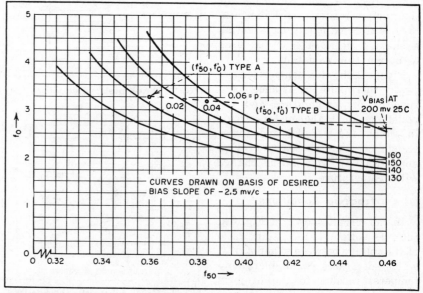

FIG. 4—Conditions necessary for three-point tracking at O, 25, and 50 C

14

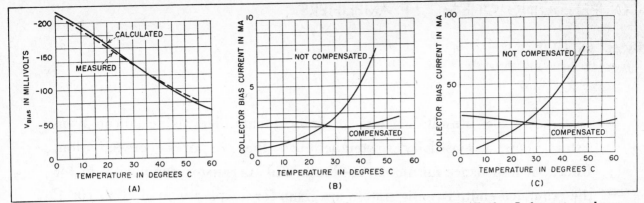

FIG. 5—Bias (A) collector current-temperature characteristic (B) when R_a is zero and characteristic (C) when R_a does not equal zero

chosen to be about twice the desired base-to-emitter bias-circuit resistance.

When R_a is not equal to zero, the resistance presented to the base-emitter terminals is nearly equal to R_b if R_a is sufficiently large. Thermistor resistance then depends on the inequality, Eq. 6.

In either case, the type of thermistor material used will depend on the magnitude of the desired base-emitter voltage at 25 C. Various types of thermistor material differ in resistance-temperature characteristics and, therefore, have different points on Fig. 4 for corresponding values of f'_{50}, f'_0. The types A and B thermistor material used as examples possess typical resistance-temperature characteristics.

Thermistor current in the final circuit design should not be high enough to produce self-heating. Otherwise the obtained bias will be lower than calculated, and tracking points will be changed.

Application

Figure 3B shows the application of this biasing method using 2N109 transistors with a supply voltage of 6 volts when R_a equals zero. The 2N109 transistors require a base-emitter bias voltage of 150 mv for an idling current of 2 ma. From consideration of input power loss and battery drain, the bias resistance between base and emitter should be about 100 ohms.

Because the point f'_0, f'_{50} (3.3, 0.36) for type-A thermistor materials lies to the left of the 150 mv bias curve as shown in Fig 4, a straight dashed line having a slope

m is drawn from this point to intersect the 150 mv curve. From Eq. 1 slope m is -3.59. Values of f'_{50} and f'_0 may be obtained from published data on thermistors. The intersection for type-A material occurs at point $f_0 = 3.25$, $f_{50} = 0.3725$. From Eq. 2 the value of p is 0.02.

Example

If a thermistor having a resistance of 250 ohms at 25 C is chosen, resistor $R_2 = 0.02 \times 250 = 5$ ohms. A value of 4.7 ohms may be chosen as the nearest value. The value of f_{50} for this combination is then

$$f_{50} = \frac{R_{t25} + R_2}{R_{t25} + R_2} = 0.371$$

The value of k_1 is determined by $k_1 = V_{B25}/V_{B50} = 1.714$, hence from Eq. 4, $K_1 = 1.37$. The value of R_b as determined from Eq. 3 is then 194. Note that V_B at $tC = V_B$ at 25 C $- \Delta t$ (2.5 mv/C), where $\Delta t = -25$ C $+ t$.

Choosing a value of 200 ohms for R_b, R_1 may be calculated from Eq. 5 and becomes 4,370 ohms. A value of 4,300 ohms is chosen for R_1.

Figure 5A shows both calculated and measured bias as a function of temperature. Figure 5B shows experimental data on collector current-temperature characteristics for 2N109 transistors.

Method

Figure 3C shows this design method used when R_a does not equal zero. A bias voltage of 200 mv is required to minimize cross-over distortion. Because the transistors used in this application have low input impedance, the bias resist-

ance between base and emitter should be approximately 5 ohms.

Figure 4 shows that either type-A or type-B material may be used because the curves for both types begin at points to the left of the 200 mv bias curve. Type-B material will be used because it permits the use of lower resistance values. The dashed line drawn from the point f'_{50} f'_0 for type-B material has a slope $m = -3.085$ (from Eq. 1).

At the intersection of this line with the 200 mv bias curve, f_{50} equals 0.4556 and f_0 equals 2.68. The value of p is then calculated from Eq. 2 and is 0.0847.

The factor k_1 is 1.4546, therefore, K_1 equals 1.62. The resistor R_3 can then be selected to satisfy the inequality, expresssion 6, and equals 7.52 R_b.

If a value of 5 ohms is used for R_b, then $R_3 \leqq 37.6$ ohms.

Available Values

An available thermistor value is 31.5 ohms at 25C. Although lower values are available, they would entail higher and undesirable values of bias-circuit current.

The value of R_2 is calculated from $R_2 = p \times R_{T25} = 2.7$ ohms.

Then $R_3 = R_2 + R_{T25} = 2.7 + 31.5 = 34.2$ ohms.

The value of N given by Eq. 8 is 11.46 and the value of R_a given by Eq. 7 is 56 ohms.

The value of R_4 is calculated from Eq. 10 and is 61 ohms. Therefore the value of R_1 as calculated from Eq. 9 is 32 ohms.

The compensation effected by the bias circuit shown in Fig. 3C is shown in Fig. 5C for an experimental high-power transistor.

COMPENSATING ONE-MC I-F AMPLIFIERS

By S. H. GORDON
Diamond Ordnance Fuze Laboratories, Washington, D. C.

Results of tests on 1-mc i-f amplifiers using silicon transistors show a sensitivity to temperature that requires compensation. Effects of impedance mismatch, feedback and thermistor compensation for the grounded-emitter configuration are given for temperatures of 20 to 100 C

SILICON TRANSISTORS were thought a cure for many of the temperature problems that exist in circuits using germanium transistors. They are rated at temperatures up to 150C and, because silicon withstands high temperature better than germanium, temperature variations should have little effect on their characteristics.

In many cases these original hopes were fulfilled. However, when the silicon transistors were tried in high-frequency i-f amplifier circuits, the gain dropped considerably as temperature increased. This decrease in gain was not expected, as it had not occurred in any previous experiences with wide-band low-frequency R-C coupled amplifiers.

Temperature Compensation

Analysis of the changes in gain caused by temperature changes in an uncompensated amplifier were made. The curves of Fig. 1 show the correlation between the calculated change in gain with temperature and the measured gain. The lower measured gain is probably caused by either losses in the output transformer or errors arising out of initial assumptions in determining equations to calculate the gain. The operating frequency in both cases is 1 mc. Further analysis that includes temperature compensation techniques is certainly possible however because of the expected complexity of the resulting equations and the time consumed to analyze each temperature compensation technique it was decided to investigate experimentally.

The three general methods of temperature compensation investigated were impedance mismatching, feedback and the use of thermistors. Only the temperature range from 25C to 100C was investigated, however the techniques developed for this temperature range can be applied at the lower temperatures. An important factor here was to obtain the desired compensation with a minimum loss of gain.

Impedance Mismatch

The input impedance of a grounded-emitter 1-mc i-f amplifier at room temperature is normally about 200 ohms. Output impedance is normally about 1,500 ohms. As the transistor is heated, the input impedance will increase and the output impedance will decrease. In an attempt to compensate for the effects of these impedance changes an approximate 8-to-1 mismatch of 1,500 ohms in series with the input was tried. An output load of 120 ohms through an 8-to-5 transformer ratio was also tried. The curves obtained using these matching values are shown in Fig. 2A. Since little or no improvement was noted and because the power loss due to mismatch was large, this method of temperature compensation appeared unsatisfactory.

Two types of feedback in the grounded-emitter amplifier were

Laboratory setup used to measure gain-temperature characteristics of silicon transistors in 1-mc grounded-emitter i-f amplifiers with feedback and thermistor compensation

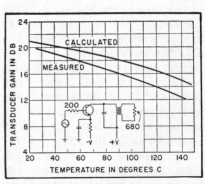

FIG. 1—Curves give calculated gain and measured gain for configuration shown

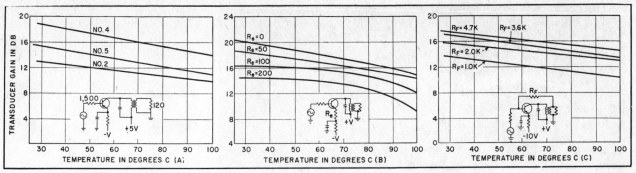

FIG. 2—Effects of compensating grounded-emitter narrow-band amplifier by mismatched impedance (A), emitter feedback (B) and negative feedback (C). Numbers identify transistors tested in circuit configuration associated with curves

investigated. They were degenerative feedback in the form of an unbypassed emitter resistance and feedback from the collector to the base. The curves of the degenerative emitter feedback are shown in Fig. 2B and curves of the collector-base feedback are shown in Fig. 2C. The input and output matching resistor was changed for each different feedback condition that was used so a matched condition could be obtained at the start of each temperature run.

No significant improvements in the temperature characteristics were noted with the amounts and types of feedback shown. Analysis of why the feedback did not help as expected was not made, however some important parameters in the equations governing feedback in a transistor amplifier are the input

tors were investigated. The first method used a thermistor in series with the input circuit to attenuate the input signal more at the low temperatures than at the high temperatures and the second used a thermistor in series with the emitter to change the d-c operating point with temperature.

The results of the first type of compensation are shown in Fig. 3A. If the input thermistor-resistor network is designed so that the amplifier is matched at the higher temperature and the input signal is attenuated at the lower temperature so the output power is the same as that obtained with the higher temperature, then the temperature-gain curve should be relatively flat. The loss in gain is the difference between the high and low-temperature gains. The trans-

thus increasing the gain.

This method has an advantage in that the increasing input impedance resulting from heating the transistor is compensated by the increase in emitter current. This also tends to lower the input impedance. All types of compensation that make use of thermistors require some matching of the thermistor to the transistor. The thermistors used in Fig. 3A were the only type of compensation tried which displayed good results with only a small loss in gain.

The author thanks R. C. Carter, W. H. Schuette and E. H. Harrison Jr. for their helpful suggestions.

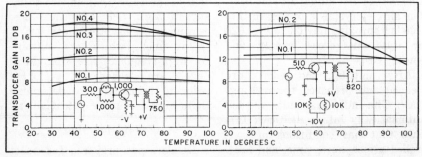

FIG. 3—Compensation resulting from thermistor in input circuit (A) and in emitter circuit (B) for silicon units of same type shows stability throughout temperature range

impedance and the amplifier gain.

Since these parameters vary with temperature, the improvement expected with feedback may not be obtained.

Thermistors

Two types of temperature compensation which employ thermis-

ducer gain shown in Fig. 3A was measured after matching at the high temperature end which results in an R_g value of 400 ohms.

Results of the second type of compensation are shown in Fig. 3B. As the emitter current is increased from low to medium the beta of the transistor increases,

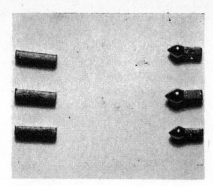

Mounted tetrode before capping (left) and complete in housing (right) after capping while enveloped in a dry inert atmosphere

Germanium pellet before (left) and after (right) meltback

HIGH-FREQUENCY CIRCUITS USING MELT-BACK TETRODES

By D. W. BAKER

Design Engineer, Semiconductor Products Department,
General Electric Company, Syracuse, New York

Application of transverse bias to *pnp* germanium transistor through second base lead enhances intermediate and high-frequency power gain. Design, construction and evaluation of tetrode unit are covered along with applications to tv receiver, pulse-amplifier and oscillator circuits

FREQUENCY LIMITATIONS of junction transistors have been a deterrent to the wide application of transistors in the vhf range. Several types of devices have shown possibilities for the improvement in high-frequency performance of transistors. The tetrode type[1] has been developed in production quantities specifically for applications up to 100 megacycles. Developmental units have produced maximum oscillating frequencies up to 1,000 mc.

Several companies have been working independently on tetrode design and development in various sections of the country. This article covers design, techniques of fabrication, evaluation and application of one of these tetrodes.

Construction

This particular tetrode is made from germanium which contains donor and acceptor impurities added in specific predetermined amounts to the single crystal melt. The single crystal is processed into pellets approximately 0.020 by 0.020 by 0.120 in. cross section and length, respectively.

After cutting, the pellets are cleaned and prepared for melting-back. The individual units are fed automatically into a meltback furnace that has conditions of controlled atmosphere, time, and heat.

The doping ratio existing within the initial pellet allows the impurities to segregate out under controlled conditions. When the melted portion of the pellet solidifies, a greater number of acceptor impurities separate out first, thus producing a *p*-type doped base. The width of this base may be varied to a considerable degree by using different doping ratios, impurities and heating and cooling combinations to obtain desired results.

The entire melting-back operation is carefully programmed to insure a high yield of pellets possessing a base width under 0.0005 of an inch, which are suitable for high-frequency use. Individual pellets are cleaned and evaluated prior to fabrication to determine their high-frequency potentialities.

Pellet Mounting

Headers are prepared for pellet mounting by welding, cutting and joining the end leads for pellet support. The melt-back pellets are attached to the end leads of the header mount, while the device is elevated to a high temperature in an inert atmosphere. The pellet after mounting on the header is step-etched. All units are thoroughly washed and dried after the etch.

Base and end leads must make

good ohmic contact to their respective areas. A poor emitter contact, for example, would seriously reduce the ratio of output to input impedance, limiting gain.

In the attachment of leads to the base region, it is important to have not only a good ohmic contact but negligible emitter and collector overlap. Base one, the active base of the tetrode, must be secured to the pellet along its entire base depth by a low-resistance ohmic contact with minimum overlap. Lack of a good ohmic contact or a small area contact would increase the base resistance.

Emitter overlap capacity C_{eb} at high frequencies might shunt the input terminals of the transistor causing a reduction in alpha. Additional collector capacity produced by overlap will also reduce gain; however, overlap on the collector is not as serious as overlap on the emitter. Another disadvantage of overlap is the reduction of barrier breakdown voltage.

At certain frequencies h_{11} may be inductive reactive; which in some cases could produce an alpha greater than unity, due to the formation of an antiresonant circuit of low Q with the input circuit of the inherent transistor.[2]

The second base-lead contact, the tetrode-bias lead, must have qualities similar to the active base; however, it is not as critical. Increased contact resistance may be overcome by a corresponding increase in tetrode bias in some cases, but the collector breakdown voltage must exceed the sum of the transverse bias voltage and collector bias voltage. Overlap capacity can seriously limit, if not prohibit, the measurement of certain high-frequency parameters.

After both base leads are attached to the pellet, the other ends are welded to the header leads.

Surface conditioning consists of two operations, final etch and surface stabilization. The tetrode is then capped while enveloped in a dry inert atmosphere. The cap is welded to the header and provides a tight hermetic seal. Header leads spaced 0.100 in off center permit easy adaption to printed circuit applications. An extra lead is pro-

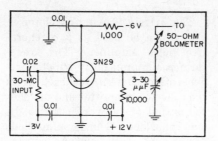

FIG. 1—Thirty-megacycle amplifier has 10-db power gain with 2-mc bandwidth

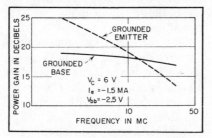

FIG. 2—Tetrode power-gain characteristics for grounded-base and grounded-emitter configurations

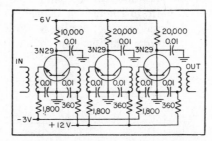

FIG. 3—Three-stage 25-mc tv i-f amplifier has 57-db gain with 4-mc bandwidth

vided for grounding the case when the device is used at high frequencies.

Evaluation

A number of device characteristics have been measured. The resulting information should not be construed as final specifications for the device. However, all units will operate under the same bias condition and may be easily substituted for one another.

While objective specifications call for a 10-db power gain at 30 mc with a 2-mc bandwidth, in a circuit similar to that of Fig. 1, the actual device exceeds this requirement quite readily. As shown in Fig. 2, grounded-base power gains average around 15 to 20 db at 30 mc and a 2-mc bandwidth, with a number of units having gains in excess of 30 db. Though grounded-emitter power gain is greater at low frequencies, grounded-base power gain begins to equal and

eventually exceed grounded-emitter power gain between 10 to 20 mc.

Useful power gain can be obtained up to and beyond frequencies of 60 mc. If the bandwidth is not restricted to 2 mc, these same 15 to 20-db power gain units will readily produce nonoscillating gains of 30 to 50 db at bandwidths around 0.5 mc at 30 mc. Power gain does not decrease 3 db over the temperature range −55 C to +85 C.

Substantial variation in bias conditions is permissible with only a slight reduction in power gain. Variations of 6 v, 2 v and 5 ma on the collector, base 2 and emitter, respectively, will not cause a drop of 3 db in grounded-base matched power gain.

Application of a transverse bias to a transistor not only enhances high-frequency power gain but intermediate-frequency gain as well. Better than 10 db of power gain can be added to transistors over the range of 445 kc to 30 mc.

Parameters

The h or hybrid parameters of the tetrode have been measured up to 50 mc. Since parameters are complex at high frequencies all quantities are presented in the conjugate form, $R + j x$ or $|Z| \angle \theta$.

At 30 mc, the short-circuit input impedance h_{11} has a resistive component which varies from 20 to 40 ohms and a reactive component which is 10 to 20 ohms. The reactive portion becomes inductive reactive in this frequency range.

Open-circuit output admittance h_{22} ($g_{22} + j\omega c_{22}$) contains a conductive component which varies from 10 to 60 ohms and a capacitive component which varies from 5 to 12 $\mu\mu$f over the frequency range of 1 to 30 mc.

Both the magnitude and phase, ($|a| \angle \theta$), of $-h_{21}$ the short-circuit current-transfer function, drop gradually with frequency. At 30 mc, alpha averages around 0.75 to 0.85 with a phase angle of approximately 40 degrees. The frequency of alpha cutoff, f_{aco}, varies from 40 to 60 mc. The units may be subjected to fairly substantial emitter currents with no change in alpha.

The application of tetrode bias produces a reduction in the $r_b' C_{22}$

product. Usually $r_b' C_{22}$ drops from several-thousand ohm-micromicro-farads to a few-hundred ohm-microfarads or less.

The noise figure of these tetrodes at 30 mc is in the 10 to 15-db range. It varies from approximately 8 db to 21 db over the frequency range from 10 to 50 mc. Only slight variations in noise figure are produced by changes in bias, the greatest change being caused by transverse bias.

Thus far the tetrode has successfully completed JAN 193, 20,000 g's centrifuge and 1,000 g's shock tests.

Video Amplifiers

In television circuitry, several of the earlier tetrodes were designed into the tv-if amplifier shown in Fig. 3[3]. This amplifier consists of six stages and operates at a center frequency of 25 mc. Gain is 57 db with a 4-mc bandwidth.

To utilize the maximum bandwidth from each stage, the tank circuit of each transistor consists of a variable inductance only. This inductance resonates with the output capacitance of the transistor plus the stray circuit capacitance.

The coupling transformer matches the input and output impedances approximately at 25 mc for all stages. The stages are tuned away from 25 mc to give three staggered pairs. Two series-resonant traps, one a sound trap at 21.9 mc and the other an adjacent-channel sound trap at 27.9 mc, were inserted in the input circuit.

No neutralization is used in the amplifier and transverse bias is adjusted to give optimum gain. The bias points used are $I_e = -1.5$ ma, $V_c = +10$ v and $V_{bb} = -2$ to 3 v.

The two-stage video amplifier in Fig. 4 was designed for low-level input preamplifier applications. The frequency response compensates for the effect of a constant-current signal source. This circuit produces a power gain of 32 db ± 0.4 db from 30 cps to 10 mc, which is equivalent to vacuum-tube performance.

Several types of oscillator circuits have used tetrodes in their operation, one an oscillator operating at 108 mc produced 10-mw output power.

The driven blocking oscillator in Fig. 5 can be used as a regenerative pulse amplifier.[4] The transformer simultaneously supplies regenerative current feedback from the collector to emitter and matches the output of the transistor to the load impedance. The damping diode across the transformer primary provides a path for the rapid discharge of the energy stored in the magnetizing inductance of the transformer.

This oscillator has a peak-pulse power gain of 32 db; peak power of the trigger pulse is 50 μw with a pulse width of 1 μsec. The repetition rate is 30 kc, with a rise and fall time of 0.3 μsec.

A variation of this circuit contained an additional feedback path from the output of the transformer to the auxiliary base. This feedback path introduced a degenerative signal at the auxiliary base so current amplification of the

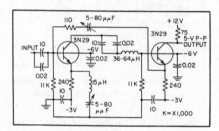

FIG. 4—Video preamplifier has 32-db power gain over range of 30 cps to 10 mc

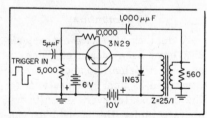

FIG. 5—Driven blocking oscillator has 0.3-μsec rise and fall times at 30-mc repetition rate

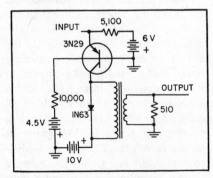

FIG. 6—Linear pulse amplifier has 10-db power gain at 1-mc pulse repetition rate

transistor was increased as the pulse built up across the output. The experimental performance of this second driven blocking oscillator circuit was similar to that of Fig. 5.

A minor variation in the processing of the tetrode controls to some degree the magnitude of internal feedback parameter h_{12}. Units possessing high values of h_{12} are ideally suited as high-frequency oscillators. Other types, designed with values of low h_{12}, operate well as high-stability amplifiers.

Pulse Amplifier

Excellent results are obtained when tetrodes are used as linear pulse amplifiers. In the circuit of Fig. 6, a pulse transformer couples the output to a resistive load. However, some distortion is produced by the transformer's magnetic properties. The diode in parallel with the transformer primary dampens the transient response of the output circuit.

Pulse-repetition rate is 1 mc with a peak pulse power gain of 10 db. Output pulse rise and fall times are 0.025 μsec and pulse width is 0.07 μsec.

Tetrodes can be used in free-running Eccles-Jordan multivibrator circuits with the same coupling as conventional transistor triode circuits.

Appreciation is expressed to associates within the General Electric Company for their helpful suggestions, with particular thanks to R. L. Pritchard, R. N. Hall, I. A. Lesk, S. O. Johnson and W. P. Barnett for their guidance and encouragement.

All work done on this project has been supported by the USAF Air Research and Development Command under contract AF 33(600) 38956.

REFERENCES

(1) R. L. Wallace, L. G. Schimpf and H. Dickton, A Junction Transistor Tetrode for High-Frequency Use, *Proc IRE*, Nov. 1952.
(2) R. L. Pritchard, Effect of Base-Contact Overlap and Parasitic Capacities on Small Signal Parameters of Junction Transistors, G. E. Research Lab. Report RL-1189, Oct. 1954.
(3) W. F. Chow, Transistorized IF and Video Amplifiers for Television Receivers, G. E. Tech. Inf. Ser. 1955ELP115.
(4) J. J. Suran, G. E. Tetrode, Transistor Applications to Pulse Circuits, G. E. Tech. Inf. Ser. DF55ELP96.

LOW-IMPEDANCE PREAMPLIFIER

By W. F. JORDAN   Teaneck, N. J.

AVERAGE low-signal general-purpose transistors in common-emitter circuits have input impedances of somewhere between 300 and 1,200 ohms. To get a voltage gain in between a microphone having 30 to 50 ohms impedance and a transistor amplifier, either matching transformers or cascading of two or more transistors is necessary. Bulk is the primary objection to these solutions.

Transistors are available, such as the CBS Hytron 2N256, that are designed for power output stages of automobile radios. The low impedance both in and out of this transistor makes it suitable for a low-impedance preamplifier.

The amplifier shown in Fig. 1 has a voltage gain of 26 db between a 50-ohm microphone and a 50-ohm input to an amplifier. It operates on very low current and with very little noise. Collector current is 5 ma with the bias, which is set by R_1. Hence a small mercury cell can be used.

Reducing the value of R_1 increases biasing voltage and gain but also increases collector current. The value was chosen mainly to keep current low, since maximum gain was not needed in the intended application.

The 2N256 transistor was designed to handle high currents. However, when operating at low current, temperature stability proved outstanding. Heating the casing to the point where it was too hot to touch resulted in no appreciable change in current gain. Neither was there any change in

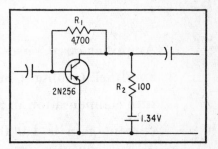

FIG. 1—Input and output impedances of 25-50 ohms make preamplifier suitable for use with low-impedance microphones

wave shape of the signal.

By direct comparison there was less noise than in a conventional vacuum-tube amplifier of the same gain.

PREAMPLIFIER MATCHES INPUT IMPEDANCE

AUTOMATIC impedance matching is accomplished in a transistorized preamplifier developed by I.D.E.A., Inc. The unusual features of the amplifier, which was designed for high-fidelity systems, may make it useful with other low-level sources.

The low-impedance input circuit shown in Fig. 1 is said to have practically constant voltage sensitivity over a range of impedances from less than 10 ohms to more than 10,000 ohms. One volt rms output is gotten from about 0.2 millivolts input.

The input is coupled directly to the base of Q_1. Resistor R_4 is common to the input load and the negative feedback circuit comprised of R_9 and C_5.

When input impedance is low, it has a shunting effect across R_4. This decreases negative feedback and increases overall gain.

As input impedance increases, it has less shunting effect, permitting more negative feedback. Amplifier gain is therefore decreased.

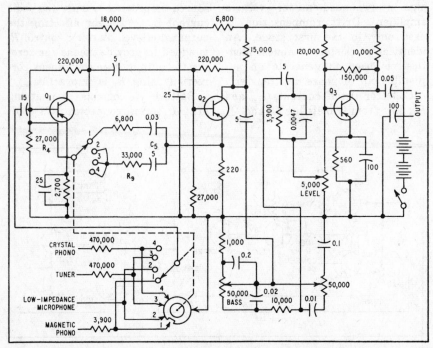

FIG. 1—Preamplifier uses controlled amounts of negative feedback to adapt to a range of low-level input impedances automatically

D-C AMPLIFIER WITH HIGH-IMPEDANCE INPUT

By DONALD SCHUSTER

Psychological Services, Inc., Los Angeles, Calif.

Amplifier circuit uses double emitter follower and grounded emitter voltage amplifier to obtain input impedance of 0.4 megohm. Adjustable temperature compensation in first stage gives good short term drift stability. Overall current gain is 1,000 and voltage gain is 40. Application is in photocell circuits where signal voltages are not large and output impedance is high

USE OF DOUBLE EMITTER followers or two cascaded grounded-collector stages to achieve an input impedance between one-half and one megohm in a-c transistor amplifiers is fairly well-known.[1] For d-c amplifiers, however, high input impedance is a problem since there is the added requirement of temperature drift compensation. The circuit discussed here maintains good short term temperature stability while maintaining high current gain.

Drift Compensation

Collector current variation with temperature is the major source of drift in a transistorized d-c amplifier. Another source of drift results from changes in current gain.

The effects of drift can be minimized, but not completely eliminated by appropriate bias stabilization or by negative voltage or current feedback. In practice, bias stabilization is difficult to maintain for d-c signals and negative voltage feedback is conducive to decreasing the input impedance. Since negative current feedback is practically feasible and increases the input impedance, it is used to compensate for drift in the circuit described here.

When using a double emitter follower minimization of the effects of drift with negative current feedback is insufficient. It is also necessary to compensate for or balance out drift resulting from the large

overall current gain in the stage.

A practical solution to the problem is to make the compensation adjustable. This is accomplished with a circuit in which one transistor is used as a drift current generator to compensate another transistor of the same type and a potentiometer is used for fine adjustment of the compensation current in a two-stage grounded-emitter amplifier.[2] A variation of this compensation method is the one employed in this amplifier.

The amplifier shown in Fig. 1A consists of a double emitter follower plus a grounded-emitter voltage amplifier. Drift compensation is used only in the first stage. An identical transistor, or almost identical, is used to generate an adjustable out of phase drift current.

Collector cutoff currents I_{co1} and I_{co2} flow from transistors Q_1 and Q_2,

respectively. Compensation control R_1 has no effect on the signal when its wiper arm is near the end connected to the emitter of Q_1. As the arm is moved towards the end connected to the base of Q_1, compensation current I_{co2} is multiplied by the current gain of Q_1 which is expressed as $\beta + 1$. In practice, Q_2 is selected to give a compensation current I_{co2} which is somewhat less than I_{co2}.

Readjustment of R_1 is necessary each time the voltage developed across bias control R_2 changes.

Any variation in the driving source impedance varies the bias current in Q_1 thereby affecting the output d-c level. Zeroing control R_3 is added to provide means for zeroing the output since continuous correction using R_1 is impractical.

Battery E_1 consists of a string of five series connected RM-400R

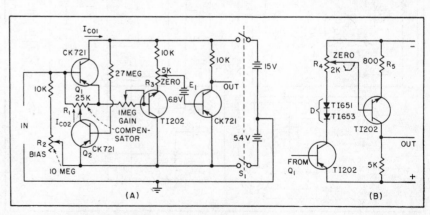

FIG. 1—Complete high-impedance d-c transistor amplifier circuit (A) has battery coupled output stage. Alternate output circuit (B) eliminates battery through use of Zener diode

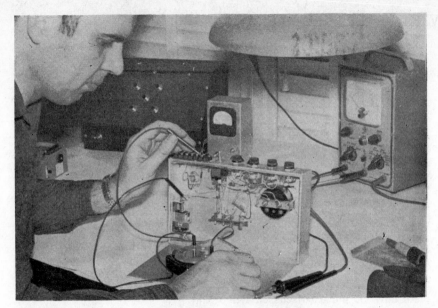

Applying signal from cadmium sulphide photocell to high-impedance d-c transistor amplifier. Output is used to operate timing circuit relay

mercury cells. When power switch S_1 is open, the current flow through E_1 is approximately three microamperes; when S_1 is closed, the current reverses and battery E_1 recharges while the amplifier is in use. If the amplifier is continuously inoperative, it takes several thousand hours to discharge E_1 completely since an RM-400R has a rated capacity of 80 milliampere hours.

An alternate output arrangement that operates without battery E_1 is shown in Fig. 1B. Additional signal losses occur in this circuit which result from the possible variable voltage drop across Zener diode D, the voltage divider action of zero control R_4 and resistance of emitter resistor R_5. To increase gain, replace R_5 with an appropriate Zener diode.

Adjustment

Since all controls interact to some extent, they should be adjusted after the preliminary zeroing procedure to establish correct current gain. Gain, zero, bias, and compensator controls are set in that order.

Zeroing of the amplifier is accomplished in two steps. First, the input terminals are shorted with a 10,000-ohm resistor and the output voltage adjusted to zero with the zero control. Second, the input short is removed and the output voltage readjusted to zero with the bias control. This zeroing procedure assures a zero output for a zero input at the desired operating characteristics.

Amplifier Characteristics

Amplifier sensitivity is sufficient to use a current change of 0.01 microampere as an input signal. Maximum usable output of approximately 1 milliampere is obtained at 0.1 microampere change.

Overall maximum current gain is 10,000 and the voltage gain is 40. Frequency response is nearly flat from d-c to 15 kc where it is down 3 db.

Direct measurement of input resistance is impractical since an ohmmeter overloads the voltage ampliying capabilities of the amplifier. Therefore, measurement is made by applying a known reference voltage across a variable resistor connected to the input terminals.

Initially the variable resistor is set at zero, and the reference volt-age gain at the output terminals measured. Then the variable resistor is advanced until the reference voltage gain becomes half its former value. The amount of ohmic increase represented by the change in the variable resistor setting is equal to the amplifier input resistance.

Using this method, the a-c input impedance for the amplifier was found to be 0.4 megohm. A dynamic check using an R-C method gave essentially the same result.

Short term temperature stability of the amplifier was measured over a fifteen minute interval and found to be five percent. Better drift characteristics are obtainable if both the second and third stage are temperature compensated in a manner similar to that used in the first stage.

Applications

This amplifier is particularly useful whenever it is used with a driving source that develops moderate voltages at high output impedances. Two typical drivers are bioelectric phenomena and solid-state photocells.

Measurements of camera shutter speeds were made using the amplifier to increase the output signal from a cadmium sulphide photocell sufficiently to operate a timing relay circuit. Since the dark resistance of the photocell is approximately 1,000 megohms and the light resistance was between 0.1 and 1 megohm. The amplifier developed an output only when the shutter opened.

For some applications, higher input impedance can be obtained by adding a resistor in series with the input. If this is done, appreciable reduction of current and voltage gains result.

REFERENCES

(1) A. Coblenz and H. L. Owens, "Transistors, Theory and Applications", McGraw-Hill Book Company, Inc., New York, 1955.

(2) J. W. Stanton, Transistorized DC Amplifiers, *Trans IRE PGCT*, CT-3, Mar. 1956.

LOW POWER DRIVES POWER AMPLIFIER

By I. DLUGATCH

Senior Engineer, Hycon Mfg. Co., Pasadena, Calif.

INCREASED use of semiconductor devices has resulted in attempts at driving an r-f power amplifier with a transistor. No problem exists normally except where large power outputs are sought. Use of the transistor implies an intended conservation of prime power, which requires high efficiency in the power-amplifier stage. The latter requires high power sensitivity if there is to be any possibility of success for the design.

Plate efficiencies in the order of 80 percent can be achieved without

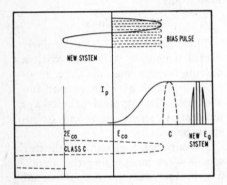

FIG. 1—Comparison of class C and new system

difficulty in a class C amplifier but higher efficiencies are desirable. The chief disadvantage of the class C amplifier is the necessity for high levels of driving power to achieve high efficiency. This is indicated by the dotted curves of Fig. 1. The transistor as a driver is generally limited by low power output particularly at frequencies above 5 mc.

Present day transistors are usable for this application by means of the system described below. The scheme involves negatively modulating the p-a grid with a frequency several times higher than that being amplified, as shown in block diagram form in Fig. 2.

This method attains the goal of reduction in driving power and increases the efficiency of the p-a stage. Evaluation of the scheme is simplified by making several as-

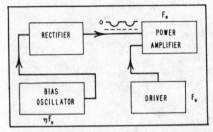

FIG. 2—Negative modulation of grid with high frequency

sumptions. First, assume that all plate current pulses discussed will have the shape of half a sine wave. Second, in the class C operation used as a basis of comparison, assume that its bias is twice cutoff so the result operating angle is 120 degrees.

In the suggested system of Fig. 2, a half-wave rectifier changes the output of the bias oscillator to produce a negative, half-sine wave pulse of sufficient amplitude to cut off the p-a. This bias pulse appears at the grid of the power amplifier with a repetition rate of n times the frequency of the driver stage. It permits the p-a to conduct during only half the time it would normally conduct.

Assume the p-a to be biased at cutoff and plate current pulse amplitude maintained at the same level as for the class C system as shown by the solid lines of Fig. 1. The mean effective operating angle is now 90 degrees, mean referring to variations with phase shifts.

If a pentode is used for the p-a, plate supply voltage can be increased to maintain the same d-c input as for the class C system without negating previous assumptions.

That is, 15 percent more power output can be realized.

The higher efficiency now possible, though extremely desirable, is not the most significant factor favoring this plan. Class C amplifiers are theoretically capable of 100-percent efficiency and can achieve practical values of 90 percent. The additional prime power

to operate the bias signal source may further decrease the advantage.

▶ **Driving Power**—The real merit of the proposed system is in the reduction in driving power required to obtain such high plate efficiencies. This is demonstrated in Fig. 1. With this method it is now possible to utilize transistors as drivers and oscillators where their power capabilities previously prevented it.

Audio modulation at the grid of a power amplifier is an obvious comparison. Such modulation will increase the efficiency of the stage just as the negative bias described above but this is only in terms of the modulated wave. The carrier efficiency is actually reduced by a factor of two. That is, twice as much driving power is necessary as with the unmodulated class C stage.

The proposed system does not preclude the modulation of the p-a by any of the known modulation methods.

The bias oscillator, since it is not required to provide substantial power, can be a transistor, if frequency limitations permit, to minimize its prime power requirements. High stability is not vital except

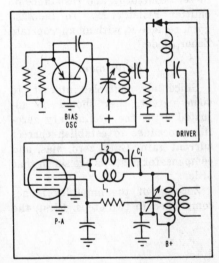

FIG. 3—Typical circuit shows the new system

that rapid variations in frequency are to be avoided to prevent possible audio-frequency modulation. Deliberate sweeping of this oscillator could provide tone modulation if so desired. In some cases, the oscillator is not necessary because a high-frequency signal is available elsewhere in the equipment.

No particular value of n is preferred but the same precautions are to be observed that would apply in any mixer application where beats may cause difficulties.

Excessive sideband generation might reduce the efficiency to a point at which this method would

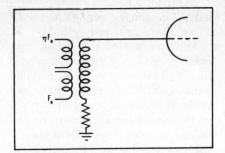

FIG. 4—Alternate circuit that eliminates rectifier

be impractical. Therefore, the Q of the p-a tank circuit must be high to increase the rejection. The use of harmonic filters is definitely

recommended as a preventive measure. The choice of as high a value for n as possible will reduce this disadvantage.

Figure 3 is a typical circuit for the system described. Components L_1, L_2 and C comprise a harmonic suppression filter. Applications are not limited to transistors, alone, either for the bias oscillator or the driver.

Figure 4 is an alternate scheme eliminating the rectifier. Other circuits will suggest themselves for applying this idea toward reduction of the operating angle without increasing the drive.

HIGH-FREQUENCY COMPLEMENTARY SYMMETRY

By YASUO TARUI

Electrotechnical Laboratory, Tokyo, Japan

FOR measurement of $r_{bb'}$, which is one of the most important high-frequency figures of merit of transistors, Giacoletto proposed a multifrequency bridge for which the schematic circuit is shown in Fig. 1 (see also ELECTRONICS, p 144, Nov. 1953). Since one of the output terminals of the square-wave generator is grounded in this case, a differential oscilloscope is required for the detector. If, however an ideal transformer that will pass square waves without distortion is obtained, the circuit can be altered to a normal bridge arrangement as shown in Fig. 2.

Complementary symmetry of high frequency pnp and npn transistors can be used for this purpose. The circuit of the bridge is shown schematically in Fig. 3. Analysis of the equivalent circuit shown in Fig. 4 shows that the

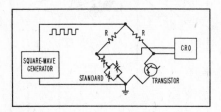

FIG. 1—Giacoletto's circuit

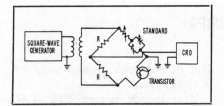

FIG. 2—Circuit with an ideal transformer

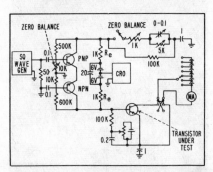

FIG. 3—Circuit with transistor complementary symmetry

potentials at A and B, denoted by V_1 and V_2 respectively, are approximately given as follows

$$V_1 = -Z_{c1} \quad A \; (1-a_1)$$
$$V_2 = Z_{c2} \quad A \; (1-a_2)$$

where

$$A = \frac{i_2}{i_1} = \frac{a_1 Z_{c1} + a_2 Z_{c2}}{Z_{c1}(1-a_1) + Z_{c2}(1-a_2)}$$

Therefore the important parameters that must be considered for

selection of transistors are Z_c, a, f_{ac}.

With high frequency pnp and npn transistors ($f_{ac} \approx 10$ mc) an

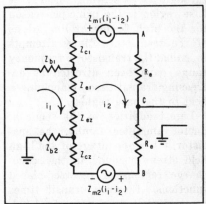

FIG. 4—Equivalent circuit

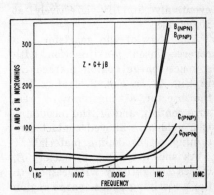

FIG. 5—Output impedances of two transistors

experimental set was constructed. Representative pulse figures at respective points are shown in the photographs.

On null condition the point C is ground potential. Hence, no error is introduced from the transistor output impedances, provided they are well balanced. Figure 5 shows the measured output conductances and susceptances of A to ground and B to ground with exclusion of R_e. The differences of those two

halves is almost negligible up to 1 mc in comparison with $1/1\ k$.

The accuracy of this bridge was checked with a dummy transistor like that shown in Fig. 6. The dummy, with accurately measured value of r_{bb}', was inserted in place of the transistor specimen and the bridge was balanced. Examples of the inserted r_{bb}' and measured r_{bb}' are given in Table I. Fairly good agreement can be seen.

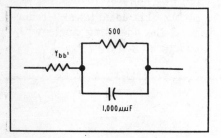

FIG. 6—Elements of a dummy transistor

Table I—Accuracy Check with Dummy

Dummy r_{bb}'	Measured r_{bb}'	Error in prtcent
99.7	100.4	+0.7
199.9	199.5	−0.2
300.6	299.0	−0.53
400.0	399.0	−0.25
500.0	498.0	−0.5
599.2	596.5	−0.52
699.2	695.0	−0.6
799.0	794.3	−0.6
901.0	894.0	+0.78
1001.0	992.0	−0.9

LOW-TRANSIT-TIME AMPLIFIER

SLOW DIFFUSION of charge carriers through an essentially field free base region is the principle reason for the high frequency limitations of the transistor. Many attempts to extend the transistor's frequency range have been directed at decreasing transit time by applying a field to the base region.

Latest addition to the semiconductor amplifier family, the spacistor, takes advantage of the high field strength found in the space-charge regions of reversed-biased junctions. Electron transit times are such that Raytheon, who developed the spacistor, predicts it will eventually amplify effectively at 10,000 mc.

The body of the spacistor is a reverse-biased p-n junction with a space-charge region marked SC in Fig. 1. In Fig. 2, contact I is the injector, a tungsten-wire pressure contact, and M, the modulator is a gold-wire alloyed contact containing p-type doping material.

As shown in Fig. 2, battery B_1 biases I negatively with respect to the underlying space-charge region SC. Contact I is still positive, however, with respect to point B. Emis-

sion of electrons from I into SC is space charge limited.

Modulator M is connected to SC

Extreme closeup of experimental spacistor assembly on transistor mount

between I and the N region of the body. Battery B_2 biases M negatively with respect to SC preventing holes from flowing from the p-type doping materials to SC. As a result, M draws practically no current.

The field produced by M affects the entire space charge region,

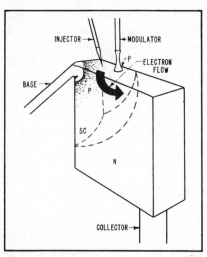

Fig. 1—Construction of present experimental spacistors

varying the emission of the injector I and thereby modulating the d-c bias with the input signal. The modulator also makes the injector bias practically independent of the base-to-collector voltage. As a result, the output impedance is greater than 30 megohms for an injected current of 0.3 ma.

Because of the wide space-charge region, the output capacitance is very small. Values less than 1 $\mu\mu f$ are feasible.

Present experimental spacistors have transconductances considerably smaller than those of good vacuum tubes. Nevertheless, they are expected to operate at over 1,000 mc. This frequency is equivalent to the inverse transit time through the space-charge regions.

Operation of the spacistor is practically independent of charge carrier lifetime. This makes it feasible to supplement germanium and silicon with other semiconductors whose short charge carrier lifetime makes them unsuitable materials for transistors. Raytheon

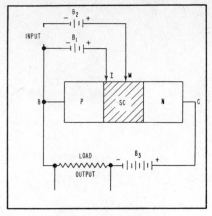

Fig. 2—Basic spacistor circuit

expects that silicon-carbide spacistors will operate up to 500 C.

The output and input circuits of the spacistor are not coupled by internal feedback, as in the case of the transistor. This, as a result, makes the spacistor well-suited for use in multistage amplifiers.

Chapter 3

DESIGN OF TRANSISTOR OSCILLATORS

COLPITTS OSCILLATOR SUPPLIES STABLE SIGNAL

By LEON H. DULBERGER
Project Engineer, Fischer & Porter Co., Hatboro, Pa.

Colpitts circuit, employing one germanium transistor and one Zener diode, operates from a laboratory regulated power supply to maintain a sine-wave voltage of precise amplitude

OPERATING AT A FREQUENCY of 10 kc, the sine-wave oscillator to be described provides 0.5-rms output at 15-ohms impedance. Amplitude stability is 0.1 percent over the temperature range of 30 C to 50 C. Checks indicate a drift of frequency under 0.25 percent.

Circuit

The Colpitts configuration, Fig. 1, operates from a 1N429 Zener reference diode rated at 0.01 percent per degree C temperature coefficient. A laboratory power supply, regulated to 0.25 percent for line and load, provides preliminary voltage control. This can be replaced by a semiconductor regulator.

The adjustable tank coil is coarse-tuned by C_1 and C_2, which also provide impedance match to the emitter. Fine frequency adjustment is obtained by the position of the core within the coil. A low L-C ratio, which allows high emitter current, provides a large voltage across R_2.

Some distortion is generated in the collector circuit, in addition to the desired positive peak clipping when providing a high signal level to the load. The output signal, taken from the emitter, is an undistorted sine wave as shown in the photograph. For a fixed supply voltage, R_1 and R_2 are adjusted for peak-to-peak collector swing just under twice the supply voltage. The final base-to-emitter bias current produces about 2 percent limiting of the peak collector swing. Final adjustment produces the collector waveform shown.

Capacitor C_3 grounds the base at the operating frequency. Under these conditions, the collector to base voltage varies directly with the collector supply voltage.

Transistor Operation

The 2N270 transistor operates at 7.6-ma collector current and 6.3-v d c supply. Best amplitude stability is obtained with 11.6 v peak-to-peak measured at the collector. At this point, output impedance is high and unsuited to driving a load. However, a 10,000-ohm load is easily driven from the emitter at 0.5-v rms without sacrificing performance.

Current-adjustment resistor R_3 with R_4, sets the idling current through the diode. For the 1N429 used, optimum regulation is obtained at 7.5 ma. Capacitor C_4 bypasses the a-c signal around the diode to provide low signal impedance.

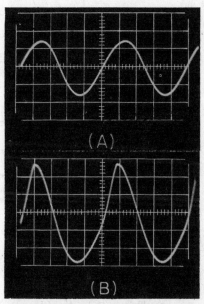

Emitter waveform (A) vertical scale: 0.5-v per cm; horizontal scale: 20-μsec per cm and Collector waveform (B) vertical scale: 2-v per cm; horizontal scale; 20-μsec per cm. Output is undistorted

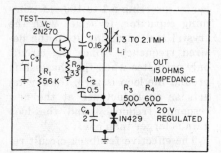

FIG. 1—Constant amplitude transistor oscillator, used as a stable amplitude carrier for a data-reduction system, replaces equipment using several electron tubes operated from a regulated power supply

Compact design of packaged oscillator is suited for minimum-space requirements

CRYSTAL OSCILLATOR WITH VARIABLE FREQUENCY

By G. A. GEDNEY and G. M. DAVIDSON

Arma Division, American Bosch Arma Corporation, Garden City, New York

Two-stage crystal feedback amplifier operates at 9.1 kc with long-term frequency stability of a few parts per million. Operating frequency can be pulled off resonance by adjustment of trimmer capacitor in series with crystal. Applications include constant-frequency power source for gyroscopic precision integrator and regulated carrier frequency for electrical resolvers, industrial generators and phase computing networks

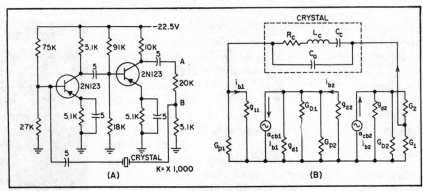

FIG. 1—Two-stage amplifier with a crystal load generates stable fixed-frequency oscillations (A); equivalent circuit (B) simplifies loop gain calculation

CONVENIENT CALIBRATION adjustment may be provided in some analog and digital systems by deviating the carrier frequency by a small but accurately known amount. This article describes the design of a fixed-frequency oscillator with long-term frequency stability of a few parts per million and provision for manually varying the output frequency a few cps in calibrated increments.

Design

Requirements of a low frequency of 9.1 kc with high stability necessitated the use of a special high-Q crystal with low variation in nominal frequency over a fairly wide temperature range.

Current degeneration type biasing compensates for the effects of leakage currents flowing from collector to base. The bias stabilization network yields a factor of two to limit the fluctuation in output voltage to less than ±5 percent over the range of −65 C to +65 C.

The basic circuit for fixed-frequency operation in shown in Fig. 1A. From the equivalent circuit of Fig. 1B, when

$$g_{22} \gg g_{d1} + G_{D1} + G_{P2}, \text{ and } g_{11} \gg G_{P1},$$

the loop current gain of the amplifier is

$$K_i \cong \left(\frac{a_{cb1}\, a_{cb2}\, G_{L2}}{g_{d2} + G_{D2} + G_{L2}} \right) \left(\frac{G_{crys}}{G_1 + G_{crys}} \right)$$

where G_D = d-c collector load conductance, g_{11} and g_{22} = a-c input conductances of the transistors, g_d = transistor output conductance, G_p = parallel conductance of the biasing network, G_{L2} = total conductance of the attenuation network and G_1 = conductance of one leg of attenuating network.

Current gain of the amplifier varies from 20 to 500 at crystal resonance depending on transistor β.

Output amplitude of the oscillator is limited only by the nonlinearities in the loop. Since the 2N123 has a Zener breakdown of 16 v that varies with age, temperature and

unit, provision must be made to prevent an output rise to 16 v. Also the maximum allowable voltage across the crystal is 1 v. The circuit can be designed to limit the output to any desired value lower than the Zener value by shaping the d-c and a-c load lines. The relatively high collector supply voltage is needed for bias stabilization.

Crystal Pulled Off Resonance

Oscillation frequency may be varied by inserting a 1 to 100-μμf trimming capacitor, in series with the crystal and adjusting it to the desired frequency. Since the circuit only oscillates when the phase shift around the loop is zero, the crystal acts as an inductance at the new resonant frequency and the loop gain is larger than one.

The effective feedback circuit resistance increases as the square of the change in frequency from resonance thereby increasing the attenuation of the feedback circuit and requiring a larger open-loop gain.

Engineer records observations of 9.1-kc oscillator performance

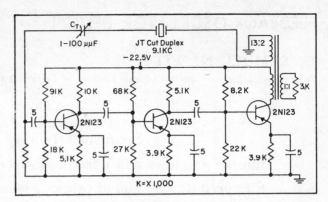

FIG. 3—Three-stage crystal oscillator provides greater power output

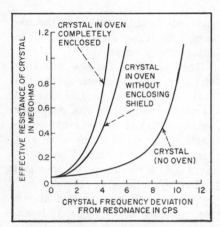

FIG. 2—Crystal resistance as function of deviation from resonant frequency of 9,100.6 cps with crystal partially or entirely enclosed within oven

The effective crystal resistance may be found from R_e where

$$R_e \cong R/[1 - (4\pi L\Delta f / X_{co})]^2$$

and effective reactance from

$$X_e \cong 4\pi L\Delta f/[1 - (4\pi L\Delta f / X_{co})]$$

where R, L and X_{co} are crystal constants at series resonance valid when

$$f_{ser} < f_{ser+\Delta f} < f_{antires}.$$

Crystal Parameters

To measure effective resistance R_e and reactance X_e of the crystal as a function of frequency deviation Δf from crystal resonance. A known value of capacitance is inserted in series with the crystal and the oscillator frequency is varied to ob-tain maximum transfer. Frequency of the oscillator is measured at maximum voltage transfer.

Measurements were performed with the crystal partially and completely oven enclosed for several values of capacitance inserted in series with the crystal. Curve 1 of Fig. 2 shows a rapid increase in feedback circuit resistance as frequency deviates from resonance indicating that the gain of the amplifier must be high to obtain a moderate frequency pull.

When the circuit requires a minimized frequency drift, the crystal should be enclosed in a constant-temperature oven to minimize temperature variation effects. This increases the parallel capacitance across the crystal and also the attenuation of the circuit as a function of frequency deviation. Therefore, for the same gain in the forward loop the crystal cannot be pulled as much. The increased gain required is illustrated in Fig. 3.

Amplifier Requirements

The amplifier must provide sufficient gain to insure a loop transmission of one, under maximum frequency pull off and yield a net phase shift around the loop of 360 deg with the crystal inserted in the circuit. Since the desired circuit frequency is a variable, each amplifier must be trimmed individually. The current flowing into the base of stage 1 is proportional to the output voltage divided by the crystal impedance. With the output voltage constant the base current swing at resonance is approximately 40 to 60 times the base current swing for a 5-cps pull off resonance.

Though all the amplifier stages tend to saturate, excessive saturation of the first stage is undesirable. The design must allow for this swing or the input current to the first stage must be attenuated by insertion of a trim resistor in series with the crystal. Value of the trim resistor is determined by the required deviation. At resonance where the loop gain is maximum the desired attenuation would be large; zero attenuation is necessary when the circuit is more than 5 cps off resonance.

The circuit of Fig. 1A was developed to yield a maximum pull off resonance of approximately 5 cps.

Impedance the external load must be about 0.3 megohm or more at point A in Fig. 1A. If the load is taken from point B, the combination of load and attenuating resistor should be equal to 5,000 ohms.

Greater Deviation

To obtain a larger frequency deviations off resonance or larger power output, a higher circuit gain is necessary. The circuit of Fig. 3 uses a transformer to obtain a phase reversal and to reflect the desired a-c load line to limit the output swing of the transistor.

31

FEEDBACK-OSCILLATOR DESIGN

By SAMUEL N. WITT, JR.

Engineering Experiment Station, Georgia Institute of Technology, Atlanta, Georgia

Frequency stability of oscillator types that can be divided into an amplifier and a feedback network is improved by design procedure that permits selection of proper geometry and elements of frequency-controlling network. Examples of oscillator circuits using transistors and designed by this method show increased stability by as much as a factor of 25

PAST APPROACHES in the design of vacuum tube and transistor oscillators have had as the principal object improved frequency and amplitude stability and/or greater output amplitude. The purpose of this article is to show how the frequency stability of most oscillators can be improved after a general oscillator type has been chosen. This method also provides for minimum harmonic distortion in the oscillator output.

This design procedure is applicable only to oscillator types which can be divided into two sections, an amplifier and a feedback network.

Stability Factors

For an oscillator with low distortion, two primary factors affect the ultimate frequency stability. The first is the change in phase shift in the amplifier portion of the oscillator when external parameters such as voltage and temperature are changed. The second is the frequency change required for equal and opposite phase shift change in feedback loop. These factors are interrelated because the total oscilla-tor loop phase shift must be zero.

The block diagram of a feedback oscillator is shown in Fig. 1A. The amplifier is considered as its equivalent Thevenin generator with voltage $K\epsilon \angle \theta$ in series with a resistance, r_o, where r_o is considered a part of the feedback network. The input impedance of the amplifier is infinite since any reactive component may be canceled at any one frequency and voltage. The resistive component may be included as a part of R. Phase angle θ is a function of all external parameters such as temperature and voltage. It is assumed that θ has been made zero for the normal static condition of the external parameters. Constant K is a real, positive constant so no distortion of the waveform occurs.

Amplitude Limiting

If the frequency-controlling network is linear, the oscillator output will be a pure sinusoidal waveform. In most actual oscillators some distortion must occur so that amplitude limiting will result. However, by careful selecting the gain k of the feedback network this distor-tion can be kept small so the assumption of no distortion will be substantially correct.

The primary factors affecting stability now becomes $\phi(f)$ and $\theta(s)$ where s is a general external parameter affecting the amplifier phase shift and $\phi(f)$ is the phase angle of the feedback network as a function of frequency. The objectives are to make $d\theta/ds$ as small and $d\theta/df$ as large as possible.

Several methods are well known for reducing $d\theta/ds$, thus, it is assumed that $d\theta/ds$ for the amplifier has already been made as small as possible.

Phase Angle

The factors influencing $\phi(f)$ are the geometry and element values of the frequency-controlling network. The choice of network geometry is limited only by practical values of K and r_o. For most amplifiers, r_o is much too large for use with practical feedback networks. Therefore, the use of transformers or coupling coils is usually necessary to obtain the desired impedances. If the values of K and r_o for the amplifier alone are compared with the values for the amplifier with the transformer, the quantity K^2/r_o will be found to remain constant. This is shown in Fig. 1B where K' and r_o' are the characteristics of the amplifier without the transformer. Either K or r_o may be selected for best performance with any particu-for feedback network, however, K^2/r_o must be held constant.

For a practical transformer-cou-

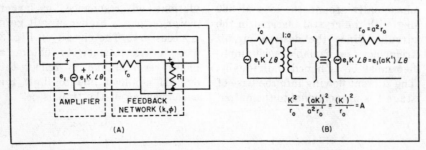

FIG. 1—Equivalennt Thevenin circuit of feedback oscillator (A) is modied in (B) to show that the quantity K^2/r_o remains constant

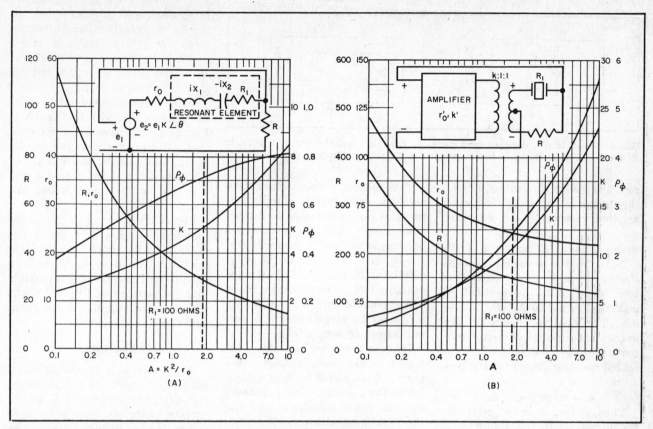

FIG. 2—Curves show the variation of circuit parameters for two different network configurations. Circuit and curves at right show what happens when the effective crystal Q is greater than its actual measured value

pled amplifier the transformer loss and phase shift should be included as part of the characteristics, $K \angle \theta$ and r_o. This is accomplished by measuring the amplifier characteristics with a representative transformer in use. The amplifier and transformer combination are then completely characterized in terms of $\theta(s)$ and $K^2/r_o = A =$ four times the available power output of the amplifier for one-volt input. Once A is known, the remaining problems are the selection of a frequency-controlling network configuration and the optimization of this network for best stability.

Network Selection

To optimize network values, consider the network shown in Fig. 2A. Here, the resonant element may be a series resonant L-C circuit or more appropriately, a quartz crystal. Any infinite amplifier input resistance may be considered as part of R. Amplifier parameters $K \angle \theta$ and r_o are both variable with the restriction that K^2/r_o is a constant.

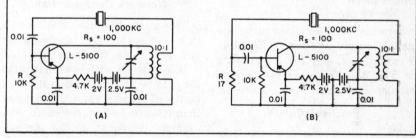

FIG. 3—Standard oscillator using surface-barrier transistor (A) has an instability of 100 parts in 10^9. Circuit in (B), designed for impedance match between input and output of transistor, has stability of 13 parts in 10^9.

For this network

$$jX = jX_1 - jX_2 \qquad (1)$$

Then,

$$e_1/e_2 = k \angle \theta = \qquad (2)$$
$$k \angle \tan^{-1} -X/(R + r_o + R_1)$$

Equation 2 shows that near series resonance $d\phi/df$ increases as $(R + r_o)$ decreases. The object is to find the minimum value of $(R + r_o)$ that will permit oscillations thus making $d\theta/df$ as large as possible.

For $X = 0$,

$$R = (r_o + R_1)/(r_o^{1/2} A^{1/2} - 1) \qquad (3)$$

from which

$$R + r_o = [(r_o + R_1)/(r_o^{1/2} A^{1/2} - 1)] + r_o. \qquad (4)$$

The quantity r_o may be selected to make $(R + r_o)$ minimum by dif-

ferentiating Eq. 4 with respect to r_o. Thus $(R + r_o)$ is minimum when

$$2 r_o^{3/2} A^{1/2} - 3r_o - R_1 = 0. \qquad (5)$$

By substituting R_1 as determined from the resonant element chosen for the oscillator and A as determined by the amplifier into Eq. 5, r_o can be found. The curves of Fig. 2A show the variation of r_o with A as aparameter and with R_1 constant at 100 ohms. For each value of r_o and A constant K is determined from $K^2/r_o = A$. Resistance R is found from Eq. 3. The variations of K and R for $R_1 = 100$ ohms are also shown in Fig. 2A.

If r_o and R could both be made

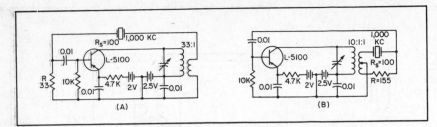

FIG. 4—Circuit in (A), designed from curves in Fig. 2A has instability of 8 parts in 10^9 while that in (B), designed from curves in Fig. 2B, has stability of 4 parts in 10^9

zero and still maintain oscillations, the change in frequency for a small amplifier phase shift, $\Delta\theta$, would be:

$$\Delta f = (\Delta\theta)\,(f_o)/2Q \qquad (6)$$

where Q is X_1/R_1 and f_o is $1/(2\pi \sqrt{LC})$ for the resonant element. For values of r_o and R other than zero, Eq. 6 must be modified by making Q equal to $X_1/(R_1 + R + r_o)$. The relative change in frequency for a practical circuit compared to the change for an unloaded resonant circuit for the same $\Delta\theta$ is

$$\rho_\phi = R_1/(R + r_o + R_1) \qquad (7)$$

Thus, ρ_ϕ is a figure of merit for the network. If the values of ρ_ϕ are compared for two different networks, then the relative stabilities, $(S_v)_1$ and $(S_v)_2$, of oscillators using the two networks can be determined since

$$(S_v)_1/(S_v)_2 = (\rho_\phi)_1/(\rho_\phi)_2 \qquad (8)$$

where the same amplifier and resonant element are used. Equation 7 expresses ρ_ϕ only for the network of Fig. 2A. For other networks, a similar expression for ρ_ϕ is needed.

Practical Circuit

To design a practical circuit to use the network of Fig. 2A, a suitable amplifier is first chosen. Parameter A is found experimentally. A suitable resonant element is chosen (crystal, series L-C circuit, or other). Resistor R_1 is determined from the resonant element. The correct value of r_o is then found from Eq. 5 and resistance R can then be found from Eq. 3. Reactance X as a function of frequency can be calculated for the resonant element. Then, ϕ as a function of frequency can be found from Eq. 2. If the amplifier phase shift θ is known as a function of voltage or temperature, then the frequency of oscillation as a function of voltage can be found by making ϕ equal to $-\theta$ for all voltages.

This method yields a theoretical stability which may be slightly better than can be obtained in a practical circuit since distortion has been neglected and R generally cannot be made as small as the value calculated in Eq. 3. If R is made exactly equal to the minimum value, when a phase shift occurs R may no longer be large enough to permit sustained oscillations.

Improvement in stability obtained by externally unloading the crystal R is found by first substituting the amplifier input resistance into Eq. 2 and calculating the stability then substituting R (minimum) into Eq. 2 and again calculating the stability. The ratio of the two stabilities is the improvement due to proper selection of R.

Network Configuration

This method is also valuable as a tool in choosing the constants for any particular feedback network. It is possible with some networks for ρ_ϕ to be greater than unity, that is, the effective crystal Q is greater than its actual measured value. For example, the equations for the network shown in Fig. 2B are

$$K = 1 + \sqrt{1 + R_1 A/2}, \qquad (9)$$
$$R = [4K^2 + R_1 A(K+1)]/A(K-1) \qquad (10)$$
and
$$\rho_\phi = 200(R + 2r_o)/ [(R - R_1)(R + R_1 + 4r_o) - 1] \qquad (11)$$

The curves of Fig. 2B show the variations of R, K, r_o and ρ_ϕ for values of A from 0.1 to 10. Resistance R_1 is again chosen as 100 ohms. In designing oscillators the curves of Fig. 2B are used in the same way as the curves of Fig. 2A.

The importance of R_1 is not directly indicated by the equations for either network described. However, it can be shown that when the inductance and capacitance of the resonator remain unchanged, the frequency stability is increased as

R_1 is decreased. For example, using Eq. 9, 10 and 11 and a typical value for A of 2 results in values for $R_1 = 1000$ ohms, $\rho_\phi = 0.795$, for $R_1 = 100$ ohms, $\rho_\phi = 2.5$ and for $R_1 = 10$ ohms, $\rho_\phi = 7.1$.

When a quartz crystal is used as the resonator with the circuit of Fig. 2B an effective Q greater than that of the crystal alone can usually be expected. With the circuit of Fig. 2A, ρ_ϕ can never exceed unity, that is, no Q multiplcation can occur.

Figure 2B makes no provision for the inclusion of a finite input impedance to the amplifier. For this condition, equations 9, 10, and 11 must be modified. Thus it is possible that the network of Fig. 2A might yield greater stability for a given amplifier and resonator.

Examples

A total of four transistor oscillators were constructed to illustrate this design procedure. An L-5100 surface barrier transistor was chosen for the amplifier because of its low power requirements and relatively high alpha cutoff frequency. The oscillator shown in Fig. 3A received no special design considerations. The component values were selected empirically and the selections were then modified as necessary to obtain oscillation.

The circuit shown in Fig. 3B was designed on the basis of an impedance match between the output and input terminals of the transistor. The only change in Fig. 3A required for this design was to replace R with a 17-ohm resistor.

The circuit in Fig. 4A was designed from the curves of Fig. 2. This required a change in the transformer turns ratio and also in the crystal terminating resistor as shown. The circuit shown in Fig. 4B was patterned after the block diagram in Fig. 2 and was designed using the curves of that figure The instability characteristics of the oscillators were 100, 13, 8, and 4 parts in 10^9 frequency change for a one percent collector supply voltage change for the circuits of Fig. 3A, 3B, 4A and Fig. 4B respectively.

This work was supported by the Signal Corps Engineering Laboratories under Contract No. DA-36-039-sc-42712.

SILICON CRYSTAL OSCILLATORS WITH HIGH-TEMPERATURE STABILITY

By E. G. HOMER
Canoga Park, Calif.

USE of silicon transistors for crystal oscillators suggests itself because of their low I_{co} and desirable temperature characteristics. However, silicon requires more initial voltage to start conduction. This characteristic requires circuits of a special nature as shown below.

Additional difficulty has occurred in using low-frequency crystals. Such crystals with little temperature variation at these frequencies are not particularly active. The loses in the low-frequency crystal from internal friction, the elastic restoring forces and the mass are such as to make it difficult to start oscillation.

The Q of a quartz crystal is very

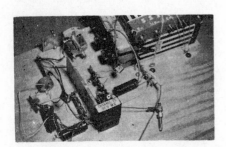

Experimental breadboard setup for testing transistor ocillator circuit

high, which makes it an ideal bandpass filter. This principle was first used as series resonant filtering between an oscillator of somewhat different design and a emitter follower.

Despite the high-frequency cutoff of Texas 904 transistors used, they did not oscillate in circuits used with germanium transistors. A dormant silicon transistor in typical oscillator circuits will have almost no current flowing since I_{co} is small. A first step is to overcome the lack of current flow by the addition of a bias circuit that will allow current to flow in the collector.

In the circuit of Fig. 1, current flowing in the tank circuit comprising choke and capacitor would allow one oscillation during the build up of current. However, the Llewellyn criteria for stability are

now satisfied and the circuit continues to oscillate.

This oscillator is sensitive to load changes. Therefore an emitter follower must be used as the next stage. In addition, this next stage will furnish power and help achieve

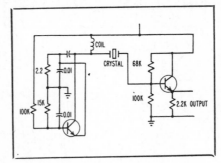

FIG. 1—Initial experimental circuit showing output stage

an impedance match for the following circuit.

The next step is to replace the coupling capacitor with a crystal and resonate the circuit. This oscillator has the advantage that only ordinary parts are required. Any toroid of satisfactory capacitance between turns and of proper core material for the frequency desired can be used. An ordinary i-f transformer or oscillator coil can also be used.

The toroid used for this application was shielded. The coil tested at 0.7 mh. The Q is 45. When this coil was substituted for one of poorer characteristics, the output voltage became 14 volts peak to peak. The oscillator was stable and measured 500,140 cycles.

Additional experiments were carried out to test the function of some of the oscillator components. When the 0.01-μf capacitor across the 2.2 kilohm resistor was removed, the output of the oscillator dropped to 4 volts. When the 0.01-μf capacitor across the 15 kilohm was removed, the oscillator ceased to operate.

Using an ordinary coupling 0.1-μf capacitor instead of a crystal the

oscillator was varied from 500 to 50 kc by the addition of series chokes and parallel capacitors.

Another oscillator employs a type of transformer common in transistor i-f and oscillator circuits that use a center-tapped primary of high impedance and a low-impedance secondary. This oscillator, shown in Fig. 2, combines the previously recommended use of proper bias for current flow in the transistor when the transistor possesses a low I_{co}. In addition, it presents a definite voltage across the crystal. This is necessary as previously explained for lower frequency crystals. It can be further substantiated by checking the relative activity in a Pierce-type electron tube oscillator.

This circuit presents the additional advantage in that oscillator frequency is controlled by the crystal and acts as more than a filter. This effect is substantiated by data taken with variation in temperature. The previous circuit tends to decrease in amplitude with small frequency drifts and would soon cut off altogether if the drift were large. In addition, precautions are taken to avoid stray capacitive coupling from bypassing the crystal and allowing false frequencies to be amplified. None of these difficulties were found in the circuit of Fig. 2.

The method of operating this oscillator is similar to the series-resonant filter. The slug or variable capacitor is adjusted until the

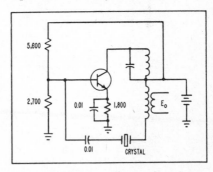

FIG. 2—Feedback oscillator using center-tapped coil

proper frequency is generated for passage through the crystal. In this case, it is more practical to short out the crystal until the oscillator is tuned to nearly the correct frequency.

Then with the crystal in the circuit, slight variations in the tank circuit are enough to allow the crystal to pass the frequency. This circuit thus still maintains the theory that a definite voltage must be present across the crystal, although in this case the voltage would only be instantaneous if the tank were improperly tuned.

The circuit, with proper bias for current flow and feedback through the crystal, assures good starting. In addition to being used with various Miller i-f transformers and oscillator coils and a GE oscillator coil, it was also extended to utilize a medium-power TI 953 transistor.

The data shown in Table I were taken in an effort to discover the effects of temperature on the oscillator. The first readings were taken with the crystal in the circuit, the second with it out. An additional fact was revealed, although not thoroughly substantiated, that the drift in frequency was not caused by the components as much by the variation in parameters of the transistors themselves. This was only checked in one way by changing the capacitor in the tank circuit.

Table I—Oscillator Behavior

Temp	Freq in kc	Output, peak-to-peak V	Current
80	521,945	1.08	5.5 ma
90	521,955	1.08	5.5 ma
100	521,975	1.08	5.6 ma
110	521,968	1.08	5.6 ma
120	521,964	1.08	5.7 ma
130	521,959	1.08	5.8 ma

Without crystal

Temp	Freq in kc	Output, peak-to-peak V	Current
80	521,994	1.22	5.6 ma
90	522,220	1.20	5.6 ma
100	522,823	1.20	5.8 ma
110	523,516	1.20	6 ma
120	524,040	1.18	6 ma
130	525,051	1.16	6 ma

DESIGN OF REGULATED POWER SUPPLIES

By MANFRED LILLIENSTEIN

Development Engineer, Components Division, Federal Telephone and Radio Co.,
Division of International Telephone and Telegraph Corp., Clifton, New Jersey

Design of transistor and diode power supplies for 60 cps provides 100-mv regulation from zero to full load with 2.5 mv ripple. Output is 70 volts at 1.5 amperes. Similar circuit for 400-cps input uses d-c amplifier in feedback loop to control 4 paralleled power-transistor regulating elements providing 150 v at 5 amperes

ADVANTAGES of a transistor power supply are lower weight and volume, the possibility of using printed wiring throughout, long life and no warm-up time.

Typical applications of such a power supply as that described here might be found in computer work, in military airborne applications or in the laboratory.

Operation

The unit, shown in block form in Fig. 1, receives a-c power from the line. This is transformed to the proper voltage and then rectified. A bridge, with a voltage-sensitive element in one leg and linear resistors in the other legs, is placed across the output of the power-supply unit. The bridge is designed to balance at the desired d-c output voltage. The bridge output is amplified and converted to an a-c signal proportional to the difference in actual d-c output and the desired d-c output.

This a-c signal, used as a carrier, is amplified and detected rectified and filtered. Further amplification may be obtained in a d-c amplifier. This carrier then is applied to the power transistor. Thus, any error between the desired d-c output voltage and the actual d-c output voltage is amplified and controls the voltage across the power transistors.

In the schematic diagram, Fig. 2, the a-c voltage is rectified by conventional selenium, germanium or silicon power rectifier and filtered by L_1 and C_1.

The voltage sensing bridge consisting of 3 fixed resistors, (R_2, R_3, R_4) one set of zener diodes and an adjustable resistor R_1 are connected across the power supply load. A pnp transistor Q_2 is placed across the bridge.

Applying a base voltage, slightly lower than emitter voltage, will cause this transistor to produce a virtual short circuit from collector to emitter. Increasing base voltage to near or above emitter voltage will cause a virtual open circuit from collector to emitter. The

voltage swing required to change states is about 0.2 volts.

Collector voltage of transistor Q_2 is essentially a chopped d-c supplied by an auxiliary power supply feeding a transistor oscillator. Thus, the collector voltage of Q_2 changes from nearly zero to twice the rectifier output voltage at oscillator frequency.

The oscillator is a modified Hartley circuit operating at approximately 15 kc. The active element of this oscillator is Q_1. The oscillator tank coil is a miniaturized transistor coupling transformer to which capacitor C_3 has been added for tuning.

Positive terminals of the auxiliary power supply are grounded to the emitter of Q_2 which, in turn, is floating in the middle of the bridge.

If Q_2 is fully conducting, all the r-f and d-c voltage is dropped in resistor R_9 since negligible voltage exists between collector and emitter. The r-f current then goes directly through Q_2 to the positive side of the auxiliary power supply.

In the nonconducting state, the chopped d-c is present on the collector of Q_2 and the a-c component is transferred to the base of Q_3 through transformer T_4. This a-c voltage is dependent on the conduction that Q_2 permits, which in turn, depends on the base-to-emitter voltage of Q_2.

have a ripple content of the carrier frequency but the output must vary with the 120-cycle power-supply ripple. Resistor R_{10} and capacitor C_7 provide a time constant between 120 cycles (twice the power-line frequency) and 30 kc (twice the carrier frequency).

more current. This causes the base-to-emitter voltage of Q_2 to increase in the negative direction. Transistor Q_2 conducts causing the bases of Q_3 and Q_4 to receive less a-c voltage. The output of the detector therefore is decreased and

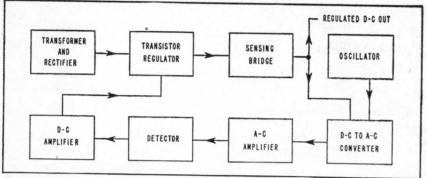

FIG. 1—Power supply regulating system in block form. The d-c amplifier between detector and regulating element is required only when oscillator frequency is above 30 kc.

Table I—Performance of Two Transistor Power Supplies

Input		
Frequency	60 cps	400 cps
No. of phases	1	3
Voltage	115 v ±10%	115 v ±5%
Output		
Voltage	70 v	150 v
Current	1.5 amp	5 amp
Regulation, zero to full load	100 mv	200 mv
Ripple	2.5 mv, rms	6 mv, rms
Maximum internal impedence from 0 to 20,000 cps	0.05 ohms	0.1 ohms

Transistor Q_3 is coupled to Q_4 through transformer T_5, to further amplify the carrier. Resistors and capacitors in the emitter circuits of the a-c amplifier sections limit d-c in the transistors. Capacitors C_4 and C_5 prevent d-c from saturating the coupling transformers.

The output of Q_4 is fed into a rectifier bridge. The d-c output of the rectifier is proportional to the a-c input. The output should not

The d-c voltage out of the rectifier is applied across the regulator element, transistors Q_5 to Q_8 which are placed in parallel to get a higher current output. Resistors in the emitter leads equalize the current through each transistor.

The transistors, in effect, regulate and filter at the same time and a large L-C combination is not required for filtering.

If the output voltage is too high, the zener diodes conduct

less current is fed into the base of the transistors Q_5 to Q_8. Therefore the voltage drop across the power transistors Q_5 to Q_8, between emitters and collectors, increases and lowers the d-c output.

A single-plate selenium rectifier D_1 is placed in the reverse direction across the transistors to protect them against excessive current.

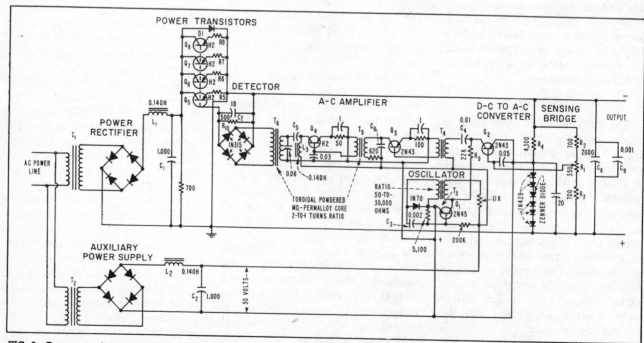

FIG. 2—Power supply uses transistors and diodes to provide 100-mv regulation at 70 volts

400-CPS Power Supplies

Where 400-cycle three-phase power rectification is used, a carrier frequency higher than 15 kc will be required. A circuit has also been designed for this purpose.

The carrier in this case is approximately 100 kc. Since the power transistors will not amplify at this frequency, the detector must be placed after Q_3 and not Q_4 and a separate d-c amplifier used after the detector. A second auxiliary power supply is also required. The 400 cps regulator circuit is shown in Fig. 3.

The detector output controls Q_9. If Q_9 is conducting, nearly all the voltage is dropped across R_{11}, with approximately 0.3 volt appearing across Q_9 from collector to emitter. In this case, the base voltage of Q_{10} approaches a value nearly equal to the positive terminal voltage of the rectifier. Furthermore, the emitter is further biased by the bleeder R_{12} causing the base voltage of Q_{10} to be positive in relation to emitter and thus to cut off Q_{10}.

The same reasoning applies to the manner in which energy is

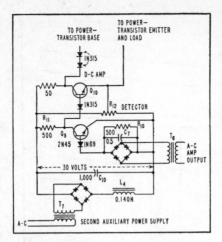

FIG. 3—Detector and d-c amplifier section used with 400-cps power supply

transferred from Q_{10} to the power amplifier except that only the germanium diodes form the biasing arrangement in this case. These diodes are virtually nonconducting unless a voltage of 0.4 volt is placed across each one, at which time practically no further increase in forward voltage results up to the rated forward current. In this arrangement, therefore, the voltage across Q_{10} collector to emitter, must

be at least 0.8 volt (2 diodes) before any change in Q_5 to Q_8 can occur. Increasing the voltage between collector and emitter of Q_{10} beyond 0.8 volt causes base current to be drawn in the power amplifiers. Therefore, the collector-to-emitter voltage of Q_5 decreases and output voltage is raised.

If a higher current output is desired, a larger number of power transistors may be paralleled. As many as 12 transistors have been paralleled.

In cases using a d-c amplifier section, a high output voltage causes the first low-power transistor (a-c to d-c converted) to conduct reducing the signal to the a-c amplifier which reduces the d-c output out of the detector and cuts off Q_9. This, in turn, causes Q_{10} to conduct, lowering the base-to-emitter potential of the series dropping transistors, Q_5 to Q_8, and thus increasing their emitter-to-collector voltage dropping the output voltage to its regulating point. Performance characteristics of both types of power supply are given in Table 1.

REGULATED POWER SUPPLY CIRCUITS

By J. WALTER KELLER, JR.
Project Engineer, Diamond Ordnance Fuze Laboratories, Washington, D. C.

Series and shunt regulator design equations provide method of obtaining low voltage power supplies with any required degree of regulation. Actual circuit derived from these equations are discussed and laboratory tests show good agreement with predicted operation

SIMPLE compact low-voltage transistor power supplies are now entirely feasible using available transistors. Virtually any degree of regulation can be obtained with only a small increase in complexity of the circuit.

The output voltage of a power supply is a function of output current and input voltage

$$E_o = E_o(E_i, I_o) \qquad (1)$$

E_o = output voltage, E_i = input or line voltage, I_o = output or load current. Since in actual practice the load is established by the choice of load resistance, or more conveniently load conductance the output voltage can be written as $E_o = E_o(E_i, G_L)$ where G_L is load conductance.

Then

$$\Delta E_o = \left(\frac{\Delta E_o}{\Delta E_i}\right)_{\Delta G_L = 0} \Delta E_i + \left(\frac{\Delta E_o}{\Delta G_L}\right)_{\Delta E_i = 0} \Delta G_L$$

Multiplying the second term numerator and denominator by $E_o + \Delta E_o$:

$$\Delta E_o = \left(\frac{\Delta E_o}{\Delta E_i}\right)_{\Delta G_L = 0} \Delta E_i +$$

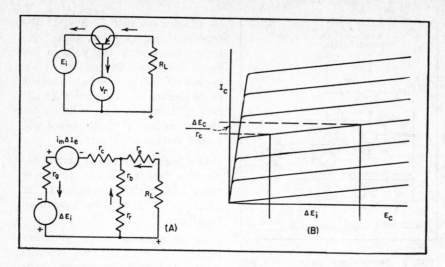

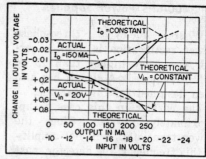

FIG. 2—Theoretical and actual characteristics of series-regulated supply

FIG. 1—Series-regulated supply and equivalent circuit (A) using transistor characteristic shown at (B)

$$\left(\frac{\Delta E_c}{(E_o+\Delta E_o)\Delta G_L}\right)_{\Delta Ei=0}(E_o+\Delta E_o)\Delta G_L$$

$(E_o + \Delta E_o)\ \Delta G_L$ differs from ΔI_o by the term $G_L\Delta E_o$. This term can be neglected because, with any reasonable regulator, it is quite small compared to ΔI_o. Thus

$$\Delta E_o = \left(\frac{\Delta E_o}{\Delta E_i}\right)_{R_{L-\text{constant}}} \Delta E_i +$$
$$\left(\frac{\Delta E_o}{\Delta I_o}\right)_{\Delta Ei=0} \Delta I_o \quad (2)$$

This equation justifies the experimental procedure of holding R_L constant instead of I_o when obtaining the voltage regulation:

This equation is the basis for the a-c type of equivalent circuit. The output voltage variation is equal to the input voltage minus an I-R voltage drop. The two terms on the right of Eq. 2 should be as small as possible. Consequently the design of a voltage regulator should concentrate on making $\Delta E_o/\Delta E_i$ and $\Delta E_o/I_o$ as small as possible.

The most simple regulator design attempted is shown in Fig. 1A. The transistor is a *pnp* power unit rated at 20 watts, with a heat sink at 25 C. The emitter of the transistor is biased in the forward direction and tends to be very close to the potential of the base which is the reference voltage. Line-voltage variations are dropped across the large dynamic collector resistance and are greatly reduced at the emitter.

To determine the coefficients of Eq. 1 assume linear operation and

* Now with Miami Shipbuilding Co. Miami, Florida.

from small-signal equivalent circuit theory

$$\left(\frac{\Delta E_o}{\Delta I_o}\right)_{\Delta Ei=0} = -r_e - (1-\alpha)(r_b+r_r) \quad (3)$$

This is the load regulation of the circuit of Fig. 1A. The input-voltage regulation employing small-signal equivalent circuit theory is

$$\left(\frac{\Delta E_o}{\Delta E_i}\right)_{\Delta Ri=0} =$$
$$\frac{r_b+r_r}{r_c\left[1+\frac{(1-\alpha)(r_b+r_r)}{R_L+r_e}\right]} \times \frac{R_L}{R_L+r_e} \quad (4)$$

This assumes $r_c >> r_b + r_r + r_g$, and R_L = load resistance.

The operation of input voltage regulation can be seen by observing that the slope of the collector characteristics, shown in Fig. 1B, is approximately the series resistance to input voltage changes.

Now Eq. 2 becomes

$$\Delta E_o = \frac{r_b+r_r}{r_c\left[1+\frac{(1-\alpha)(r_b+r_r)}{R_L+r_e}\right]} \times$$
$$\frac{R_L}{R_L+r_e}\Delta E_i - [r_e+(1-\alpha)(r_b+r_r)]\Delta I_o \quad (5)$$

For a power transistor, substituting typical parameter values $r_c = 10K$, $r_e = 1.5$, $r_b = 50$ and $\alpha = 0.96$, $R_L = 75$, $r_r = 0$, in Eq. 5 the following equation is obtained:

$$\Delta E_o = 0.004\Delta E_i - 3.5\Delta I_o \quad (6)$$

Using the circuit shown in Fig. 1A the curves of Fig. 2 were obtained. The nonlinearity of the experimental points and their departure from the theoretical value were expected in view of the large variation of transistor parameters with operating point. However, this does not limit the utility of the regulation equations in most practi-

cal cases because power supplies are usually employed in a narrow range making the assumption of a single operating point a good approximation.

Equation 5 appears to contain sufficiently accurate relationships upon which further design considerations can be based. It can be predicted that better regulation would result from a higher alpha and r_c or lower r_b, r_e and/or r_r.

One major drawback of the circuit of Fig. 1A is that it draws too much reference current. The circuit of Fig. 3 gives about the same regulation, but with substantially less reference current drain. Since $I_o = (1 - \alpha)\ I_e$, the reference current assuming identical transistors is approximately $0.002\ I_o$. This is equivalent to having a larger α.

Employing the equivalent circuit of Fig. 3 regulation will then be:

$$\frac{\Delta E_o}{\Delta I_o} = -[r_{e1}+(1-\alpha_1)(r_{b1}+r_{e2})+$$
$$(1-\alpha_1)(1-\alpha_2)(r_{b2}+r_r)] \quad (7)$$

$$\frac{\Delta E_o}{\Delta E_i} = \frac{\frac{r_{b1}+r_{e2}+(1-\alpha_2)(r_{b2}+r_r)}{r_{c1}}\frac{r_{b2}+r_r}{r_{c2}}}{1+\frac{(1-\alpha_1)}{R_L+r_{e1}}[r_{b1}+r_{e2}+(1-\alpha_2)(r_{b2}+r_r)]}$$
$$\times \frac{R_L}{R_L+r_{e1}} \quad (8)$$

These equations show that the reference resistance can be much larger than in the single transistor circuit before it becomes significant. Since Q_1 will be carrying the full-load current and since the circuit will be required to deliver currents close to the peak ratings of Q_1 the alpha and r_c of this tran-

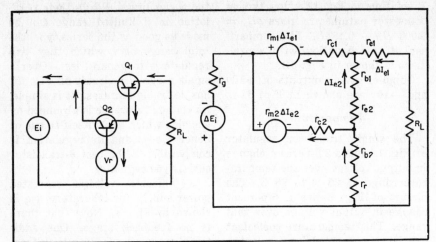

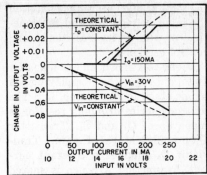

FIG. 4—Operating characteristics of two-transistor regulator circuit

FIG. 3—Two transistor regulator characteristics are determined by equivalent circuit as described in text

sistor may be much smaller than for small-current bias. It would not be wise to assume equal parameters for the two transistors when using Eq. 7 and 8. The values of the parameters used in the single-transistor regulator were large bias values. The proper values for Q_2 are $r_c = 90K$, $r_b = 60$, $r_e = 2.5$ and $r_r = 50$ ohms and $a = 0.97$.

The equation for regulation can be shown to be

$$\Delta E_o = 0.0067 E_i - 3.7 \Delta I_o \qquad (9)$$

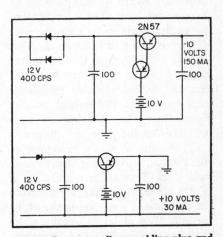

FIG. 6—Power supplies providing plus and minus 10 volts designed for use in ordnance field equipment

age between the load and the base, or reference, less the drop in voltage caused by current in the emitter, base and reference resistances. The regulation is a function of the I-R drop in the circuit path between output voltage and reference. This drop can be made small by making the resistance of this path small or by making the change in current small in this circuit. Clearly, the size of the reference current change is dependent upon the amplification in the loop that tries to

correct for output voltage changes.

When a single transistor is used to sense the difference in reference voltage and output voltage, regulation can be expressed as

$$\frac{\Delta E_o}{\Delta I_o} = \frac{\left[\dfrac{r_{ef}}{1-\alpha_f}+r_{bf}+r_r\right]}{G_{cs}} \qquad (10)$$

$$\frac{\Delta E_o}{\Delta E_i} = \frac{\left[\dfrac{r_{ef}}{1-\alpha_f}+r_{bf}+r_r\right]}{G_c R} \qquad (11)$$

where the subscript f, refers to the transistor parameter of the first amplifying stage, or the stage where the reference is compared with the output voltage. The G_c is total current gain in both the control and series resistance sections of the power supply and R is

This equation is compared with experiment in Fig. 4. The agreement is good since parameter values used in the equations were average values of a sample of ten transistors.

Both circuits discussed are basic series-type regulators employing the collector resistance for dropping line voltage to output voltage. The size of the dropping resistor is dependent upon the difference in volt-

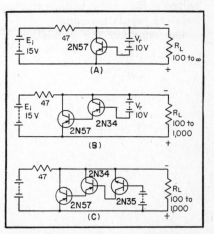

FIG. 7—Three shunt-type transistor regulators for low-voltage power supplies

FIG. 5—Regulator using four transistors including a single-stage feedback amplifier (A) shows good correlation between actual and theoretical results (B)

the total parallel resistance between the input voltage line and the junction of the series resistance and control sections.

Figure 5A shows an example of a voltage regulator employing a single-stage feedback amplifier. The dynamic series dropping resistor is made up of Q_1 and Q_2. Transistor Q_4 provides stage of gain. Transistor Q_3 supplies a constant current to junction X which is shared by the base of Q_2 and collector of Q_4.

If the output voltage decreases as the result of a greater load current the base voltage of Q_4 will decrease toward ground causing Q_4 to conduct less. The current that normally would have gone through Q_4 now is diverted through Q_2. This causes both Q_2 and Q_1 to conduct more, which is equivalent to reducing the series-regulator resistance and results in the output voltage increasing, thus tending to compensate for the drop in voltage.

Transistor Q_3 should be connected backwards, using the emitter as collector. Temperature stability of a transistor with large resistance in the base is poor unless the alpha is low. The alpha is usually quite low in the back direction. Furthermore, the quality of this particular connection as a constant current source is independent of alpha.

Under certain conditions, better regulation could be obtained by employing transistor Q_3 as an additional gain stage in the amplifier and using a resistor in the input voltage line to the base of Q_2 to supply the needed current. This would depend mainly on how much larger the input voltage is as compared to the output voltage and the required regulation range.

The equation for regulation is

$$\Delta E_o = \frac{r_{e4} + (1 - \alpha_4)(r_{b4} + r_r)}{\alpha_4 R \frac{1 - \alpha_3}{\alpha_3}} \Delta E_i -$$

$$\frac{(1 - \alpha_1)(1 - \alpha_3)}{\alpha_3 \alpha_4} \times$$

$$[r_{e4} + (1 - \alpha_4)(r_{b4} + r_r)] \Delta I_o \quad (12)$$

A solution to Eq. 12 using typical transistor parameters gives $\Delta E_o = 0.004 E_i - 0.13 \Delta I_o$. The comparison of this equation with experiment is shown in Fig. 5B.

Supplies giving outputs of +10v and −10v are shown in Fig. 6.

Temperature

The single transistor regulator of Fig. 1A has a 2.5-percent change in output voltage over the temperature range of 25 C to 80 C. The circuit of Fig. 3 has a 5-percent change in output voltage over that range. This temperature coefficient can be compensated by employing a reference with a negative temperature coefficient.

In more complicated regulators employing high-gain servo amplifiers the major temperature problems are found in the amplifier, and principally, in the first stage where the comparison between reference and output voltage is made. Silicon components employed here may alleviate the problem but there will still be a need for compensation of the emitter to base voltage change in the first or comparison stage.

Shunt Types

The shunt types of power supply shown in Fig. 7 are subject to the same general type of analysis as the series units. While their regulation in a limited range can be made as good as the series type, the total range over which they will regulate is in general less. Over a small range the circuit of Fig. 7C has desirable features. It is simple, its voltage reference is grounded on one side which lends itself to use of Zener diodes and its regulation is comparable to the best 3-transistor series-type regulator.

A variable voltage regulated power supply for laboratory use is shown in Fig. 8. Note that there is no feedback stage. One additional transistor feeding back from the reference could improve the regulation by a factor nearly equal to the beta of the added transistor.

Thanks are extended Stanley Gordon who experimentally checked the formulas, Charles Durieu for recommendations of mathematical rigor and Morris Brenner for supplying transistor parameters.

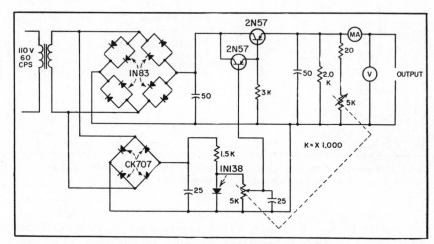

FIG. 8—Variable voltage regulated supply for laboratory use employs two transistors

VOLTAGE REGULATOR USING MULTIVIBRATORS

By WILLIAM A. SCISM

Senior Engineer, Convair Astronautics, San Diego, Calif.

> Pulse-type transistor voltage regulator uses voltage-controlled pulse generators. Astable multivibrator has square-wave output over frequency range of 100 to 3,600 cps; frequency is determined by magnitude of control voltage over 12-to-one range. Pulse width of monostable multivibrator is varied over 10-to-280 microsecond range

LOW SATURATION RESISTANCE of presently available power transistors suggests their use in pulse-type power voltage regulators.

In the pulse-type regulator shown in Fig. 1, reduction of the input voltage is accomplished by turning a gate transistor off at a repetition rate determined by the regulation requirements. The gate is controlled by a series of drivers which, in turn, are controlled by a variable-width one-shot transistor multivibrator. The pulse of no conduction at the gate is smoothed by the low-pass filter which delivers the average value of the wave to the load. This load voltage is attenuated and compared with a Zener reference diode.

The difference voltage is amplified and used to control the frequency of a voltage-controlled multivibrator as well as the width of the one-shot pulse. Since the one-shot is triggered by the variable-frequency multivibrator, the net result is that increasing line voltage increases the width and repetition rate of the pulse cut out of the line by the gate. The average value of the load voltage is thus kept constant.

Voltage-Controlled Multivibrator

Figure 2 is the block diagram of the variable-frequency multivibrator and Fig. 3 the circuit diagram. The chief difference between transistors and vacuum tubes in this application is that the transistor requires power into the base to turn it on or off, whereas the tube will bottom if no power is applied to the grid.

A transistor multivibrator will oscillate if the bases are returned to ground, but the waveform is most unsatisfactory and the period is subject to considerable variation with temperature and transistor interchange.

Returning the bases to a negative

FIG. 1—Block diagram of regulator

FIG. 2—Basic of monostable multivibrator

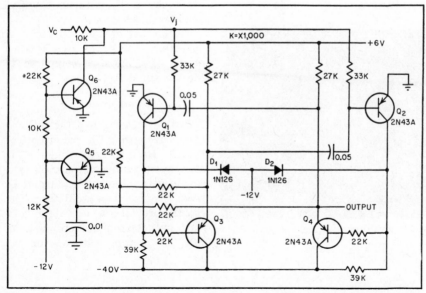

FIG. 3—Frequency of multivibrator is determined by varying return voltage of Q_1 and Q_2

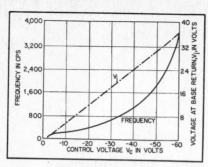

FIG. 4—Characteristics of variable-frequency circuit shown in Fig. 3

supply (in the case of *pnp* transistors) through a resistor will permit sufficient current to be drawn to hold the transistor in the saturated condition. The voltage return for the base resistors can then be varied so the frequency of the multivibrator will be proportional to the base return voltage.

The collectors of multivibrator transistors Q_1 and Q_2 in Fig. 3 are returned to a large voltage and are clamped at a small voltage by diodes D_1 and D_2, utilizing only the first position of the rise time which is, by nature, fast. Emitter followers are also used between the collectors and the bases of the opposite transistors to permit the collector of the cutoff transistor to rise against the base resistor of its emitter follower for a rapid rise of voltage. The associated timing capacitor is driven by the low output impedance of the emitter fol-

lower, which also provides a convenient output takeoff point.

The collector resistor can thus be a fairly high value, as shown in Fig. 3. This offers the advantage of a short-base base which is comparable to the grid base of a tube. The transition region of operation is thus made short compared to the voltage swing of the timing capacitor.

Starting Characteristics

The base resistors could be returned directly to a variable voltage, but the circuit will not start when the control voltage is large. Application of power will bottom both transistors when there is a large value of base return current present. Advantage is taken of the fact that the circuit will always start when the base is returned to zero volts; the base resistors are returned to the collector of Q_6 which is also connected to the control voltage.

When the circuit is oscillating properly, there is a square wave at the emitters of both emitter followers, Q_3 and Q_4. Summation of both these square waves through the 22,000-ohm resistors at the base of Q_5 keeps Q_5 in the saturated condition. This in turn biases Q_6 so it draws no current.

In this condition, the control voltage is applied to the base-return resistors through the 10,000-

ohm resistor connected to the collector of Q_6. Failure of the circuit to oscillate, either upon application of power or due to excess loading in operation, will result in the emitters of Q_3 and Q_4 remaining at essentially zero volts; Q_5 is cut off and Q_6 is turned on, effectively grounding the base resistor return.

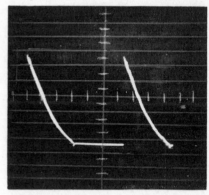

FIG. 6—Waveform at base of Q_1 in Fig. 3 has 6,500-μsec period

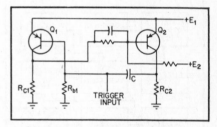

FIG. 7—Basic one-shot multivibrator

With zero volts on the return, the circuit starts oscillating and control voltage V_c governs the frequency of operation.

Calculation of the frequency of operation at $V_c = -20$ v gives a period of 2,750 μsec compared to the measured 2,500 μsec.

Performance

Figure 4 is a plot of voltage versus frequency for the circuit of Fig. 3. Base return voltage V_j is also plotted against the control voltage to determine the loading of Q_6 when the circuit is oscillating. The circuit provides a useful square-wave output from 100 cps to 3,600 cps. The output voltage over the

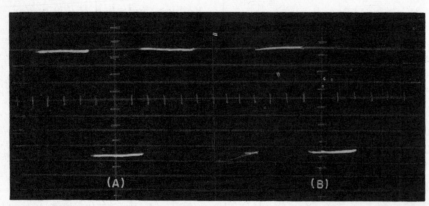

FIG. 5—Control voltages of 5.5 and 52 v applied to multivibrator of Fig. 3 produce 6.500 (A) and 600-μsec (B) outputs respectively

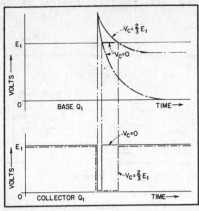

FIG. 8—Effect of change in return voltage on pulse width

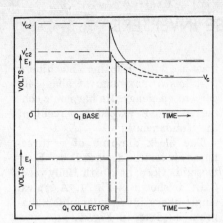

FIG. 9—Effect of change of Q_2 collector clamp voltage on pulse width

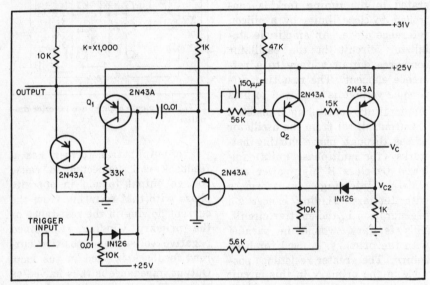

FIG. 10—Variable pulse-width monostable multivibrator with compound control

operating range is fixed at a constant level of 12 v by the diodes.

Figure 5 shows the output waveform with the control voltage near the low and high ends of the range. Figure 6 shows the base waveform. Note that the timing exponential still has considerable slope as it enters the transition region, resulting in greater stability of the half-period.

One-Shot Multivibrator

Development of the regulator called for a one-shot multivibrator whose pulse width is a function of a d-c control voltage.

A conventional transistor one-shot multivibrator is shown in Fig. 7. Transistor Q_1 is held on in the absence of a trigger by the current through R_{b1}. The collector of Q_1 is at a potential of $+E_1$ and the current from positive return $+E_2$ keeps Q_2 nonconducting.

A positive pulse applied to the base of Q_1 cuts it off; it's collector voltage drops toward ground causing it to draw current through the base of Q_2, turning it on and holding it in the saturated condition. As the collector voltage of Q_2 rises toward $+E_1$, the voltage across C drives the base of Q_1 positive, cutting it off.

If R_{b1} is returned to some voltage V_c other than ground, the pulse width can be varied. Figure 8 shows the effect of changing V_c.

The range of pulse widths is limited by this arrangement because the current drawn through R_{b1} becomes insufficient to keep Q_1 on in the no-pulse condition as V_c approaches E_1. This condition can be partially remedied by increasing R_{c1} so less base current is required to keep Q_1 saturated. However, this necessitates an increase in R_{c2} and the net result is a lower limit for the value of R_{b1}.

By using emitter followers in the coupling networks, R_{c1} and R_{c2} can be made quite large and low-impedance outputs are made available at the emitters of the followers. To extend the degree of control, the collector of Q_2 can be clamped to variable voltage V_{c2}. Figure 9 illustrates the effect on pulse width of different values of this clamp voltage with timing-capacitor return voltage V_c held constant.

The clamp voltage is derived from a grounded-emitter amplifier as in Fig. 10, whose base is driven from control point V_c. This amplifier provides the necessary phase inversion for the control voltage and compounds the effect of V_c on the pulse width.

Figure 11 shows pulse width with variation in the timing capacitor return voltage, for constant clamp voltage, and the pulse width for the compound-control circuit shown in Fig. 10. There is a noticeable improvement in slope and range due to the clamp voltage.

Figure 12 shows regulation action of complete circuit.

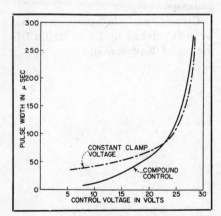

FIG. 11—Operating characteristics of multivibrator of Fig. 10

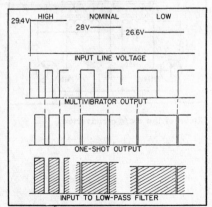

FIG. 12—Overall regulation action for different input line levels

SINGLE- AND THREE-PHASE INVERTERS

Aircraft power inverter contains no moving parts

INCORPORATION of high-power transistors into power-supply design has made possible a 300-va lightweight inverter for use with servos in the second stage rocket for the Vanguard Project.

The unit illustrated produces three-phase, 400-cps power when energized by 28 v d-c and driven by an appropriate tuning-fork oscillator. It weighs 5.5 pounds, has an estimated life in excess of 10,000 hours and operates at 60 to 70 percent efficiency—as con-trasted with rotating machinery that has a maximum of about 40-percent efficiency. Its harmonic content does not exceed 5 percent of the fundamental.

The block diagram of a three-phase inverter designed by Electrosolids Corp. of North Hollywood, Calif. is shown in Fig. 1. A typical single-phase inverter amplifier diagram is given in Fig. 2.

▶ Oscillator—The sine wave generated in the tuning fork is controlled to close limits by a silicon reference diode. An amplitude stabilizing circuit in the oscillator compares output voltage to a reference element. The resulting difference voltage is employed to bias an amplitude controlling transistor.

Output signal from the oscillator passes through phase-shifting networks (for multiphase units) followed by class B high-power amplifiers. Interstage transformers with tertiary winding cancel a-c degeneration in the emitter circuit. This tertiary, wound in parallel with the primary, is used for base return. The greater resistance possible in the primary in this circuit thus increases temperature stability.

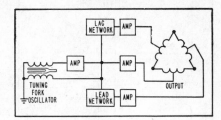

FIG. 1—Three-phase inverter

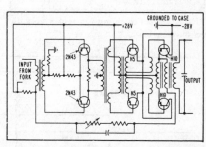

FIG. 2—Typical single-phase inverter amplifier

The output transformer has a feedback coil connected to cause voltage output to add in opposite phase with that resulting from the current flowing in the resistance of the primary winding. Combined negative voltage and positive current feedback is used in the loop. Output impedance is thus halved as compared with that obtained using voltage feedback alone.

D-C TO D-C TRANSFORMER

VOLTAGE step-up from storage battery to potential suitable for average electron-tube plate supply is generally accomplished using a vibrator power pack. An interesting function for certain types of transistors is that of generating an alternating voltage that can be stepped up and subsequently rectified by semiconductor diodes.

Low saturation voltage of the type shown in the circuit diagram reduces the internal power dis-

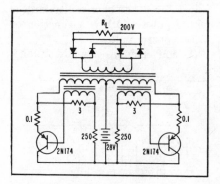

Circuit of the d-c/d-c converter

sipation in the converter application. The resultant small amount of self-heating permits a small heat sink and small size of the package.

Circuit and information have been furnished by Delco Radio Division of Kokomo, Ind.

Chapter 5

DESIGN OF PULSE CIRCUITS

NEON-LAMP INDICATING COUNTER

By H. C. CHISHOLM
Development Engineer, Berkeley Division of Beckman Instruments, Inc.,
Richmond, California

Cascaded silicon-junction transistor binary stages energize neon-lamp indicators for digital frequency meter at counting rates up to 100,000 per second. Step-by-step calculations are shown for design of binary stages. Counter has logged 14,600 hours of continuous operation with no component failures or waveform deterioration

TRANSISTOR COUNTING CIRCUITS are readily designed, but systems for count indication have led to considerable complexity. However, owing to the unique transistor characteristic of minute saturation resistance evident in switching applications, it is practical to cascade binary circuits to obtain a decade counter. Such a system, as described in this article, provides potentials sufficient to reliably operate neon lamps by direct near-conventional methods.

Readout Indicators

Experience has shown that long-term repeated operation of the NE-2A neon lamp requires ionization potentials of at least 85 v and extinguishing potentials must be less than 55 v. For a decade counter, it is possible to operate the lamps with an effective 50-v bias and a superposed 40-v swing.

The basic counter is shown in Fig. 1. Input is at binary 1, which triggers binary 2 once for every two input pulses. In turn, binary 2 triggers binary 3 once for every two of its own input pulses and similarly, binary 3 triggers binary 4.

The resistor matrix for only four lamps is shown to simplify the illustration. The matrix terminations at the blocks are, by implication, directed to the left and right-hand collectors of the two transistors composing each binary.

The upper voltage indicates d-c collector potential at zero count, while the lower voltage indicates the potential of the second stable state of the binary. The output voltage of each binary swings 40 v. The neon lamps are operated at 0.2 ma by potentials which vary at both of their terminations.

Binary 1 operates to select the odd or even numbered lamp, while the other binaries select pairs of lamps through the resistor matrix. The system can best be understood by considering the situation at zero count.

Binary 1 applies a five-volt potential to the even-numbered lamps and a 45-v potential to the odd-numbered lamps. The potential at

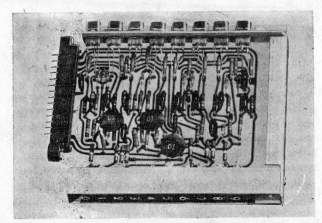
Printed-circuit techniques permit compact packaging of counter

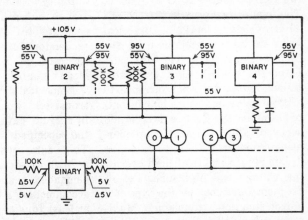
FIG. 1—Indicator system for counter uses NE-2A neon lamps

the junction between pairs of lamps is an average of the potential existing at the binary ends of the resistors. The resistors connecting the zero-one lamp pair are each terminated at potentials of 95 v. Therefore, before ionization occurs the zero lamp has 90 v applied and the one lamp has 50 v applied; the zero lamp will ionize while the one lamp cannot.

Current flow through the zero lamp is limited by the series resistors. The resistor matrix is arranged such that for the indicated voltages, a 90-v potential cannot exist across any of the other lamps.

A single pulse input to the counter triggers binary 1 to its second stable state and reverses the potentials applied to the even and odd-numbered lamps. The zero lamp is extinguished while the one lamp is ionized. Succeeding pulses cause the binaries to assume combinations of steady state voltages which ionize the lamp corresponding to the count stored in the decade counter.

Transistor Requirements

As the counting rate expressed as a frequency is 100 kc, it is desirable to have pulse rise times on the order of 1 μsec. In order that the transistor not be a limiting factor, its alpha cut-off frequency must then be 2 or 3 mc.

High current gain is not necessary and, in fact, would require larger binary crossover resistances to limit base current. This would increase waveform decay time and limit maximum counting frequencies.

Operating temperatures include the range from -20 C to $+50$ C. The former requires that the selected transistor retain a practical value of large signal gain at the coldest temperature. The high temperature requires that transistor leakage current I_{co} not increase to a large value and cause loss in output amplitude. Such a loss would seriously affect the neon lamp operating potentials.

The requirements of high operating voltage and high alpha cut-off frequency are met by the type 903 silicon-junction transistor. Maximum rated voltage for this unit is

30 v, but the collector junction has a breakdown voltage in excess of 50 v.

Binary Design

Since the binaries are of the saturating type, it is particularly important to know the base current requirements for the conditions of lowest gain and poorest operating conditions. The 903 transistor has a beta range of 9 to 19. Of more significance in binary circuit design is the large signal gain, B, which varies for this type from 7.5 to approximately 17. For complete interchangeability of transistors the binary is required to function with all values of B; therefore, it was necessary to design with the lowest value.

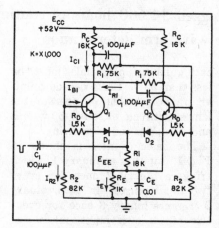

FIG. 2—Basic binary counter circuit

From the actual characteristic curves for many 903 transistors operated at -20 C, it was determined that the base current must be increased by 50 percent over the current value at 25 C to insure stability. Under this condition, a transistor with a B of 7.5 would saturate to a minor degree and transistors of higher gain would saturate in proportion to their large signal B.

Emperical data indicated that the transistors would exhibit a change in current gain of two to one over the specified temperature range. This data, when related to the normal two-to-one spread of beta, demanded that the binary circuit accommodate a total gain change of at least four to one.

The binary circuit is illustrated

in Fig. 2. The design calculations are based on the following values: $V_o = 40$v; $E_{BB} = 3$v; $I_c = 2.5$ ma (assumed); $B = 7.5$ min.

The design procedure begins with a required output voltage, V_o, and establishes a supply voltage as the last step. Collector load $R_c = V_o/I_c = 16,000$ ohms.

Assuming transistor Q_1 on and Q_2 off: $I_{B1} = I_{C1}/B_{min} = 333$ μa at 25 C ambient. At -20 C, $I_{B1} = (333)$ $(150\%) = 500$ μa.

Assuming $I_{R2} = 50$ μa, $I_{R1} = I_{B1} + I_{R2} = 550$ μa; $R_1 = V_o/I_{R1} \cong 75,000$ (assumes $V_{B1} = V_{C1}$ on); $R_2 = V_{B1}/I_{R2} = 4/50 \cong 82,000$ where $V_{B1} = E_{BB} + V_{CE}$ (sat) and V_{CE} at current saturation is 1 v. Neglecting minor leakage currents of the cut-off transistor: $I_E = I_{C1} + I_{B1} = 3.0$ ma; $R_E = E_{BB}/I_E = 1,000$ ohms and R_E is adequately bypassed when $C_E = 0.01$ μf.

Resistor R_2 establishes a reverse bias at the base of the transistor in a cut-off condition. Neglecting leakage currents, the reverse bias potential can be computed as follows: $V_{B2} = R_2$ $(E_{BB} + V_{CE})/(R_1 + R_2) = 2.1$ v; reverse bias $= V_{B2} - E_{BB} = 2.1 - 3 = -0.9$v.

The supply voltage is computed as a summation of the common emitter level, the output voltage swing, and the product of the collector load and the base driving current. The crossover capacitances improve the rise time characteristics of the binary waveforms. Larger values sharpen the rise time, but also increase the time constant for the decay characteristic.

With $C_1 = 100$ $\mu\mu$f rise times on the order of 1 μsec were obtained with fall times of approximately 3 μsec.

Complete Circuit

The complete schematic of the decade counter is given in Fig. 3. The grouping of the four binaries is the same as that indicated in Fig. 1; the binaries are identical with only minor differences existing in their drive circuits. The resistors of the indicator lamp matrix shunt the 18-kilohm collector load resistors to an effective value of 16 kilohms.

The upper value of potential for each collector is for the zero count

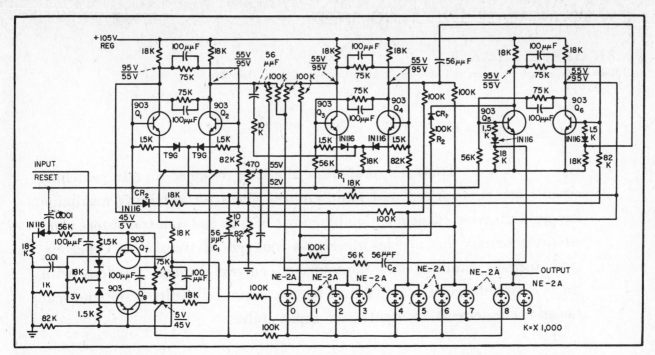

FIG. 3—Decade counter for digital frequency meter has four cascaded stages of two-transistor binary counters

state and the lower potential is the value to which the binary switches in proper sequence.

A basic four-binary counter scales to a count of 16. To scale to a count of 10, the counter is gated after the eighth input pulse and then, after the tenth pulse, returned to the zero count conditions. The first three binaries scale in a normal manner up to the eighth input pulse.

After the eighth input pulse binary 4 is triggered to its second stable state. The collector potential of transistor Q_6 rises to 95 v and is applied through R_1 to the juncture of the steering diodes in binary No. 2. The diodes are hereafter reverse biased by approximately 40 v and input pulses from binary 1 through coupling capacitor C_1 are effectively blocked. With these conditions, lamp 8 is ionized.

The ninth counter input triggers binary 1 and switches the ionizing potential from lamp 8 to lamp 9. The tenth input pulse resets binary 1 to its normal steady state. During this regenerative switching, a pulse is applied through capacitor C_2 to the base drive circuit of transistor Q_5 in the fourth binary, which returns to its normal steady state and completes the cycle of 10 counts. An output pulse may be taken from the fourth binary to

drive a second decade counter which will then indicate the tens count.

D-C Gating

The d-c gating method of the counter is simple and reliable. The two steering diodes in binary 2 are special only in that their reverse impedance and breakdown voltage are sufficiently high to withstand the 40-v reverse bias.

Relatively large leakage current here is sufficient, under certain counting conditions, to trigger binary 2 and nullify the gating function. Silicon junction diodes at this circuit point are unsatisfactory because the diode-junction capacitance is large enough to transfer triggering pulses under certain operating conditions.

At zero count the even numbered transistors, or right-hand units, are in the on condition and lamp 0 is ionized. On count 8, the right-hand transistors of the first three binaries return to the on condition. Two high potentials now exist attempting to ionize lamps 0 and 8. Lamp 0 is prevented from ionizing by the application of a 55-v potential at transistor Q_5 through diode CR_1 and R_2 to the 0-1 lamp junction.

In addition, a current-limiting resistor is omitted in the path

from the collector of Q_6 to the 8-9 lamp junction. These two lamps then operate at a 300-μa current level, as limited by the matrix resistors connected to binary 1. The result is a rise in the potentials applied to the lower side of the odd and even numbered lamps and an assist in preventing lamp 0 from ionizing.

Reset

The reset circuit is composed of two networks performing identical functions. The desired condition at reset is 0 count in which the right-hand transistors are conducting. This is accomplished by application of a pulse which will switch-off the left-hand transistors should they happen to be conducting. In the base return circuit corresponding to R_2 of the computations, a diode is inserted.

A negative pulse applied to the anode of the diode drives the transistor to a nonconducting state. The base return resistors of binaries 2, 3 and 4 are returned together through common diode CR_2 to the common bias potential. The reset pulse must be of 50-v amplitude and 40-μsec duration.

The developments were supported by contract with the Frequency Control Branch of the Army Signal Supply Agency.

HOW TO INCREASE SWITCHING SPEED

By RICHARD H. BAKER

Staff Member, Massachusetts Institute of Technology, Lincoln Laboratory, Cambridge, Mass.

Transistor properties affecting response time in switching circuits are summarized and basic circuits given for obtaining maximum energy conversion efficiency. Combined use of pnp and npn transistors gives circuit symmetry that utilizes inherent advantages of transistors. Other circuits include saturated and nonsaturated current-demand flip-flops with single or double triggering, designed for maximum reliability despite normal variations in circuit constants and input pulses

THE NORMAL three-region junction transistor (excluding graded-base or drift types) is a slow device when compared to a vacuum-tube triode. In a tube, the movement of electrons from cathode to plate is aided by strong electric fields, whereas in a transistor the transport of carriers (electrons or holes) is only by diffusion.

In designing transistor circuits for high-speed switching, the designer must consider normal integrative effects due o shunt capacitances as well as the delay or carrier transit time between emitter and collector. When the transistor is operated in the saturated mode, there exists an additional effect, that of hole storage or saturation delay.

Response Times

There may be as many as three separate response times, depending upon the mode of operation, associated with a single-stage transistor network. These are rise time, storage or saturation delay and fall time, all shown in Fig. 1. If R_L, R_e, V_{cc} and V_T are chosen so that the voltage polarity across the collector junction maintains the collector junction under reverse bias at the peak of the output pulse, the saturation delay vanishes.

The magnitudes of response times T_1, T_2 and T_3 are different for each of the three basic connec-

tions. In all modes of operation, however, the transistor switching time is dependent on the constants of the device and the amount of overdrive supplied at the input.

The single most important factor affecting the switching time is the frequency response of the device itself. Also, minimum response time occurs when current gain a_N is 1.

There is promise of obtaining high-frequency transistors by using graded-base structures and other configurations. However, the interim solution of transistor manufacturers has been to build transistors with very narrow base widths to increase the frequency response. This approach is fruitful to a degree, but there is an op-

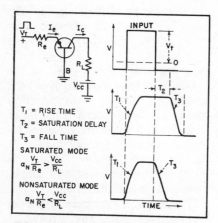

FIG. 1—Response time with storage delay (middle curve) and no delay (lower curve)

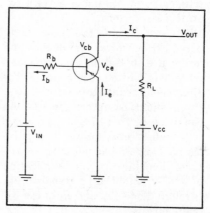

FIG. 2—Back-clamping technique giving voltage gain along with good efficiency

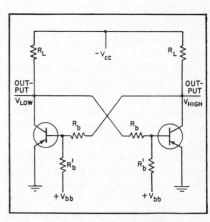

FIG. 3—Bistable saturated circuit is less efficient at low output power

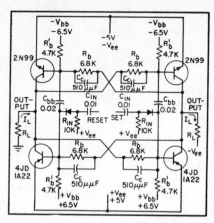

FIG. 4—Saturated current-demand single-triggering circuit with 1-mc transistors

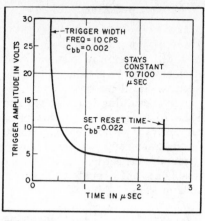

FIG. 5—Characteristics of single-triggering circuit of Fig. 4

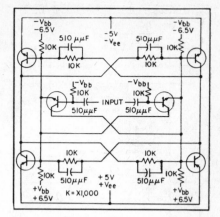

FIG. 6—Saturated current-demand flip-flop double-triggering with 1-mc transistors

timum base width that yields minimum switching time for practical switching circuits. This optimum base width is generally different for each of the three basic connections.

In transistor circuits the transit time of the carrier across the base region imposes an absolute minimum input pulse width. This in turn sets rather large minimum capacitance values in a given circuit, creating recovery time problems that may be more serious than actual rise time considerations.

Signal Levels

Because transistors are extremely efficient voltagewise, the system levels are usually set by a combination of system and transistor considerations.

The low voltage limit is automatically set if the transistors are allowed to saturate, this being primarily determined by speed considerations.

The upper voltage limit is set by the total power consumption of the system and by the punch-through and avalanche phenomena in the transistor.

The signal voltage swing in an all-transistor system is usually chosen as a compromise between two inherent opposing effects. As the signal level increases (total swing), the amount of energy dissipated in charging and discharging capacitance increases. This effect indicates that the signal level should be low. On the other hand, for convenience of circuit design the signal level should be large

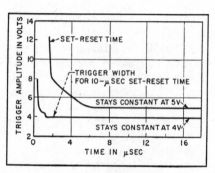

FIG. 7—Characteristics of double-triggering circuit of Fig. 6

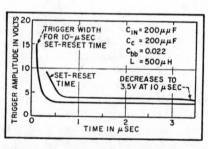

FIG. 8—Characteristics of double-triggering circuit using 5-mc transistors

compared with the transistor off-on uncertainty region, which is about 0.2 volt for germanium transistors and about 1 volt for silicon transistors.

Energy Conversion Efficiency

Fundamentally the transistor, like the vacuum tube, has gain by virtue of dissipation changes. Unlike the vacuum tube, the input impedance is much lower than the output impedance. In the design of realistic transistor systems,

then, a serious problem arises in the available power to drive succeeding stages. This situation is aggravated still further in the design of high-speed systems, since it is necessary in the transient state to overdrive the stages to obtain fast switching. This fact, more than any other, accounts for the large number of transistors required to build transistor systems compared with equivalent vacuum-tube systems.

These considerations indicate that circuitry should be designed to deliver maximum output power and that a high percentage of the available output power should be available to drive other transistors. Further, since currently available high-frequency transistors are ex-

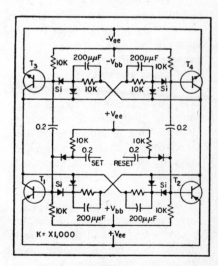

FIG. 9—Nonsaturated current-demand single-triggering flip-flop

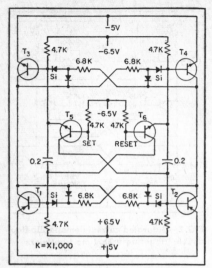

FIG. 10—High-speed nonsaturated current-demand double-triggering flip-flop

tremely low-power devices (on the order of 0.5 mc-watt as a figure of merit for SBT-100 at 50 mc and 10 mw), circuitry should be designed to give highest possible energy conversion efficiency. The ratio n_1 of useful signal output power to transistor dissipation should therefore approach infinity.

For optimum circuit design of a minimum-power-drain system, the ratio n_2 of useful signal output power to power supply drain should approach 1.

The product of n_1 and n_2 should be made as large as possible. The ratio represented by n_1 can be made large by allowing the transistor to saturate or by controlling the voltage from collector to base through the use of clamping diodes. However, minimum-power systems can be built only by making n_2 close to unity (this must be true if there exists a minimum power level to process intelligence).

In most present transistor circuit designs, a high percentage of useful output power from the transistor is dissipated in the load resistors. This is especially true for direct-coupled logic. Therefore, the value of n_2 may be increased significantly by removing the standby power dissipated in this area.

The circuit design techniques in the following sections show how the values of n_1 and n_2 may be increased to give minimum power dissipation, maximum speed and minimum sensitivity to component and transistor drift circuits.

Maximum-Efficiency Circuits

Aside from eliminating the power dissipated in load resistors, an additional gain in system power efficiency may be obtained by using circuits that draw power from the supplies according to the power and demand at the output. This process always involves feedback. Cathode-follower and emitter-follower (grounded-collector) circuits do this, but unfortunately have no voltage gain.

A transistor circuit involving voltage gain, along with an ability to convert d-c power into signal power as required by the load, is shown in Fig. 2. The transistor dissipation is low and the output power is high for collector currents less than the maximum output current. The major portion of the power drawn from the supplies is available at the output for dissipation in the load resistor, so that n_2 approaches 1 and the circuit draws from the supplies only the power dissipated in R_L and R_b (neglecting transistor dissipation). The only transistor parameter of importance in the conducting state is the minimum base-to-collector current gain β_N.

To illustrate the design of this circuit, assume that β_N is 20, input d-c voltage V_{in} is 5 volts, transistor saturated base resistance r_b is 50 ohms, V_{cc} is 10 volts, R_b is 10,000 ohms and R_L is 1,000 ohms. Then I_b is about 0.5 ma, $I_{c\,max}$ is 10 ma and I_c is the sum of these or 10.5 ma. Useful signal output power is $\frac{1}{2} V_{cc} I_c$ or 50 mw and transistor dissipation is 10.5×0.2 mw, so that n_1 is about 24. Power supply drain is 50 mw + 2.5 mw, so that n_2 is about 0.95.

Two of the circuits of Fig. 2 may be coupled together, with only slight modification, to form the bistable circuit of Fig. 3. This has several drawbacks, however. The low-voltage level is not fixed, being dependent on I_{co} and other factors. The power dissipated in internal load resistor R_L (in shunt with the actual load) may be an appreciable percentage, particularly at low output power levels. For fast fall time (when the transistor is turned off), R_L must be made small.

Current-Demand Circuit

A circuit that circumvents these disadvantages is shown in Fig. 4. Here essentially all of the output current (collector current) is available to drive load R_L.

Standby power is low; when there is no load, the power taken from the supplies is approximately equal to the dissipation $2 I_b^2 R_b$ in the base resistors. Both the high and low voltages are clamped (the transistors saturate). The

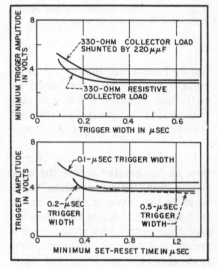

FIG. 11—Triggering characteristics of flip-flop of Fig. 10

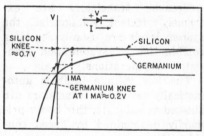

FIG. 12—Diode characteristic curves

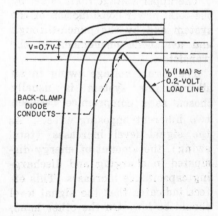

FIG. 13—Family of collector curves for silicon transistor

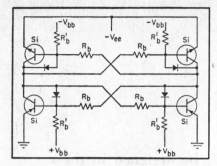

FIG. 14—Basic silicon transistor circuit for nonsaturated current-demand flip-flop

tolerance on all resistors may be large (on the order of 50 percent).

Circuit operation is substantially independent of transistor parameters. The stability of the configuration is insensitive to supply voltages. The configuration leads to fast rise and fall time since large transistor overdrive is inherent. The last three advantages accrue because the configuration allows the transistors to set their own levels. Some of the operating characteristics of the circuit are shown in Fig. 5.

One difficulty with the circuit of Fig. 4 is that there is an appreciable delay around the loop because the conducting transistors are saturated. This difficulty may be minimized by double triggering (triggering all four transistors simultaneously). The current-demand flip-flop circuit of Fig. 6, employing this feature, gives the characteristics shown in Fig. 7. By using 5-mc transistors in this circuit and changing all 510-$\mu\mu$f capacitors to 200 $\mu\mu$f, the characteristics of Fig. 8 can be obtained.

The circuit techniques described may be extended to nonsaturating circuits. The primary gain in designing the circuits to operate in the nonsaturating mode is decreased switching time. Figure 9 shows a typical design using the nonsaturated configuration with single triggering. Figure 10 shows a higher-speed version using double triggering, and Fig. 11 gives triggering conditions for the circuit. The diode characteristics in Fig. 12 show why these back-clamped circuits do not allow the transistors to be saturated.

If silicon transistors are used, the nonsaturated circuits do not require the four silicon diodes. This may be seen from the silicon collector curves in Fig. 13. The basic circuit using silicon transistors is shown in Fig. 14.

The salient features of the saturated back-clamping current-demand technique are low transistor dissipation, high conversion efficiency, insensitivity to component and transistor parameters (standby load resistors not needed), insensitivity to voltage supply drift, maximum system efficiency (power drawn from supplies according to needs of load), fast rise and fall time (inherent overdrive) and loop delay (caused by saturation time). Nonsaturated circuits give increased operating speed because they have no saturation delay, but are otherwise identical.

Gating Circuits

The design of maximum-reliability switching systems depends heavily upon the reliability of the voltage-pulse voltage-level gate. To assure maximum system reliability (assure positive action and suppress superfluous triggering), the gate circuits should be independent of pulse width, pulse amplitude, pulse repetition frequency and pulse level (within given limits), and should have fast response to pulse and level changes. The circuit design should also be insensitive to component values and transistor parameters, require minimum standby power, have high output power, present a constant load to the pulse source (driver) and deliver standardized output pulse and level amplitudes.

A circuit configuration that fulfills to a high degree the above reliability characteristics is shown in Fig. 15 along with its gating waveforms.

Conclusions

The reliability of transistor switching systems is closely related to the design of circuits. The circuit designer must consider the drift of operating points caused by aging and ambient self-generated temperature changes. For high-speed networks, due to the lack of high-speed transistors, overdrive must be used to speed up the circuit response.

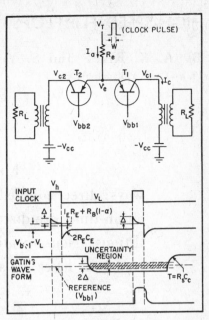

FIG. 15—Circuit and waveforms of pulse-level gating circuit

Transistors are inherently efficient devices (both voltagewise and powerwise). This, along with the fact that two types of transistors are available (*npn* and *pnp*), allows circuit design that is extremely efficient in terms of power supply drain for a given signal power output.

The transistor, being an efficient, reliable and small device, may be soldered into systems much as are ordinary resistors and capacitors. This, plus the fact that it is basically a three-terminal passive device which can produce power gain, makes its use attractive in networks where feedback techniques are widely employed.

The research work herein described was supported jointly by the Army, Navy and Air Force under contract with Massachusetts Institute of Technology.

HIGH-SPEED FLIP-FLOPS

By A. K. RAPP and S. Y. WONG
Philco Corp., Philadelphia, Pa.

Flip-flops for digital computers may be designed in three ways: direct coupled, emitter coupled or resistance coupled. Delay and transition times are discussed for these circuits under loaded and unloaded conditions. Nonsaturating circuits and the effects of loading with zero to four transistor bases are also given

TRANSISTORS HAVE made possible electronic computers requiring only a small fraction of the space and power required by the vacuum tube equivalent.

This article compares maximum speeds measured for changes of state for three types of circuits using surface-barrier transistors.

Basic Pulse Amplifiers

Shown in Fig. 1, the three basic types of pulse amplifier circuits are the direct-coupled, the resistance-coupled and the emitter-follower-coupled cascades.

Figure 1A illustrates the simplest method of coupling cascaded common-emitter stages by connecting the output of one stage directly to the base of the following stage.[1]

The circuit modification in Fig. 1B overcomes one of the principal limitations in the switching speed of transistors, the storage of minority carriers in the base region of a saturated transistor. This storage produces a time delay at the collector on application of a cutoff signal.

One convenient means of reducing turnoff time delay is to limit the base current by inserting resistance in the base leads. Capacitors are shunted across the base resistors to help speed up the circuit.

Figure 1C illustrates the use of an emitter follower as an active coupling network. The emitter follower does not suffer from hole storage and its low source impedance permits high charging currents to be supplied to the stray and internal capacitances, usually the limiting factors in fast pulse operation.

Figure 2 shows flip-flops using

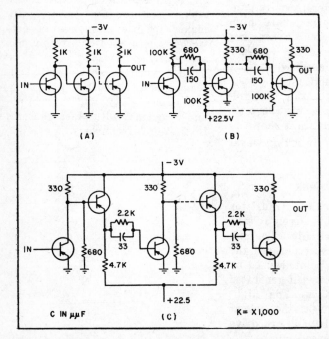

FIG. 1—Direct-coupled (A), resistance-coupled (B) and emitter-follower-coupled (C) pulse amplifiers coupled in cascade

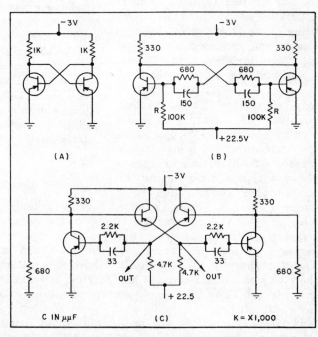

FIG. 2—Flip-flops derived from circuit configurations of Fig. 1. Circuit (A) contains a minimum number of components

the three types of coupling given in Fig. 1. The circuit of Fig. 2A is a simple flip-flop circuit containing a minimum of components.

The flip-flop of Fig. 2B is analogous to the Eccles-Jordon circuit. It may be operated without the bias network consisting of +22.5 v and resistors, R, since the network has no effect on speed but increases reliability.

The emitter-follower stages of Fig. 2C produce buffering action and consequently provide convenient output terminals.

Switching Times

The direct-coupled flip-flop of Fig. 2A is redrawn in Fig. 3 to illustrate the switching times employed as criteria of high-speed performance. The delay time, δ is the time interval between the point where the collector potential of the gating transistor changes by 10 percent of its total voltage excursion and the point where the collector potential of the off-going transistor falls 10 percent of its total voltage excursion.

The total transition time X is the delay plus the fall time of the collector voltage of the off-going transistor. Fall time is the time required for the collector voltage of the off-going transistor to move from 10 percent to 90 percent of its voltage excursion.

Time Measurement

Delay and transition times were measured with the flip-flops loaded symmetrically with from zero to four pairs of transistor bases. Owing to its higher ratio of collector to base current, the resistance-coupled circuit has considerable less delay than the direct-coupled circuit. Transition time is more indicative of operational speed since it represents the total time required for the flip-flop to acquire a new state. Using transition time as the criterion, the resistance-coupled circuit requires about 18 percent less time than

the direct-coupled, while the added complexity of emitter-follower coupling results in a 70 percent reduction in transition time.

Speed

The speed of the direct-coupled circuit increases to double its no-load value when it is loaded with four pairs of bases, while the resistance-coupled circuit is slowed down by the addition of load. The buffer action of emitter followers allows the emitter-follower-coupled flip-flop to remain relatively unaffected by load.

The nonsaturating, resistance-coupled flip-flop[2] shown in Fig. 4 makes use of small resistors connected from each base to ground to form a voltage divider which limits the voltage swing of the base. The insertion of resistance into the circuit common to both emitters provides d-c feedback which forces the emitter to follow about 0.3 v more positive than the base of the conducting transistor. Since this base-to-emitter potential is independent of the emitter resistance, the choice of resistance controls the emitter current. Once the base voltage and emitter current are fixed, the collector current is also determined. A small collector resistance then holds the collector voltage of the conducting transistor sufficiently negative with

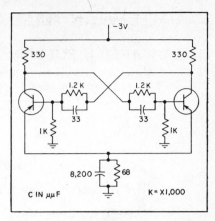

FIG. 4—Nonsaturating resistance-coupled flip-flop has d-c feedback

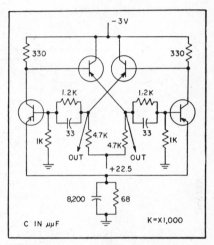

FIG. 5—Nonsaturating emitter-follower-coupled flip-flop uses emitter biasing

respect to the base to prevent saturation.

The nonsaturating, emitter-follower-coupled flip-flop shown in Fig. 5 also applies the emitter-biasing method of preventing saturation to the emitter-follower-coupled flip-flop.

The avoidance of saturation produces a reduction of more than 60 percent in the transition time of the resistance-coupled circuit. Prevention of saturation in the emitter-follower circuit results in an increase in speed to the extremely fast transition time of 22 millimicroseconds.

REFERENCES

(1) R. H. Beter, W. E. Bradley, R. H. Brown and M. Rubinoff, Directly Coupled Transistor Circuits, ELECTRONICS, p 132, June 1955.
(2) R. E. McMahon, Designing Transistor Flip-Flop, ELECTRONIC DESIGN, p 24, Oct. 1955.

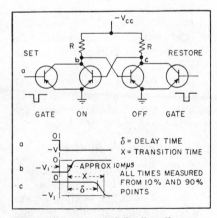

FIG. 3—Direct-coupled flip-flop illustrates switching time criteria for high-speed performance with symmetrical loads

VARIABLE-WIDTH PULSE GENERATOR

By EDWARD J. FULLER

Microwave Electronics Division, Sperry Gyroscope Company,
Division of Sperry Rand Corporation, Great Neck, New York

All-transistor pulse generator achieves pulse amplitudes of 28-v negative, 50-v positive with widths continuously variable from 1.0 to 10 μsec and rise-decay time of 0.3 μsec. Internally generated repetition rates vary in steps from 50 to 5,000 pulses per second with external triggering by either positive or negative pulses, and 1 to 100 μsec time delay internally

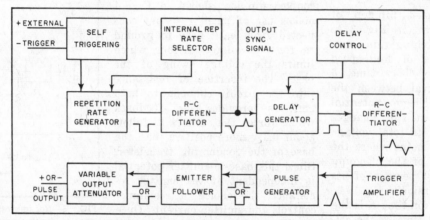

FIG. 1—All-transitor pulse generator employs repetition-rate generator, both free-running and externally triggered. Delay is achieved by using a variable-width monostable multivibrator in conjunction with a RC differentiator

DURING the course of the design and development of a completely transistorized pulse generator, it was soon apparent that presently available junction and surface barrier type transistors imposed certain limitations upon the desired performance. The completed unit, presented here, is one approach to the problem and particularly emphasizes how these limitations were overcome and what refinements were necessary to fulfill certain definite requirements.

The use of transistors offers several advantages: excellent adapt-6ability to printed circuitry; economy of power consumption; reliability over an extended period of time, and small size.

Basic Circuit

Figure 1 shows the basic circuit in block form. The overall schematic diagram is shown in Fig. 2.

The repetition rate generator is an astable multivibrator with a provision for switching to monostable operation so that it can be externally triggered. The common-emitter circuit used allows for phase reversal between base and collector.

The 904 npn silicon-junction transistor used allows a maximum collector potential of 30 volts and a collector current of 25 ma.

Common Emitter

A typical measured set of common-emitter characteristics is shown in Fig. 3. To achieve an amplitude of about 22 volts, the load line is drawn from $I_c = 0$ ma, $V_c = 30$ v to $I_c = 10$ ma, $V_c = 0$ v. The calculated load resistor $= 22$ v/ 0.010 ampere $= 2,200$ ohms.

The load line is intersected by base current $I_b = 300$ μa at $V_c = 8$ volts which allows for the 22-volt

collector swing. Therefore, the value of R_b must be determined for $I_b = 300$ μa and is calculated at 100k ohms.

The values of the coupling capacitors are determined to give the proper R-C time constant to generate the designated repetition rates. Because the inherent resistance between the base and the emitter is considerably smaller than the value of R_b, timing capacitor C_c will discharge through the transistor between base and emitter to ground, thus limiting the value of the R-C time constant.

To insure that the R-C time constant is determined solely by R_b and C_c, a diode is placed between the emitter and ground. The back resistance of the diode is so large that the capacitor C_c must discharge through R_b.

The circuit is unsymmetrical to free-run over a wide range of repe-

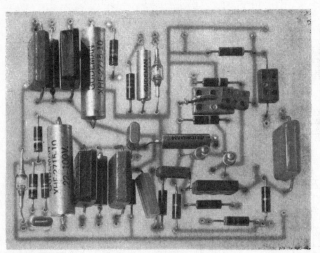

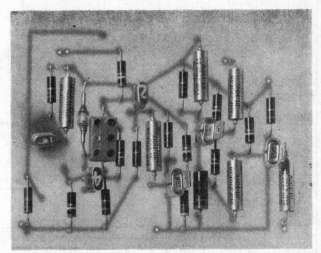

Repetition-rate generator and time-delay circuits excluding controls. Back-lighting shows the printed circuitry

Wide-band amplifier, pulse generator, emitter follower and amplifier circuit show adaptability to printed-circuit techniques

tition rates, especially at values below 500 pulses per sec (where R_{L2} should be approximately twice the value of R_{L1}). To achieve a large variation in the repetition rate, the value of C_c is varied from 0.018 to 0.22 μf, as changing the value of R_b changes the d-c operating point of the transistor, seriously affecting the voltage waveform.

Triggering

For monostable operation, a negative bias is applied to the base of Q_2. The circuit permits self-triggering or external triggering either positively or negatively. An external signal of $\pm$ 5.5 volts with a width greater than 0.7 μsec triggers the generator during monostable operation. When the emitter of Q_2 is triggered positively, it is grounded directly instead of being connected to ground through a diode, as in the free-running position. This allows the multivibrator to trigger on a smaller signal than when the diode is in the circuit. The output waveforms are shown in Fig. 2.

Delay

To achieve delay, a variable-width, monostable multivibrator is used with an R-C differentiating

circuit. To attain a pulse width of less than 1 μsec, a transistor with a high β cutoff frequency is used.

In the common emitter configuration, the β cutoff frequency ($f_{c\beta}$) = ($f_{c\alpha}(1-\alpha)$), where α and $f_{c\alpha}$ are the common-base configuration gain and bandwidth parameters.

The 904 A transistor was considered. Of those measured, $F_{c\alpha}$ varied from 7.5 mc to 15 mc and α from 0.981 to 0.964, giving values of $f_{c\beta}$ from 0.171 to 0.480 mc. The other transistor considered was the SB-100 surface-barrier type. The measured $f_{c\alpha}$ was in excess of 50 mc for all of the transistors tested and α varied from 0.944 to 0.978, giving values of $f_{c\beta}$ greater than 1.1 mc.

The 4.5-v max collector voltage and 5-ma max collector current of the SB-100 limit the magnitude of the output signal to a value of less than 4.5 volts, but this value is sufficient for a delay generator.

The surface-barrier transistor gave a much narrower pulse in a monostable multivibrator circuit than any of the other available types of junction transistors. Pulse width of 0.3 μsec were obtained with rise times of 0.03 μsec and decay times of 0.05 μsec.

The value of R_{L1} was determined

to give a collector voltage swing of 4.2 volts and the base resistance value was taken from the load line in the same manner described for the repetition-rate generator.

Stage Q_1 is normally on and Q_2 is held off by a positive bias on the base.

To vary pulse widths from less than one μsec up to 100 μsec, the R-C time constant must be varied. The conventional R-C circuit used in the repetition-rate generator allows only C to be varied. Over a range of one to 100 μsec, the dynamic range of the capacitor would be excessively large. Therefore, the pulse width variation is brought about by a special R-C circuit.

Pulse Shape

The base resistance for Q_1 equals 27,300 ohms and the value of C is determined by the values of C_{T1} and C_{T2} in series with each other. C_{T1} isolates the collector of Q_2 from the base resistance of Q_2 and the 25,000-ohm variable resistor allows a pulse width variation from about 0.8 μsec up to over 100 μsec. Although pulse widths narrower than 0.8 μsec can be obtained with the SB-100, the R-C values are not practical when widths of 100 μsec are to be achieved with the same

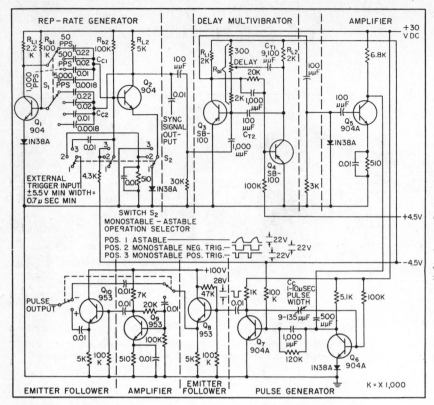

FIG. 2—Pulse generator circuit uses 904 junction transistors and produces pulses with amplitude of 100-v width, continuously variable from 0.1 to 10 μsec

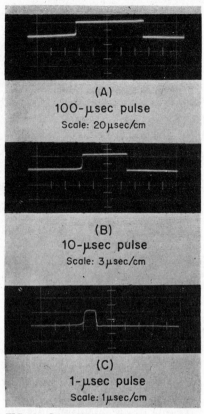

(A)
100-μsec pulse
Scale: 20 μsec/cm

(B)
10-μsec pulse
Scale: 3 μsec/cm

(C)
1-μsec pulse
Scale: 1 μsec/cm

FIG. 4—Output pulse shapes of the delay multivibrator are shown for widths of 1, 10 and 100 μsec at 4.2-v amplitudes

vibrator is not loaded.

The transistor finally chosen for the pulse generator was the 904-A npn junction type which has a f_{co}

type of circuit.

The optimum pulse shape is achieved by driving the transistors into saturation, which gives a flat top. Decay time can be improved by lowering the value of R_{L2}, so that the transistor is just into saturation. Values lower than 2,000 ohms are feasible for narrow pulses, but attempts to achieve widths greater than about 10 μsec cause the pulse amplitude to droop excessively.

The delay multivibrator output pulse shapes are shown in Fig. 4 for widths of 1, 10 and 100 μsec. The pulse amplitude is 4.2 volts in a positive-going direction. The differentiated pulse gives a delayable positive voltage spike of 1.1 volts.

Pulse Forming

The pulse waveform generator is a monostable multivibrator. As the surface-barrier SB-100 allows a maximum voltage swing of only about 4.3 volts, the output pulse must be greatly amplified making their use in this application impractical.

Transistors capable of swinging a signal of 100 volts are the 953 and 970 *npn* silicon-junction transistors, whose collector potentials are rated at 120-v maximum. The 953 transistors were tried out in a monostable multivibrator circuit.

Transistor Selected

Though the output pulses had amplitudes greater than 100 v, the minimum pulse width obtainable was about three to four μsec, with poor rise and delay times. In addition, the output impedance of a common-emitter configuration is high, necessitating the use of an emitter follower when pulsing a low impedance oscillator so that the output stage of the multi-

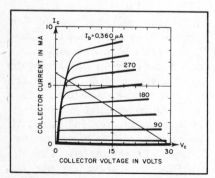

FIG. 3—Common-emitter characteristics

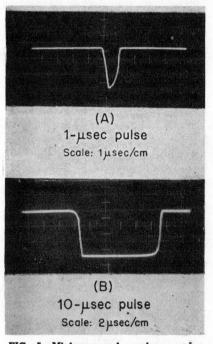

(A)
1-μsec pulse
Scale: 1 μsec/cm

(B)
10-μsec pulse
Scale: 2 μsec/cm

FIG. 5—Minimum and maximum pulse width output wave forms of the pulse generator circuit

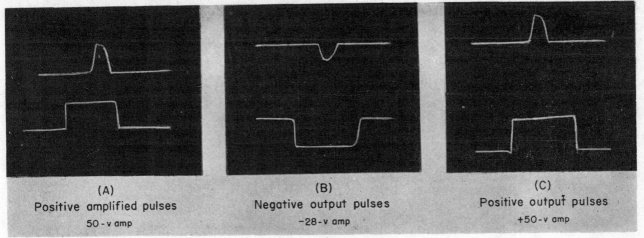

(A)	(B)	(C)
Positive amplified pulses	Negative output pulses	Positive output pulses
50-v amp	–28-v amp	+50-v amp

FIG. 6—Amplifier output waveforms (A) for 1 and 10-μsec pulses. Negative (B) and positive (C) output pulses from the bases of the emitter followers Q_{10} and Q_8 respectively

greater than 8 mc and a maximum collector potential of + 30 volts. This transistor possessed the best high frequency characteristics of any of the higher-voltage transistors investigated (maximum collector potentials of at least 25 volts) and proved to be the best compromise, giving consideration to both collector potential and bandwidth.

Varying capacitor C_o from 9 to 135 μμf gives a continuously variable pulse width ranging from 1 to 10 μsec. This width may also be varied as was shown for the delay generator. The minimum and maximum width output waveforms are shown in Fig. 5. The amplitude is 29 v in the negative-going direction.

Trigger Amplifier

The negative-going spike of the differentiated pulse from the collector of Q_4 has an amplitude of 1.1 volts, which is not sufficient to trigger the pulse generator, and requires amplification. Turning off the normally conducting transistor of the pulse generator gave a better waveform than turning on the nonconducting transistor. Therefore, the triggering waveform was taken from the collector of Q_3. It is important that this spike remain as narrow as possible since its width affects the minimum width of the pulse generator.

The amplifier utilizes a 904A transistor. A voltage gain of six was achieved with no noticeable loss of fidelity. This signal triggers the pulse generator for a pulse as narrow as one μsec.

Emitter Follower

The pulse output taken from the collector of Q_7 is negative-going. The positive-going waveform at the collector of Q_6 has poor shape and is not suitable for use as the positive output pulse.

To get a good positive-going pulse, it is necessary to invert the negative pulse through a common-emitter amplifier circuit, the only configuration that gives phase inversion.

Amplifying Input

The common-emitter circuit also has a low input impedance, which would load the multivibrator output. Thus, to generate a positive pulse, the negative pulse must be passed through the emitter follower and then amplified.

The 953 and 970 transistors are the only ones capable of passing and amplifying an input signal of 28 volts. Emitter follower Q_8 presents a higher impedance to the collector Q_7 and prevents any loading effects by the low impedance inputs. The 47,000-ohm resistor from the collector to the base of Q_8 puts the necessary positive bias on the base so that the large negative-going signal can be passed properly.

Amplifier Q_9 inverts this negative-going signal and Q_{10} is an emitter follower which provides a low impedance output.

The amplifier output waveforms at both one and 10 μsec are shown in Fig. 6A. The negative and positive pulses from the bases of emitter followers Q_{10} and Q_8 respectively are shown in Fig. 6 B and C.

Performance

The achieved results were as follows: 1) pulse amplitudes 28-v negative, 50-v positive; 2) pulse widths continuously variable from 1.0 to 10 μsec; 3) rise and decay times of 0.3 μsec; 4) internally-generated repetition rates varied in steps from 50 to 5,000 pulses per second; 5) external triggering by either a positive or negative pulse at any repetition rate from 50 to 5,000 pulses per second; 6) internally generated time delay from 1 to 100 μsec.

Available Transistors

These results bring out the two inherent weaknesses of the present transistors: low collector potential and limited high-frequency response. However, the use of recently developed transistors will increase the pulse amplitude and allow for the generation of a much narrower pulse.

The author thanks Thomas O'Brien for his encouragement and comments, and George Pate and William Shephard for their suggestions and assistance.

BIBLIOGRAPHY

R. F. Shea, "Principles of Transistor Circuits," John Wiley and Sons, Inc., New York, 1953.
L. Krugman, "Fundamentals of Transistors," John F. Rider Publisher, Inc., New York.

UNIJUNCTION TRANSISTOR FORMS FLIP-FLOP

By E. KEONJIAN and J. J. SURAN

Electronics Laboratory, General Electric Co., Syracuse, N. Y.

Two-to-one economy in circuit components over conventional transistor multivibrators is afforded by multivibrator consisting of three resistors, one capacitor, one diode and one unijunction transistor. Circuit can be astable or monostable in operation and has particular application in digital computers and counters where component cost and network complexity can be restrictive

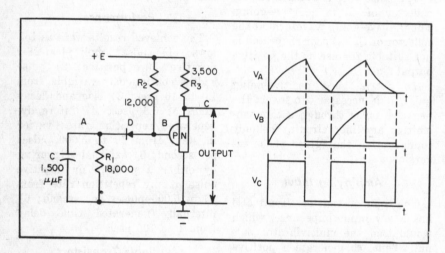

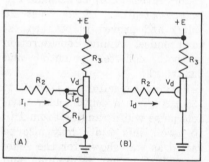

FIG. 2—Direct-current (A) and steady-state (B) equivalent circuits

FIG. 1—Basic unijunction transistor multivibrator (left) and waveforms for astable operation (right)

MOST FUNDAMENTAL of digital-type circuits is the multivibrator. The semiconductor multivibrator to be described may be considered as a diode flip-flop. Its active element is a unijunction transistor, which is a three-terminal, single-junction, negative-resistance device.[1,2,3]

An almost two-to-one reduction in circuit components required by the unijunction transistor multivibrator, compared to conventional transistor configurations, affords a higher degree of circuit simplicity, miniaturization and economy.

Operation

The basic circuit configuration of the unijunction transistor multivibrator is illustrated in Fig. 1.

During astable operation, C is charged from the battery supply through R_2 and diode D. During the charging cycle of the capacitor, D is conducting but the unijunction transistor is in the cut-off state. When the potential across the capacitor becomes equal to or greater than the peak-point potential of the unijunction transistor, the latter becomes unstable and switches into the conducting state. The junction potential at point B is then clamped almost to ground potential, causing D to become cut off.

When D is in its nonconducting state, point A is virtually isolated from point B. The capacitor then discharges through R_1 until the potential at A is approximately equal to the junction potential of the unijunction transistor. At this instant, the diode becomes conducting again.

When the diode reverts to its conduction state, the current through the junction of the unijunction transistor decreases and the latter is driven into its cut-off state. Capacitor C will then recharge and the cycle will be repetitive.

Waveforms

Since capacitor C alternately charges and discharges through R_2 and R_1 respectively, the waveform at A consists of a periodic ex-

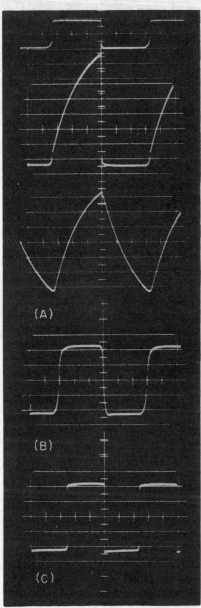

Waveforms for circuit of Fig. 1 operating as 10-kc astable mvbr (A); output waveforms for 55 kc (B) and 7 kc (C)

ponential rise and decay.

During the time D is conducting, the waveform at B is almost identical to that at A. However, when the unijunction transistor becomes conducting, the potential at B is clamped to a near-ground value until the capacitor has completed its discharge cycle.

When the unijunction transistor is in its cut-off state, the current through R_s is comparatively low. However, when the unijunction transistor switches on, its bar resistance drops by an order of magnitude and the current through R_s increases. Thus, the current

through R_s is either high or low, depending upon the operating state of the unijunction transistor and the waveform across R_s at point C is a square wave. Frequency and symmetry of this square wave are dependent upon the time constants associated with resistors R_1, R_2 and capacitor C.

Graphical Analysis

A better understanding of the operation of the multivibrator may be obtained by considering its equivalent circuit. Figure 2A illustrates the d-c equivalent circuit of the unijunction transistor when D is conducting. The capacitor is omitted and the diode is assumed to have negligible forward resistance.

The loop equations for this circuit are

$$E = (R_1 + R_2)I_1 - R_1I_d \quad (1A)$$
$$0 = -R_1I_1 + R_1I_d + V_d. \quad (1B)$$

In Eq. 1B, $V_d = f(I_d, E, R_s)$, which represents the input characteristics of the unijunction transistor for a battery supply E and load resistance R_s.

Solving for V_d as a function of I_d

$$V_d = \frac{R_1}{R_1 + R_2}E - \frac{R_1R_2}{R_1 + R_2}I_d. \quad (2)$$

When D in Fig. 1 is nonconducting, the steady-state equivalent circuit of Fig. 2B is obtained. It is assumed that R_1 is effectively isolated from the unijunction transistor by the high back resistance of the diode. For this circuit

$$V_d = E - R_2I_d. \quad (3)$$

The application of Eq. 2 and 3 to the operating characteristics of the unijunction transistor permits the graphical load-line analysis of Fig. 3.

For the condition that the diode conducts, the steady-state input load line is determined by Eq. 2 and is represented by the dashed line. The intersection of the load line with the ordinate axis is at a point $V_d = (E)\,[R_1/(R_1 + R_2)]$ and the slope of the load line is the parallel combination of R_1 and R_2.

For the condition that the diode is nonconducting, the load-line characteristic is determined by Eq.

3 and is represented by the solid load line (slope $= R_2$) in Fig. 3.

Astable Operation

For the diode multivibrator to be astable, or free running, the input load line should not intersect the unijunction transistor characteristic in the cut-off region when the diode is conducting.

When the diode is nonconducting, the input load line must intersect the unijunction transistor operating characteristic in the transition, or negative-resistance, region. Circuit conditions are

$$R_1E/(R_1 + R_2) > V_p \quad (4A)$$
$$E/R_2 \leq I_v. \quad (4B)$$

In Eq. 4A and 4B, V_p is the peak-point potential of the double-base diode and I_v is the input current corresponding to its valley point. The operating path of the multivibrator, in relation to the input characteristics of the unijunction transistor, may be approximately determined from the graphical analysis, as indicated in Fig. 3.

10-KC Generator

Figure 4 shows the characteristics of an experimental unijunc-

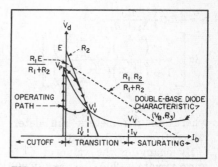

FIG. 3—Graphical load-line analysis of unijunction transistor characteristics

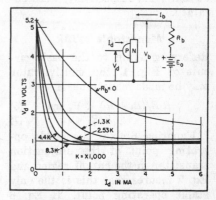

FIG. 4—Characteristics of experimental transistor similar to 4JD5SA1

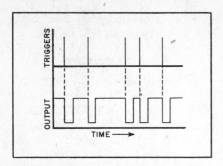

FIG. 5—Input and output of circuit of Fig. 1 for monostable operation

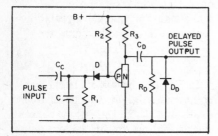

FIG. 6—Basic delayed-pulse generator

tion transistor, similar to the 4JD5A1, that was used in the circuit of Fig. 1 to obtain a 10-kc symmetrical waveform multivibrator with a maximum output of 3.7 v.

The period of oscillation can be determined from

$$t_T = -R_1C \left\{ \frac{R_2}{R_1 + R_2} \right.$$
$$\ln \left[\frac{1 - \left(\frac{V_p}{E}\right)\left(\frac{R_1 + R_2}{R_1}\right)}{1 - \left(\frac{V'_v}{E}\right)\left(\frac{R_1 + R_2}{R_1}\right)} \right] +$$
$$\left. \ln \frac{V'_v}{V_p} \right\} \qquad (5)$$

where all parameters are determined from Fig. 1 and 3.

By changing the value of C to approximately 300 $\mu\mu$f and 2,000 $\mu\mu$f, operating frequencies of 55 kc and 7 kc, respectively, were obtained.

Monostable MVBR

Referring again to the basic circuit of Fig. 1, the multivibrator may be made monostable if

$$R_1E/(R_1 + R_2) < V_p \text{ and} \qquad (6)$$
$$E/R_2 < I_v \qquad (7)$$

Equation 6 fixes the stable operating point of the unijunction transistor in the cut-off region and Eq. 7 insures that this is the only stable operating point. If Eq. 6 and 7 are satisfied, a positive pulse

will trigger the unijunction transistor from the off to the on state. The unijunction transistor will then remain conductive until the capacitor discharges through resistor R_1.

When the diode reverses at the end of the capacitor discharge cycle, the transistor becomes nonconductive. Since it is stable in cut-off state, the multivibrator circuit remains stable until the next positive trigger pulse is applied. Thus, the regenerated output waveform duration is

$$t_D = -R_1C \ln (V'_v/V_p) . \qquad (8)$$

Figure 5 shows the waveform generated by a monostable multivibrator. Minimum spacing of the trigger pulses is limited by the circuit's time constants.

On the other hand, if

$$R_1E/(R_1 + R_2) > V_p \qquad (9)$$
$$E/R_2 > I_v \qquad (10)$$

a monostable circuit, having a stable operating point associated with the conductive state of the unijunction transistor is obtained. Negative pulses may then be used to trigger the circuit into its regenerative cycle.

Delayed-Pulse Generator

Use of the monostable multivibrator as a delayed-pulse generator is illustrated in Fig. 6. Here R_D and C_D are used as a differentiating network while D_D filters out the pulses of unwanted polarity. The output waveform consists of pulses which are generated by differentiating the trailing edge of

the multivibrator output.

Diode D_D filters out the pulses which are generated by the leading edge of the multivibrator waveform. Hence, the output of the delayed-pulse generator consists of a train of pulses which have the same polarity and repetition rate as the input pulses but which are delayed in time by an interval t_D determined by the time constants of the monostable circuit.

Figure 7 illustrates the relationship between the pulse delay t_d and the magnitude of the multivibrator capacitor C for the experimental delayed-pulse generator circuit shown. In this circuit the conditions defined by Eq. 6 and 7 are required for operation. Time delays from 50 μsec to 2 millisec have been obtained for pulse repetition rates from 0 to 5 kc. The time delay is related to the magnitude of C in a linear manner. This relationship is convenient in design and facilitates constructing simple variable-delay pulse generators.

The advantages of this new circuit should be particularly significant in complex systems such as digital computers and counters where component cost and network complexity can be restrictive.

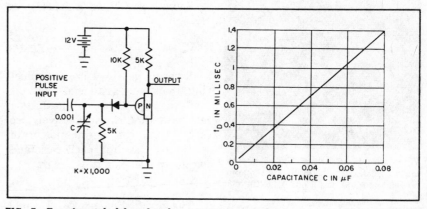

FIG. 7—Experimental delayed pulse generator and delay characteristics. Transistor used is similar to 4JD5A1

REFERENCES

(1) J. A. Leak and V. P. Mathis, The Double-Base-Diode—A New Semiconductor Device, 1953, *IRE Conv Rec*, p 2.
(2) R. F. Shea, "Principles of Transistor Circuits", p 466, John Wiley & Sons, Inc., New York, 1953.
(3) J. J. Suran, The Double-Base Diode—A Semiconductor Thyratron, *Electronics*, p 198, Mar. 1955.
(4) J. J. Suran and E. Keonjian, A Semiconductor Diode Multivibrator, *Proc IRE*, p 814, July 1955.

TRIGGER AND DELAY GENERATORS

By H. L. ARMSTRONG

Pacific Semiconductors, Inc., Culver City, Calif.

PULSE TECHNIQUES are used in a variety of modern electronic equipment, including digital computers, pulse code modulation systems, telemetering and radar. The advantages of transistors are being exploited more and more in these pulse circuits. The blocking oscillator type trigger generator and the multivibrator delay generator described here are examples.

Trigger Generator

The circuit shown in Fig. 1 is a trigger generator that generates relatively narrow pulses at an adjustable repetition rate. The trans-

Formerly with National Research Council, Ottawa, Canada

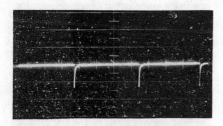

Vertical scale of 2 volts per cm and horizontal scale of 3 milliseconds per cm to show trigger generator output

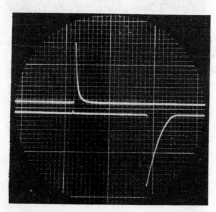

Double-beam oscilloscope was used to show positive trigger in and negative trigger out of delay generator

former for this blocking oscillator is not a regular pulse transformer

but a miniature audio type giving a three-to-one step down from collector to base. The 2.5 megohm variable resistor controls repetition rate.

Delay Generator

It is often desired to introduce a controllable delay somewhere in the train of pulse circuitry. For example, to see the leading edge of a multivibrator pulse on an oscilloscope, the scope would have to be triggered before the multivibrator. This can be done with the delay generator shown in Fig. 2.

Output of the trigger generator can be applied directly to start the oscilloscope sweep and at the same time applied to the delay generator. After a set time, the delay generator will produce a pulse that is used to trigger the multivibrator.

The delay generator is a transistorized monostable multivibrator. The trailing edge of the pulse at the collector of T_1 is sharpened and used as the delayed trigger. The delay is controlled by the 10,000-ohm potentiometer.

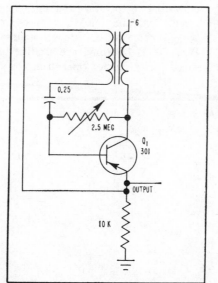

FIG. 1—Audio transformer is used to provide positive feedback in blocking oscillator type trigger generator

In some cases, it may be possible to dispense with the potentiometer,

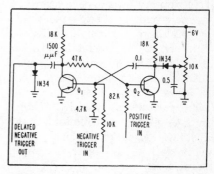

FIG. 2—Positive or negative delayed triggers can be gotten from transistorized monostable multivibrator

as well as the associated diode and capacitor. The delay may be controlled by switching in other capacitors in place of the 0.1 μf capacitor that couples T_2 to T_1. A variable resistor of up to a few thousand ohms in series with this capacitor will also give a certain amount of control of the delay.

These circuits are quite economical of battery power. The delay generator draws about one milliampere, of which half goes to the 10,000-ohm potentiometer for the delay control. The trigger generator, in typical use, drew only 15 microamperes.

Although *pnp* transistors were used in these circuits, *npn* types would work as well. Then all polarities would be reversed, and the diodes used in the delay generator would have to be reversed.

MULTIVIBRATOR SWEEP GENERATOR

By H. L. ARMSTRONG

Pacific Semiconductors, Inc., Culver City, Calif.

TRANSISTOR multivibrators can be operated either as triggered or free-running sweep generators.

The circuit of a triggered sweep generator is shown in Fig. 1. Initially, Q_2 is conducting rather heavily while R_2 provides enough bias current to almost bottom the collector of Q_2. Transistor Q_1 is near cut off. A positive trigger cuts Q_2 off and its collector voltage instantly rises in absolute value. This rise, which causes the initial step in the waveform of Fig. 2 is coupled to Q_1, turning it on. Then the signal at the collector of Q_1 is fed back to Q_2, cutting it off.

The current drawn by Q_1 through R_1 makes the voltage available at point A for Q_2 less, in absolute value, than it would otherwise be. This voltage begins to charge capacitor C, and the charging current keeps Q_1 turned on. As C becomes charged, the rate of charge, (slope of the sweep), tends to decrease. This causes Q_1 to draw less current. Thus the voltage at point A increases in absolute value, tending to maintain the sweep rate. This feedback improves the linearity of the sweep.

At the end of the sweep, Q_2 begins to conduct. The change which this produces at the collector is coupled to Q_1, cutting it off.

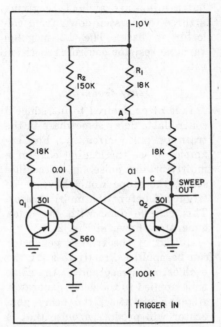

FIG. 1—Triggered sweep generator can be made free-running by increasing value of resistance R_2

This feeds back a signal to Q_2, causing it to conduct very heavily discharging C. Thus the flyback is achieved at the end of the sweep.

If *npn* transistors were used all polarities would be reversed, thus obtaining a positive sweep. The circuit can be made to free-run by altering R_2 to larger values. Waveforms obtained are similar to those for triggered operation.

Figure 2 shows a typical waveform for triggered operation. The trigger, shown on the lower trace, represents about 80 microamperes peak signal. Duration is not critical. The output sweep waveform is shown in the upper trace. Its linear part represents about 0.5 volt, and duration is about two milliseconds.

The output of the multivibrator can be loaded heavily enough to drive an ordinary grounded-emitter transistor amplifier stage. In one application it was possible to connect the output directly to the base of the amplifier stage through a 10,000-ohm resistor without harming the waveform.

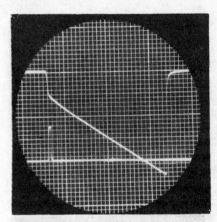

FIG. 2—Trigger pulse and sweep waveform for simple transistor sweep generator

Chapter 6

HOME-ENTERTAINMENT CIRCUITS

FOUR-TRANSISTOR REFLEX RADIO

By ERICH GOTTLIEB Application Engineer, General Electric Company, Syracuse, New York

THROUGH REFLEX circuitry a 6-v 4-transistor radio can duplicate at lower cost the performance of a 5-transistor radio having a 200-μv/m sensitivity and a power output of 50 mw at 10 percent distortion. The radio uses standard transistors and has no more playthrough and distortion than the average receiver.

Electron-tube reflex radios were fairly common in the United States and Australia in the pre-World War II era and are still in commercial use in Australia. These receivers suffer from serious distortion and high playthrough, although the advent of remote-cutoff tubes and the use of low a-f plate load resistors have provided a considerable improvement. Now the development of transistors has reawakened the industry's interest in such a receiver since there is no longer a filament feedthrough problem, and the saving of one transistor is still incentive enough to warrant a closer look.

Characteristics

The diagram of one such receiver is shown in Fig. 1. The output stage is a single-ended class-A circuit operating at 23 ma collector current with a collector-to-base voltage of about 4.5 v. The d-c dissipation is about 100 mw. Using a commercial output transformer this circuit will give 50 mw maximum power output at 10 percent distortion and 40 mw with less than 5 percent distortion. The bias and temperature stability is adequate to permit the 2N241A to perform acceptably up to 55 C ambient temperature without danger of damaging the transistor.

The driver is part of the reflex system, but it can be treated separately since there is little interaction between audio and i-f loads. Transistors are essentially inde-

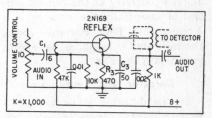

FIG. 2—Audio function of reflex circuit

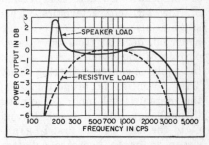

FIG. 3—Overall audio response curve

pendent of collector voltage within the 3 to 12-v range and a relatively large collector resistor therefore affects i-f gain little from a d-c standpoint. This audio load must be bypassed for i-f signals. The i-f load is essentially a short circuit at audio frequencies, thus permitting the use of a split input and output load.

The audio signal is taken off at the volume control by C_1 and applied at the base of the 2N169 which is used as a combination audio driver and second i-f amplifier. The audio function alone is detailed in Fig. 2. The additional gain of this reflex circuit over a conventional circuit is basically the audio power gain of the R-C coupled amplifier, which is approximately: $P_G = h_{fe} (R_L/h_{ib})$.

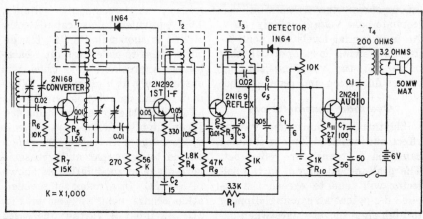

FIG. 1—Schematic of four-transistor reflex portable radio using six-volt supply

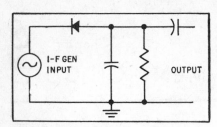

FIG. 4—Equivalent circuit of detector

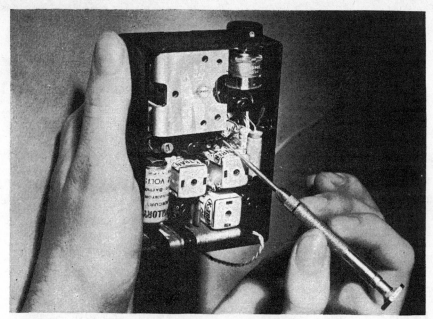

Compact superheterodyne receiver made by Westbury Electronics uses four transistor reflex circuit to get five transistor performance

Since both R_L and h_{ib} are somewhat fixed by i-f overload considerations the best gain can be obtained with high h_{fe}, which is 50-200 in the 2N169. This transistor also has a narrow h_{fe} spread.[1] The audio sensitivity at the volume control is essentially the same as in a 5-transistor R-C coupled radio. Overall audio response is shown graphically in Fig. 3.

Detector and Second I-F

In the detector stage a slightly forward-biased diode operates out of the square-law detection portion of the I-E characteristics. This stage is also used as source of agc potential, derived from the filtered portion of the signal as seen across the volume-control detector load. This potential is proportional to the signal level and is applied through agc filter network R_1 and C_2 to the base of the first i-f transistor so as to decrease collector current at increasing signal levels. The operating point of the first i-f stage is chosen at 0.6 ma to obtain almost optimum gain at a point where it takes little power to get maximum avc action.

The second i-f stage is conventional, with the operating point at 1 ma collector current. This represents about maximum gain for relatively small supply current. Two important changes, however, stem from its use as a reflex stage. The first is the large emitter bypass capacitor C_3. This bypasses R_3 in the emitter both for i-f and audio and must therefore have low impedance at both frequencies. Second and more important is the choice of i-f transformer. With a supply voltage of 6 v, an operating point of 1 ma and a total emitter and collector load resistance of

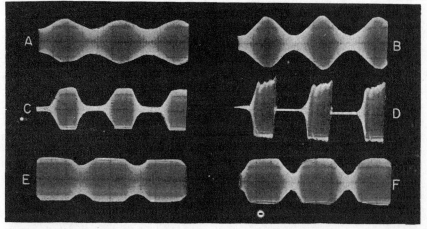

FIG. 5—Detector waveforms at various levels of signal and modulation. No distortion occurs at medium signals with 30-percent modulation (A) and 60-percent modulation (B). Other patterns show clipping or regeneration

1,500 ohms, the remaining collector-to-base voltage is only 4.5 v. At high signal levels the peak-to-peak a-c swing may approach this value and the resulting clipping causes distortion and regeneration.

Overload Considerations

Playthrough, minimum volume effect and overload have a common cause in a transistor reflex set. The peak-to-peak signal in the collector will tend to exceed the applied d-c potential, causing clipping in high signal level stages when op-

erating at low supply voltages. The equivalent classical detector circuit is shown in Fig. 4. The diode (base-to-collector) is back-biased appreciably and thus will operate only at high signals. When it does, the modulation envelope is affected in the following sequence. When clipping first occurs there is a squaring of the envelope accompanied by an apparent increase in percentage of modulation as shown in Fig. 5C. In a reflex set regeneration occurs next as seen in Fig. 5D. As input is further increased

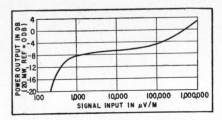

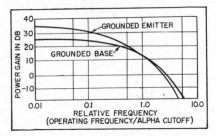

FIG 6—Severe bend in avc characteristic is necessary to avoid signal clipping

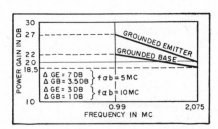

FIG. 7—Grounded-emitter configuration offers greater gain in broadcast band

the modulation is affected in a compression-like manner giving the appearance of a decreased modulation percentage as in Fig. 5E and 5F.

As the input signal is increased still further the envelope is compressed nearly to the point of complete elimination of the modulation. As a result of such severe overloading little or no audio at all is seen at the detector. This condition is not a result of reflexing but can be found in any transistorized radio without auxiliary or multiple stage avc. In reflex circuits, however, another problem arises from this clipping.

Referring again to Fig. 4, the clipped and thereby detected signal appears across the same a-f load resistor as the normal feedback signal. Since both are in phase, regeneration occurs at high signal strength as in Fig. 5D.

The limiting factors in terms of clipping are the supply voltage and the i-f load impedance. If the latter were low, more power would be received at the output stage before i-f clipping, thus delaying this clipping to a larger signal input.

Desirable design criteria in this radio thus require that no i-f clipping occur before the audio stage obtains enough drive to produce maximum output. At that moment the avc should take over radically and keep the signal from becoming large enough to cause clipping in the output i-f transformer. Thus high audio system gain, a supply

voltage as high as possible, a low-impedance i-f transformer and an excellent avc system are the ingredients of the successful application of reflex circuits to transistor radios.

The audio gain in this radio is the gain of the reflex driver plus the gain of the class-A output stage. It is quite adequate and about 60 db. The supply voltage is fixed at 6 v, the reflected impedance of the i-f transformer is only about 3,600 ohms and auxiliary avc gives adequate performance.

First I-F, AVC and Converter

The first i-f stage is conventional. In the operation of the auxiliary avc a diode is connected to the primary tap of the first i-f transformer, where the d-c potential is fixed at 5.75 v by the stable operating point of the converter transistor. When tied to the d-c load of the first i-f stage at the top of R_4, this diode has about 0.75-v reverse bias and will thus appear as a high impedance. As the signal level at the detector creates an agc potential the collector current of the first i-f stage drops and the voltage at R_4 rises. At high signal level this d-c voltage approaches 6 v, thus forward-biasing the diode. The resulting low impedance shunts T_1 and thereby reduces the gain at this point. The avc characteristic is shown in Fig. 6.

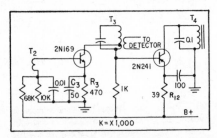

FIG. 8—Lesser slope of grounded-base power-gain characteristic assures better linearity over operating frequency range

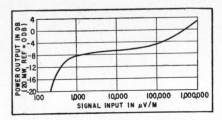

FIG. 9—Circuit modifications for direct coupling of output audio amplifier

In the converter stage the mixer operates in the grounded-emitter configuration for optimum gain. The oscillator operates in the grounded-base configuration since more linear gain over the frequency band can be obtained. The reasons for this choice can be seen in Fig. 7 and Fig. 8. Gain superiority of the common-emitter (over the common-base) configuration in the broadcast frequency range is apparent. The common-base configuration has a lesser slope, however, and will give more linear gain over the band.

The bias stability factor is expressed as $(1/R_5)/(1/R_6 + 1/R_7)$ and in this case is 4. This permits the replacement of the converter transistor without great variation of operating point. The oscillator will function down to one-half the supply voltage without appreciable frequency shift.

Direct Coupling

The advantages of the reflex radio are savings in size and cost since three resistors and one transistor have been eliminated. Further economies can be effected by the use of an *npn* transistor driving a *pnp* unit, which lends itself well to the application of direct coupling. As shown in Fig. 9, C_5, R_{10} and R_{11} have been completely eliminated and the base bias voltage of the output transistor derived from the drop across R_9. This voltage is fixed since the relatively good stability figure of the driver stage fixes the collector current. By lowering R_{12} and changing the stability figure of the driver stage, varying degrees of stability can be obtained for the output circuit.

Using this circuit the radiated sensitivity at 24 in. is increased to 120 μv/m at 1,000 kc. This improvement results from the removal of the initial coupling and bias circuit which introduced a 1-db loss.

In addition to reducing the set drain by the 1 to 2 ma formerly required by the bias circuit of the output stage, the set size and cost is additionally reduced by one electrolytic capacitor and two resistors.

REFERENCE
(1) T. P. Sylvan, Conversion Formulas For Hybrid Parameters, ELECTRONICS. p 188, Apr. 1, 1957.

MINIATURE FERRITE TUNER COVERS BROADCAST BAND

By E. A. ABBOT and **M. LAFER**
Chief, Communications Department   Chief, Mechanical Design Department
Emerson Radio and Phonograph Corporation, Jersey City, New Jersey

Rotary-axial tuner consists of two pairs of ferrite cups with ground D-shaped center cores ganged to produce linear frequency variation from 500 to 1,600 kc with mechanical motion. Operating frequencies can be extended to 15 mc. Tuning sensitivity is reduced as each band is covered in 270-degree rotation rather than in the 180 degree of normal capacitor tuning

GROWTH IN TRANSISTOR use has spurred the search for miniature electronic devices.

Miniaturization of an r-f tuner operating between 0.5 and 50 mc can be achieved with permeability rather than capacitance tuning. Further, permeability tuning is free from vibration and shock troubles. A linear, permeability type tuner is described for which a 3 to 1 frequency range, maximum Q and minimum variation of Q with frequency are assumed to be desirable.

Gap Tuner

Present tuners may be classified as slug tuners, gap tuners, variometers and those that vary the number of turns on a core.

A permeability gap tuner is the best compromise between electrical and mechanical considerations. It uses the relative motion of two ferrite cups or a ferrite cup and cover plate or two C-shaped cores. Although the core material is always within the coil, the inductance is varied by changing the size of a gap in the magnetic-field path.

A gap tuner is shown in Fig. 1A. The tuning technique permits miniaturization since the cover movement does not exceed ⅛ in. for a 3 to 1 frequency range.

Frequency variation with cover movement is nonlinear. A 250-kc change occurs in the first 10 mils of travel, and at the upper end of the band, a 10-mil change may cause a 25-kc shift. A precise mechanical drive system is required to control the complete frequency range in a ⅛-in. travel.

Another gap tuner is shown in Fig. 1B. The center section of the cups has the shape of a D. Rotation of one cup with respect to the other changes the effective gap length. Frequency changes of 2 to 1 are feasible with this device. Furthermore, since the complete frequency range covers a long rotary path the mechanical drive system is not critical.

The rotary-axial gap tuner takes advantage of the slow frequency variation of the D-type tuner and the wide frequency variation of the cup and cover-type tuner. It consists of two pairs of ferrite cups with ground D-shaped center cores.

Tuner operation is described in the curves of Fig. 2. Curve 1 is the frequency response obtained upon rotating two ferrite-cup cores with D-shaped center cores without gap separation. Frequency initially varies slowly as the cups are rotated, increases to a maximum when the D figures are mirror images at 180 deg and then de-

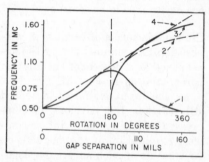

FIG. 2—Curve 4 is the linear frequency response of the rotary-axial tuner obtained by combining responses of curve 1 for D-shaped center cores without gap separation and curve 3 for a tapered coil within the core

creases slowly. The curve is bell shaped.

When separating two ferrite-cup cores without rotation, curve 2 is the frequency response obtained. The frequency varies rapidly and then reaches a point where increased gap separation has no effect.

When a tapered coil is used inside the core, the frequency response of curve 3 is obtained. Saturation frequency is 100 kc higher than in curve 2.

Composite curve 4 is obtained when both rotation curve 1 and axial-movement curve 3 are combined by a cam. Since the gap separation is small, rotation of the cup initially exerts the greater control over frequency. Therefore, the frequency varies comparatively slowly.

During the first 250-kc change there is a long mechanical path of

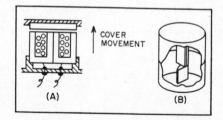

FIG. 1—In gap tuner (A) inductance is varied by changing gap size in magnetic-field path. Rotation of one cup of D-type tuner (B) with respect to the other changes length of the gap

Rotary-axial tuner size is indicated by comparison with transistor

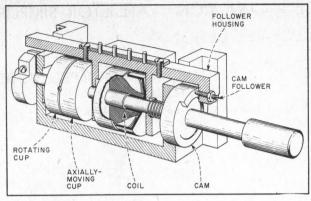

FIG. 3—Cutaway view of rotary-axial tuner shows parts

rotation rather than a short mechanical path of axial movement. As the cups are rotated 180 deg with respect to each other, the gap separation increases until the D-shaped center-core separation no longer affects the frequency change. Thus, the right half of the bell-shaped curve has no effect on the frequency. Beyond 180 deg the frequency change results from gap separation, although the cups continue to rotate with respect to each other. The resulting curve 4 has a linear frequency variation from 500 kc to 1,600 kc.

By further tapering the coil the frequency range may be extended linearly to 1,700 kc. Maximum possible Q is 151.

Mechanical Description

Axial motion of the cups is caused by the cam shown in the tuner drawing of Fig. 3.

A cam follower is in contact with the cam. After adjusting the cam

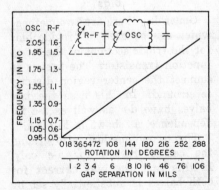

FIG. 4—Straight-line tracking is achieved with rotary-axial tuner

follower to provide initial positioning, it becomes a fixed element of the follower housing and provides linear movement of the axial cups in accordance with the cam form.

Linear motion of the axial cups without backlash is provided by a spring.

The portion of the shaft which is in the immediate area of both rotating and axial cups is made of a dielectric material.

Three-point tracking of the r-f coil is obtained by placing shunt and series coils, together with a new tuning capacitance, in parallel with the r-f coil. A 2 to 1 oscillator-frequency range results. The oscillator tuning follows the r-f tuning curve as shown in Fig. 4.

To show the application of the tuner to transistorized circuits, the tuner r-f coil is connected by a capacitance divider to the input of the transistorized mixer circuit shown in Fig. 5. Mixer output is fixed tuned with a coil resonant at 455 kc. The oscillator coil of the tuner is connected through two trimmer coils to the collector of transistor Q_2 which acts as a Clapp oscillator. The oscillator signal is capacitance-coupled to the emitter of the mixer stage. After making adjustments for stray capacitance, the tuner operates with the linearity and tracking characteristics of Fig. 4.

Tuners covering the ranges 1.5 to 5 mc and 5 to 15 mc have the same frequency slope as the broadcast-band tuner, and therefore can be used with the same cam. When

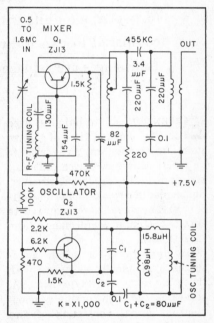

FIG. 5—Transistorized circuit checks r-f tuner whose linearity and tracking characteristics are shown in Fig. 4

the coils are successively tapped down after each rotation of the cam, a semicontinuous tuner can be constructed. A coil wound with several taps is placed in ferrite cups and tested in an r-f tuning jig. The bands from 0.5 mc to 1.5 mc and from 1.5 mc to 5 mc are easily covered by tapping down the coil. The highest band shows an upper-frequency limit of 7.3 mc. The total frequency ratio obtained is 14.6 to 1 with the same tuner.

The aid of Capt. C. Green and G. Tarrants in this work, done under contract No. AF 33(600)-31464 is acknowledged.

TETRAJUNCTION TRANSISTOR SIMPLIFIES RECEIVER DESIGN

By R. J. FARBER, A. PROUDFIT, K. M. ST. JOHN
and C. R. WILHELMSEN
Hazeltine Research Corporation, Little Neck, N. Y.

Dual-triode transistor, with emitter of one unit and collector of the other section of same germanium region, provides performance of two triode units with considerable reduction of circuit components when compared to two individual units. Superheterodyne receiver in which first four stages are replaced with two tetrajunction units is described

EXTENSION of techniques used in the construction of grown junction transistors gives rise to an interesting new device that may be applied to a broadcast receiver. Some modifications to normal circuit arrangements are required to obtain optimum performance from these units.

An *npn* transistor, often represented symbolically as a bar with three zones as shown in Fig. 1A, serves as the base for the development of a bar consisting of two transistors formed in a single unit as shown in Fig. 1B.

When work on this device was initiated, it was based on the thought that the new approach might lead to significant economy in the cost of a broadcast receiver because of reduced transistor cost. Early indications lent support to this expectation. At present, manufacturing techniques have apparently not advanced to the point of immediate production.

In the tetrajunction transistor, the emitter of one unit and the collector of the other are parts of the same germanium region. This limitation on freedom of connection results in applications of the new transistor being somewhat more involved than for a dual-triode tube.

General

Since use of a transistor in the common-emitter configuration provides the most power gain, it is generally desirable to operate both sections of the tetrajunction transistor in this fashion. If the input to a transistor is connected between base and emitter and the output is connected between collector and emitter, the emitter is common to input and output, hence the transistor provides maximum power gain. This is irrespective of the ground point in the circuit, as shown in Fig. 2A. All of the circuits in this figure have the same power gain.

As shown in Fig. 2B, operation of the two triodes can be independent. Ground can be placed at any point. In the receiver to be described the central collector-emitter region was chosen, as shown.

The operation of the two halves of the tetrajunction transistor is then as follows: the upper half is connected in a conventional grounded-emitter common-emitter circuit using a bootstrapping technique. In this way the two halves are operating with complete independence. There has been no evidence to date of any interaction between the two sections because of the common use of the single "N" layer.

Bias

Generally, there are two methods which can be used to provide bias for the triode sections of the tetrajunction transistor. The first is to connect the center region directly to ground. In this case the two halves have d-c as well as a-c independence, as shown in Fig. 3A.

The second method is to ground the center region for a-c only. In this case the bias current for both halves is substantially set by the biasing elements for the bottom half alone as shown in Fig. 3B. The choice of use of the two methods will depend, to a large extent, on the type of circuit being implemented. The second method, how-

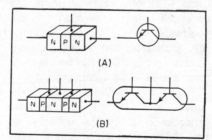

FIG. 1—Normal characterization of triode transistor (A) and that of tetrajunction unit (B)

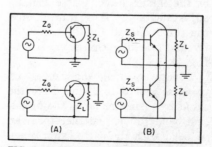

FIG. 2—Configurations shown give maximum power gain regardless of ground point. Central collector-emitter region has not produced interaction between the sections so far

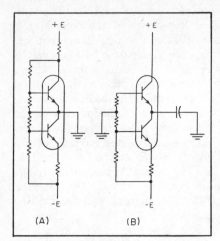

FIG. 3—Tetrajunction unit can be biased by connecting collector-emitter region directly to ground (A) or by grounding center region for a-c only (B)

Component layout of tetrajunction transistor receiver. Two tetrajunction units are visible at right

ever, generally can result in a component saving.

Thus the tetrajunction transistor is a device that potentially can be used to replace the transistors in any existing receiver design, two at a time. The tetrajunction transistors studied to date have been of small crosssection, suitable for low-level r-f and audio applications.

Two of these units plus a diode and normal circuit elements can accomplish all the functions of a receiver up through the audio driver. No restriction is placed on the audio output stage and any of the popular class A or class B arrangements may be employed.

Problems

When an attempt is made to use the tetrajunction device in the circuit of an existing receiver, several problems arise because of the common connection between the emitter and collector which is generally grounded, at least for signal frequencies.

Since the two halves of the device are in series, to achieve the same collector-to-base voltage as when using two separate triodes in a conventional manner, a battery of

a higher voltage is required. The power requirements, in both cases, however, will be the same. In addition, for some applications, it is desirable that the battery be center-tapped. This is done to advantage in the receiver described The push-pull *pnp* audio stage can take advantage of the full battery potential regardless of the connections elsewhere in the receiver.

Stabilization

The stabilization circuits require some review. A typical method of biasing a triode transistor amplifier

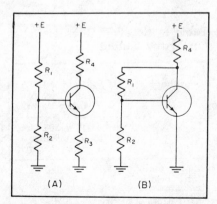

FIG. 4—Typical biasing configuration for temperature stability (A) is modified to provide required degree of stabilization for top triode of tetrajunction unit (B)

is shown in Fig. 4A. Here, d-c degeneration is introduced by R_3 to stabilize the operating point against changes in temperature and/or variation in transistor parameters from unit to unit. In some tetrajunction transistor circuits however, the center region is grounded. Therefore no stabilization resistor can be inserted between the emitter of the top triode and ground. It is possible to get the required degree of stabilization for the top triode by using the circuit of Fig. 4B, in which R_4 introduces the necessary degeneration.

However, for a given degree of stabilization, the resistor in the collector circuit is generally considerably larger than one which would be used in the emitter circuit. Thus with a given size battery and current in the transistor, there will be a larger voltage drop and hence less voltage applied to the transistor.

Using the bootstrap technique for the bottom half of the transistor presents some additional problems. In a transformer-coupled stage, both sides of the winding feeding the base-emitter input are hot with respect to ground as shown in Fig. 2B. This can at times result in undesirable coupling to other circuits. Direct-current biasing elements for the base generally constitute shunt paths to ground and may result in a compromise between stabilization and stage gain. Furthermore, capacitance between the base and ground, such as stray capacitance or capacitance from the high side of the input coupling network to ground, constitutes feedback between collector and base, since the collector is grounded. This may require neutralization.

Receiver Description

A block diagram of a typical transistor receiver which also applies to a receiver using tetrajunction transistors is shown in Fig. 5A. The set consists of an autodyne converter, two i-f stages, a diode detector which also supplies the agc voltage, and a class A audio amplifier driving the push-pull class B output stage.

Figure 5B shows a modified block diagram of the configuration with each half of the two tetrajunction transistors, as well as two 2N241 audio output transistors. The present arrangement was chosen because it provides an optimum solution to the problems described, as well as offering some chance for additional economy. As pointed out before, the receiver uses only two tetrajunction units, two audio units and a diode, while giving the performance of a six-transistor plus diode set. Figure 6 shows the schematic diagram of the complete receiver.

Conversion

The autodyne converter utilizes the bottom half of the first tetrajunction transistor in a grounded-collector common-emitter bootstrap circuit. The oscillator uses a common emitter circuit with feedback from collector to base.

A 1000-ohm resistor bypassed by a 0.01-μf capacitor is inserted between the emitter and the point common to the input and output circuits. This R-C combination provides low-frequency degeneration in the emitter-base circuit, reduces squegging tendencies and provides some degree of stabilization.

The loop antenna consists of a ferrite core as large as is consistent with the desired packaging of the receiver. A spaced winding of Litz wire contributes to a high unloaded Q. The maximum input to the converter is obtained with the resistive component of the input impedance of the converter reflected across the tuned primary equals the equivalent unloaded tuned antenna circuit resistance. This must be consistent with having a loaded Q sufficiently large to obtain satisfactory image rejection, but not so large as to make tracking a problem, or to narrow excessively the overall bandwidth.

Inasmuch as the quality of the antenna is affected by its proximity to the chassis, the antenna becomes a compromise between the packaging problem, sufficient sensitivity, adequate image rejection, ease of tracking, and a location that reduces undesired coupling.

The distributed primary winding tunes the desired range of 535 kc to 1610 kc with a 220-$\mu\mu$f variable capacitor. The secondary winding transfers the energy to the base-emitter circuit of the transistor, using the bootstrap technique. This is done by a direct connection to the base and through the oscillator tickler and a 0.01-μf capacitor to the point common to the input and output circuits.

The primary of the i-f transformer is connected between the emitter and the tap on the oscillator tank. Since there is little impedance at i-f between this latter point and the grounded collector, the transformer serves as an i-f load between collector and emitter.

The first i-f amplifier uses the

Table I—Receiver Characteristics For 50-mw Output

Sensitivity	350 to 500 μv/m 66 to 69db below 1 v/m
ENSI	15 to 20 μv/m 94 to 97db below 1 v/m
1.0-mc, 6-db bandwidth	6.0kc
AGC figure of merit	32db
Total battery drain	23ma

top half of the first tetrajunction transistor in a grounded-emitter common-emitter circuit. The output impedance of the converter is matched to the 500-ohm input impedance of the first i-f amplifier by the input i-f transformer.

Correct stabilization is obtained by returning the 33,000-ohm base bleeder to the collector side of the 2,200-ohm collector resistor. Any change in collector current resulting from variations from unit to unit or of temperature will change the voltage across the bleeder. Thus the current will tend to be maintained at a value of 1 ma.

The base bleeder and associated circuitry is designed to bias the first i-f transistor at 1 ma, place the correct amount of forward bias across the diode (about 0.10 v), provide satisfactory agc action and filtering without undue audio attenuation and provide satisfactory stabilization and a satisfactory load for the diode detector. The internal feedback capacitance in the individual triode units is sufficiently small so no neutralization network is required in most cases.

Interstage

The 15,000-ohm output impedance of the first i-f is matched to the 500-ohm input impedance of the second i-f stage by the interstage i-f transformer. The secondary of this transformer is layer wound rather than bifilar wound, as in the other transformers, to minimize capacitance to ground.

The 150,000 and 33-000-ohm resistors in the base bleeder circuit are a compromise between satisfactory stabilization and shunting of the load for the stage. The by-passed 1,000-ohm emitter resistor provides the necessary d-c degeneration for stabilization.

The second i-f stage is also biased to draw 1 ma. Since it and the audio driver are in series, with no d-c connection between the central collector emitter region and ground, the latter stage is also stabilized at 1 ma d-c.

The load for this i-f stage is connected between emitter and ground, with the emitter connected to the bottom of the primary rather than the tap so that a voltage of the correct phase may be fed back to

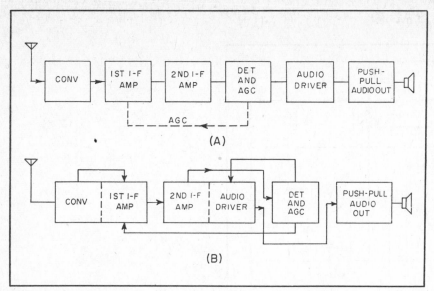

FIG. 5—Block diagram of standard superheterodyne receiver using triode transistors (A) and same receiver using tetrajunction transistors (B)

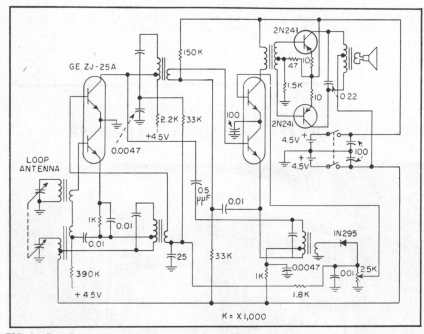

FIG. 6—Complete schematic of tetrajunction superheterodyne

through the combination of the 1,800-ohm resistor and 25-μf capacitor, a time constant of 0.045 second.

The bias current is set by the current in the second i-f. The base is essentially at ground potential, as previously described. The driver transformer provides an output load of approximately 40,000 ohms to the driver. This impedance is chosen so that the driver collector runs out of voltage at the same point as do the collectors.

Audio Output

The output stage is a common-emitter push-pull class B amplifier using two 2N241 *pnp* transistors. A slight amount of forward bias is provided by the 1,500-ohm and 47-ohm resistors to reduce crossover distortion. The unbypassed 10-ohm resistors in the emitter circuits provide d-c stabilization and some audio degeneration that reduces the overall distortion. The audio frequency response is limited by the 0.22-μf capacitor across the primary of the output transformer.

The output stage is placed across the entire 9-v battery to reduce the current drain, as well as to minimize distortion by using higher voltage—lower current operation. When loaded with a 12-ohm loudspeaker, the output transformer presents a load of approximately 1,100 ohms, collector to collector, which constitutes a compromise between gain and maximum audio output. One may be exchanged for the other depending on the desired receiver characteristics. In this receiver, maximum power output of 200 milliwatts and undistorted power output of 150 milliwatts are obtained.

Performance characteristics of the set are shown in Table I.

There has been some thought given to the possibility of constructing the tetrajunction transistors from bars having a larger crosssection than the present units. While such transistors will have large capacitances, they should also have high peak current capability and thus be useful for high-level audio applications. Acknowledgment is made to C. J. Hirsch, B. D. Loughlin and A. V. Loughren for their guidance and assistance.

the tap on the primary of the interstage transformer to effect correct neutralization. The 0.5-μμf capacitor provides the feedback coupling to neutralize the additional collector to base feedback resulting from stray capacitances between base and ground. The collector is bypassed by a 100-μf capacitor, thus effectively isolating the second i-f and audio stages. The collector is essentially at ground potential because the base of the upper half is connected to ground through the 2,500-ohm volume control, the voltage drops across the

volume control and emitter junction being small.

The second detector uses a point-contact diode (1N295 or equivalent) as a rectifier. The diode is slightly forward biased for maximum sensitivity on weak signals. The 15,000-ohm output impedance of the second i-f amplifier is matched to the approximately 1,000-ohm input impedance of the detector by the i-f output transformer. The 2,500-ohm volume control serves as the detector load, with the 0.01-μf capacitor acting as an i-f bypass. The agc bias is fed

TWO-TRANSISTOR REFLEX RADIO

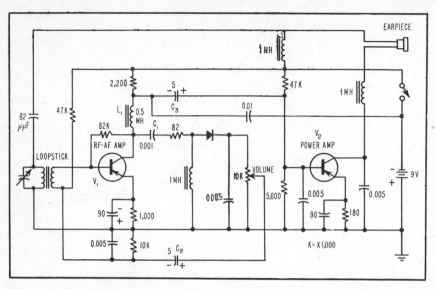

Two transistor receiver circuit using reflex r-f and a-f amplifier. The r-f signal is amplified by V_1 and coupled through C_1 to crystal detector. The detected audio is then returned to transistor V_1 through C_2 for amplification. Audio output of V_1 is coupled to to power amplifier V_2 through L_1 and C_3. Circuit is used in kit made by Allied Radio Corp.

PUSH-PULL OUTPUT BROADCAST RECEIVER

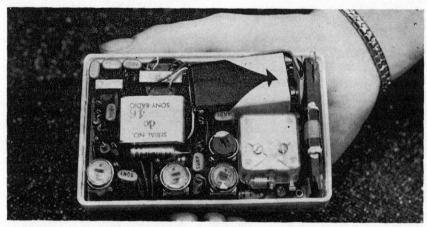

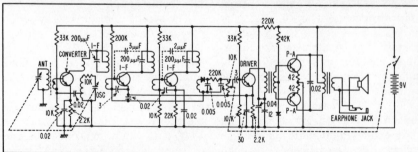

Latest in a line of compact Japanese broadcast receivers is the Sony TR-63 that employs six transistors and a varistor to avoid damage in case the battery is reversed. The varistor also compensates for temperature effects. Sensitiviy of the receiver is a millivolt and it has a maximum undistorted output of 20 mw. Selectivity is approximately 15 db 10 kc off. Set is made by Tokyo Tsushin Kogyo, Ltd., Western Electric Transistor licensees

FIVE-TRANSISTOR BROADCAST RECEIVER

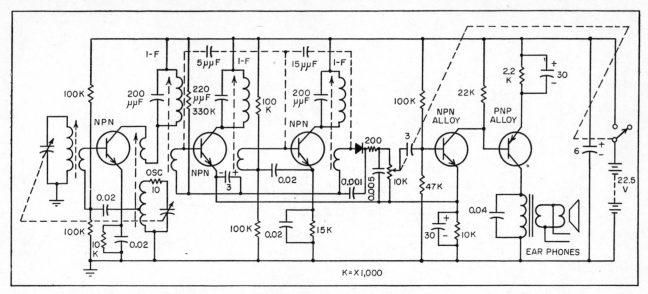

K = X 1,000

Similar to the Sony TR-55 receiver described on p 174 of the Dec. 1955 issue, the TR-57 shows evidences of simplification, having fewer parts, although it uses a 22.5-volt battery. Sensitivity is improved, being about 1 microvolt per meter. Selectivity is said to be 15 db and maximum undistorted output 20 milliwatts. The *npn* audio amplifier is a special alloy type

---

SIMPLIFIED REFLEX-TYPE RECEIVER

By S. A. SULLIVAN

Sonoma, Calif.

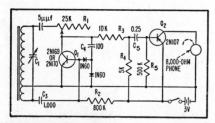

FIG. 1—Transistor Q_1 is used regeneratively as an r-f amplifier and reflexively as the first a-f amplifier while Q_2 functions as the power amplifier

SENSITIVITY and selectivity are said to be quite good for a transistor radio that uses a minimum of parts. The circuit is a descendant of the reflex-type of the early twenties.

In Fig. 1, r-f energy applied to the base of Q_1 through C_3 is ampli-fied and coupled to the diodes through C_4. Resistor R_3 acts as an r-f choke. The audio output of the diodes is applied directly back to

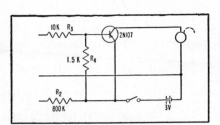

FIG. 2—Alternate output stage eliminates C_5 and R_5 but R_2 and R_4 must be determined by trial and error

the base of Q_1, which now functions as the first audio amplifier. The amplified audio is coupled to the output stage by C_5. Regeneration through C_2 is controlled by R_1. An 800,000-ohm resistor is nominally selected for R_2, but the correct value must be chosen empirically.

Performance can be improved even further by replacing R_3 and R_4 with r-f and audio chokes, but size and cost will be increased.

Using the direct-coupled alternate output circuit shown in Fig. 2, C_5 and R_5 can be eliminated. However, R_2 and R_4 become interdependent and require more juggling to determine optimum values. In addition some loss of output is experienced because of the shunting effect of R_4 on the input of Q_2. Occasionally a transistor will be found for Q_2 the characteristics of which also allow R_4 to be eliminated.

ELEVEN-TRANSISTOR AUTOMOBILE RADIO

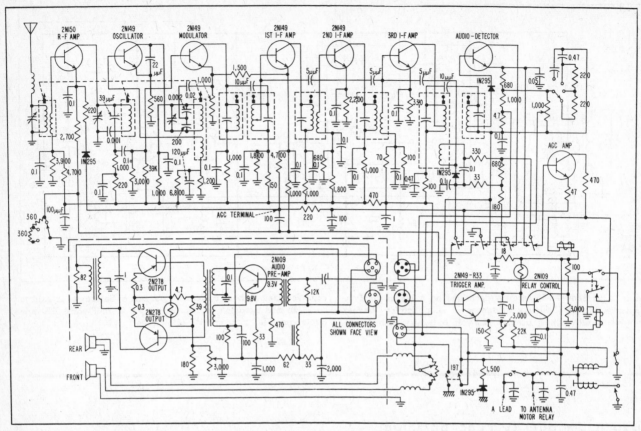

The Delco transistor radio used in the Cadillac El Dora brougham employs 11 transistors together with two more in the relay control circuits. Push-pull output stage supplies 10 watts to **loudspeakers**.

HIGH-FREQUENCY CRT DEFLECTION CIRCUIT

By WALTER B. GUGGI
Stanford Research Institute, Menlo Park, California

Single power transistor operating as switch provides efficient cathode-ray-tube deflection with five-percent linearity for standard television horizontal-sweep applications. Sweep amplitude may be adjusted continuously without affecting linearity or general circuit performance

POWER CONSUMED in conventional television sets in the production of the horizontal deflection has long been a problem since the total useful deflection work output is zero. The entire input of (typically) 45 watts, or even more in the case of color television, is used up in replacing circuit losses.

The completely transistorized deflection system to be described promises to reduce these losses, reducing the power input by a factor of as much as three or four. In addition, most of the transient difficulties that give rise to ringing are eliminated. The simplest embodiment yields a deflection linearity of ±5 percent or better. Addition of simple passive elements corrects linearity to ±2 percent or better.

The basic principle, shown in Fig. 1 in symbolic form, is one of circulating energy among various reactive elements by low-dissipation on-off switches. While lost energy is replaced during the early part of the sweep cycle, the circulating energy is actually switched into the deflection coil during the retrace period.

Basic Circuit

Figure 2 shows the circuit conditions together with plots of the important voltages and currents, for three different portions of a complete sweep period. In Fig. 2A, the voltage is shown across the sweep coil, starting at the instant S_1 is closed, Fig. 2B shows the corresponding sweep current and Fig. 2C shows the current flowing through the switch.

Assuming that retrace capacitor C_2 has reached its final charge with the polarity shown and at this instant (t_1) the switch is closed, the full capacitor voltage appears across L_1. The closed switch completes a tuned circuit, L_1C_2 initiating currents i_1 and i_2. Within one quarter cycle of the resonant frequency of L_1C_2, all energy is transferred from C_2 to L_1. Diode D_1 blocks current flow through branch C_1D_1 during this initial quarter cycle.

At t_2, C_2 is completely discharged. However, two independent current components continue to flow within two branches of the circuit. During period t_2t_3, the sweep current i_1 flows through $L_1C_1D_1$ and loop current i_2, which is but a small fraction of the peak sweep current, flows through $L_2C_1S_1$ and continues

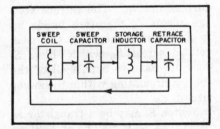

FIG. 1—Pictorial representation of circuit operation

buildup has occurred. This may be noticed in Fig. 2B, in interval t_1t_2.

It may be observed in Fig. 2C that at turn-off point t_3 a relatively small current flows through the transistor, followed by a delayed voltage buildup providing at least several microseconds of switching time for the transistor to reach its

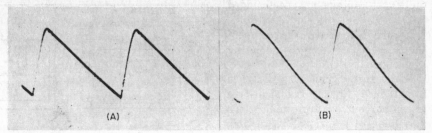

FIG. 5—Sweep current of circuit of Fig. 3B (A) and current for wide-angle deflection (B)

highest impedance before any appreciable voltage buildup has occurred (Fig. 2D). Therefore, the switching operation in either direction is supported by the circuit itself and takes place with relatively low instantaneous peak transistor power dissipation during either switching transition.

For horizontal-sweep operation, this feature becomes quite important because of appreciable inherent delays in the switching characteristic of power transistors at those frequencies.

Switching Operation

The switching performance of a commercially available power transistor applied to this circuit is shown in Fig. 4, where it is compared with its performance switching into a purely resistive load. At 15 kc, switching into a resistive load at the same power level would be prohibitive due to the relatively slow transition of the transistor through an area of excessive power dissipation. The operating condition in this sweep circuit, however, is such that maximum peak currents and peak voltages may be reached without reaching the maximum dissipation limit.

The peak deflection current flows through the transistor for only a small portion of the retrace time, at the most a few microseconds, as is shown in Fig. 2C. The high-voltage peak also occurs for only a very short time, as indicated in Fig. 2D. These two factors permit increase of the transistor peak collector current and voltage without exceeding dissipation ratings. No spurious high-voltage transients are present to limit the operation to uneconomically low voltage levels, as they would be in usual sweep circuits. Power transistors may therefore be used quite efficiently

in this application.

If a lossless sweep coil and lossless diode were available, the sweep current would follow a perfectly linear function, provided a large sweep capacitor or constant-voltage source were used. In practical circuits, however, a logarithmic decay takes place during each sweep cycle causing deflection distortion, which is further aggravated by the relative increase of diode resistance with decreasing sweep current. The result is a steadily increasing sweep compression which limits the sweep linearity to a value of approximately ±5 percent if standard commercial components are used throughout.

Compensated Circuit

A method has been found to compensate for this sweep compression simply by adding to the sweep voltage a suitable compensating component in the form of a

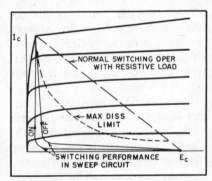

FIG. 4—Switching performance of typical power transistor

saw-tooth voltage which is readily available from another portion of this sweep circuit. Figure 3B shows the improved circuit.

During time t_3t_1, a saw-tooth voltage buildup occurs across L_2 similar to the buildup across C_2. A low-impedance secondary winding which has the purpose of supplying sufficient compensating signal to the sweep voltage has been added to L_2.

Figure 5A shows the compensated sweep current, indicating high linearity.

While this circuit was designed to produce linear, transient-free sweep current, certain applications may require additional correction, particularly for sweep distortion caused by the geometrical arrangement of wide-angle deflection systems applied to flat-face cathode-ray tubes.

This circuit can be compensated by introducing electrical sweep compression. The early portion of the sweep current may be compensated by proper selection of the sweep capacitor which permits shifting the sweep current as much as may be desired from the linear portion into the curved portion of a sinusoid. The negative portion of the sweep current may be corrected by reducing, omitting or reversing the compensation obtained by the system shown in Fig. 3B or by adding resistance to the sweep coil.

Figure 5B shows the corrected sweep current for compensating sweep distortion due to flat-face cathode-ray-tube geometry. The bar pattern of Fig. 6 shows the degree of linearity possible with this improved circuit.

Operating Performance

The deflection angle is 60 degrees at 16-kv accelerating voltage obtained with a 90-degree yoke, which does not give the highest efficiency at this angle. Input power required is six watts.

The transistor used in this circuit dissipates approximately three watts. Peak voltage amounts to 180 v and peak current is five amperes. Driving power of approximately one watt is supplied to the transistor base by transformer coupling from a synchronized pulse generator. Transformer coupling keeps until S_1 is opened at t_3.

Current i_2 transfers energy from C_1 into L_2. When S_1 is opened, at t_3 i_2 continues to flow in L_2; however, it changes its path and flows within loop $L_2D_1C_2$, transferring energy from L_2 to C_2. This is indicated by the voltage buildup which occurs across S_1 as shown in Fig. 2D, and across C_2, as shown in Fig. 2E.

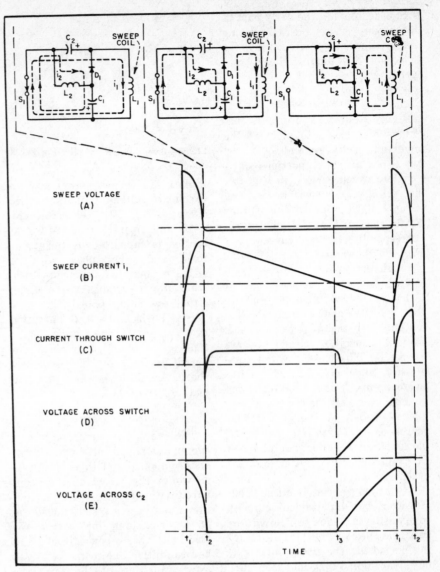

FIG. 2—Current and voltage relationships in equivalent circuit of basic sweep generator

Capacitor C_2 is chosen to have a relatively small value and, since considerable energy has been stored in L_2, time period t_3t_1 permits the buildup of sufficient voltage across C_2 to obtain proper retrace performance when S_1 is closed again.

Going back to Fig. 2B, the sweep current actually crosses the zero line and turns negative. This is made possible during the last portion of the sweep cycle (t_3t_1) by i_2, which flows through $L_2D_1C_2$ and essentially permits i_1 to flow backwards through D_1 by subtraction from i_2.

Sweep-current flow stops suddenly when its increasing value reaches the value of the decreasing loop current i_2. At this instant S_1 is closed so C_2 may again be discharged into L_1. However, a portion of the sweep energy has already been returned from C_1 to L_1 because of sweep current reversal during time t_3t_1. The retrace voltage across L_1 therefore builds up to a higher value than the peak voltage across S_1, which shows S_1 does not have to handle all of the energy circulating within the sweep circuit.

Switch S_1 may be a transistor as in Fig. 3A or it may be an electron tube or other suitable switching element.

Performance Analysis

Certain desirable performance characteristics become apparent upon further analysis of this circuit. When S_1 is closed the current within the switching element (in this case a transistor) is not a step function, but follows a sinusoidal function, giving the transistor suf-

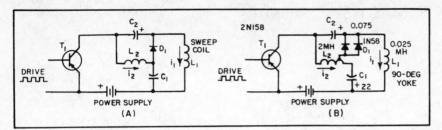

FIG. 3—Basic transistor sweep circuit (A) and circuit (B) compensated for sweep distortion

ficient time to reach its lowest impedance before appreciable current the base positive during cutoff without an additional bias source. The duty cycle of the pulse is approximately 60-percent on and 40-percent off.

It is estimated by results obtained from this experimental circuit, running at nearly one-half the power required for a 90-degree deflection system, that about 15 w of d-c input power will be required to deflect 90 degrees at 16-kv accelerating voltage, using the highly efficient transistor circuit. The d-c supply voltage will be approximately 20 v, depending on the circuit impedance which again is a function of transistor characteristics. It is necessary to correct for the d-c component in this sweep circuit by using either an output transformer, capacitive bypassing or permanent magnets.

It should be possible to derive high voltage from this circuit by replacing storage inductance L_2 with a step-up autotransformer and a following rectifier. For low-impedance transistor operation, it may be necessary to use voltage-multiplier circuits to avoid difficulties in obtaining the proper step-up ratio. Tests show that a relatively large amount of power may be taken from such a transformer without any sweep interference or increase of peak sweep current.

Design Equations

For most practical design purposes, it can be assumed that the circuit is lossless.

The basic equations are

$$I = \sqrt{2W/L} \qquad (1)$$

$$E = \sqrt{2WL/T} \qquad (2)$$

where W is energy stored in inductance, in joules; L is inductance, in henries; I is peak sweep current, in amperes; E is applied constant voltage, in volts (retrace voltage) and T is time duration of applied voltage, in seconds (retrace time).

These equations give an expression for maximum current and voltage conditions in an inductance having a linear rate of change of current. They are also valid for determining the peak voltage and peak current a switching device a sweep circuit of the type des-would have to be able to handle in scribed.

Switching Characteristic

The characteristic of the switching device can be determined from

$$IE = 2W/T \qquad (3)$$

where T is retrace time in seconds, I is peak sweep or retrace current, E is peak retrace voltage across the sweep coil and W is the energy required to deflect the beam. This product is equivalent to the peak volt-ampere requirement of the switching device and the diode.

After a switching device has been selected, peak current and peak voltage are known; yoke impedance L_1 may be calculated from Eq. 1 or Eq. 2.

The value of C_1 may be determined as a function of the sweep waveshape desired; the linearity improves with increasing capacitance. Physical size becomes the limiting factor in most cases.

The value of C_2 may be calculated from

$$C_2 = 1/(\omega^2 L) \qquad (4)$$

where the value of ω is determined by $\omega = \pi/(2T)$ and retrace time T is a quarter cycle of resonance frequency.

Storage inductor L_2 should have several times the inductance of L_1. Diode D_1 is preferably of the semiconductor type because of the low circuit impedance and the relatively large sweep currents. The supply voltage is a function of circuit losses and may be kept relatively low.

The on-to-off ratio of the switching device may be determined experimentally. Its calculation is laborious.

Since the retrace waveform is sinusoidal, the retrace voltage actually increases by a factor of $\pi/2$ over the value calculated for a constant voltage pulse, for which Eq. 3 is correct. At the same time, however, the peak voltage requirement is reduced by a factor of approximately two-thirds by the amount of energy that returns into the sweep coil during the later part of the sweep cycle. Therefore, Eq. 3 remains quite adequate for practical calculations, without the necessity of any correction factors, since $(\pi/2)(2/3) \cong 1$.

PHONO PREAMPLIFIER FOR MAGNETIC PICKUPS

By R. PAGE BURR

Burr-Brown Research Corp., Cold Spring Harbor, N. Y.

DESIGNED for use with low-impedance magnetic phonograph pickup cartridges of which the variable-reluctance type is a popular example, the battery-powered transistor amplifier described below is comparable to vacuum-tube amplifiers now widely used for this purpose in high-fidelity systems.

To improve linearity of the amplifier and to stabilize the operating gain negative feedback is used. A midband voltage gain in the order of 40 db is a reasonable figure for a phonograph pickup amplifier. Stage gain of approximately 40 db is readily obtainable from junction transistors with base-to-collector current multiplication of about 40 times. Use of liberal feedback in a two-stage amplifier is therefore feasible. The two junction transistors developed a voltage gain of approximately 78 db with feedback inoperative. When the loop is closed, the gain at approximately 1 kc is close to 40 db.

When voltage feedback is taken from the output anode circuit of an electron tube and is returned to the cathode of the input tube, the advantages of simultaneously raising the input impedance and lowering the output or source impedance of the system result. Low output impedance is desirable because preamplifiers of this type are often used to feed long shielded cables having high total shunt capacitance.

▶ **Amplification**—The gain needed for a phonograph preamplifier may be calculated using representative input and output signal levels. Most consumer-goods high-fidelity apparatus is satisfactorily operated by a signal level in the order of 1 volt rms. A magnetic pickup cartridge, such as the widely used General Electric variable-reluctance unit, will produce an output voltage of approximately 10 millivolts for a lateral stylus velocity of 4.8 cm per

* Work described was performed while the author was employed by Hazeltine Corp., Little Neck, N. Y.

sec. This velocity corresponds roughly to the average recorded velocity of many commercial long-playing $33\frac{1}{3}$ rpm recordings. Instantaneous program peak velocities are approximately 10 db higher than average. Accordingly, a pre-amplifier voltage gain of 40 db (1 volt/10 millivolts) appears to be a reasonable choice.

Two identical grounded-emitter stages shown in the circuit diagram are connected in cascade and op-

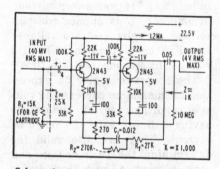

Schematic circuit diagram of the transistor preamplifier with RIAA equalization

erate from a collector supply potential of −22.5 volts. A substantial portion of the battery voltage is dissipated in the large emitter supply resistors. This practice insures good d-c stability of the circuit against variations in temperature and a high degree of circuit immunity to varying transistor parameters. Almost any junction transistor having a base-to-collector current multiplication of 30 or greater will operate satisfactorily in the circuit. Suitable types include 2N43, 2N104, 2N105 and 2N175.

Battery voltage limits the output voltage swing of the amplifier. For all units tested, limiting occurs at an output level of approximately 4 v rms. Since the gain is 40 db, the maximum allowable input signal at 1 kc is 40 mv rms, or 56 mv peak. Total battery drain is 1.2 ma.

▶ **Equalization**—The feedback cir-

cuit consists of the elements shown in the lower part of the diagram. If the amplifier were designed for uniform frequency response the feedback impedance would comprise a pure resistor. This amplifier is intended to reproduce phonograph recordings from a magnetic pickup that is essentially a velocity sensitive device.

▶ **Standard**—Under these conditions amplifier transmission as a function of frequency should correspond to the present standard playback curve for lateral disc recordings as specified by the Record Industry Association of America. This nonuniform frequency characteristic compensates the pre-emphasis employed by the record manufacturer in the original recording. The curve is specified by three time constants (3, 180, 318 and 75 μsec) affecting the low, middle, and high frequency regions of the audio spectrum. It can be synthesized in the feedback amplifier by proper arrangement of the components in the feedback path.

The 318-μsec time constant is provided by C_1 and R_t. The 3,180 μsec time constant is the product of C_1 and R_2. The high-frequency time constant of 75 μsec is obtained by placing the appropriate value of resistive loading R_1 across the input connection to the amplifier so the series inductance of the pickup causes the desired amount of high-frequency roll-off. For a GE cartridge this resistance value should be 15,000 ohms as shown.

▶ **Performance**—Two preamplifier characteristics of interest are signal-to-noise ratio and harmonic distortion.

In audio-frequency apparatus it is customary to specify noise output with respect to some reference output level appropriate to the particular equipment. For this amplifier the reference level is taken as 3 v rms. This is an output signal approximately equivalent to the peak instantaneous program level from an average long-playing recording reproduced by the typical pickup cartridge.

Relative to this 3-v reference level the average rms noise voltage for a sampling of 16 transistors of various types was 80 db down or 0.3 mv at the output of the preamplifier. The noise output of the pre-amplifier was therefore 70 db below the 1-v average signal level from a long-playing record. Such dynamic range is quite satisfactory for the reproduction of the usual recordings whose range is limited by surface noise to approximately 50 db. It is particularly true because no hum voltage need be generated in the preamplifier.

Harmonic distortion data were taken for the preamplifier at four frequencies and at three levels of signal output. In every case, distortion was less than 1 percent, ranging from 0.95 at 40 cps to a low of 0.12 at 400 cps with 1 v output.

PREAMPLIFIER FOR CERAMIC PICKUPS

By WILLIAM NEWITT
Senior Engineer, Electro-Voice, Inc., Buchanan, Mich.

IT is desirable that ceramic phono pickups have low source impedance, with an output voltage comparable to that of a magnetic pickup. Some power gain should also be provided so that the ceramic unit may compare favorably with the high power sensitivity of a magnetic pickup.

Power gain may be provided by a transistor amplifier. By placing the amplifier close to the pickup, minimum hum and noise pickup as well as minimum loss due to shunt capacitance can be obtained.

The highest input impedance obtainable from a transistor may be realized from the common-collector configuration. Since the input impedance depends also to a considerable extent on the output load of such a circuit, satisfactory results have been obtained by a tandem type of circuit. Two transistors are used in a cascade-common-collector configuration with the input impedance of one stage comprising the load of the preceding stage.

An experimental circuit, shown in Fig. 1, utilizing 2N184 transistors can handle two volts rms input without introducing distortion. With simulated ceramic pickup, having 500 $\mu\mu$f internal capacitance, coupled to the input, open-circuit output response is within $\frac{1}{4}$ db from 35 to 40,000 cycles. Output impedance is 120 ohms at 1,000 cycles.

The transistor assembly was built into a small phenolic tube $\frac{5}{16}$-inch in diameter and $\frac{3}{4}$-inch long. Plugs fitted at each end permit attachment directly to the pickup. Actual performance verified the results obtained with the simulated ceramic pickup. Total loss in gain resulting from insertion of the transistor unit was $\frac{1}{4}$ db.

At 1,000 cycles, the impedance of the 500 $\mu\mu$f pickup would be about 32,000 ohms. Increase in power sensitivity of the pickup with the attached transistor unit is about 300 times. Thus, a power gain of about 25 db has been provided, and the ceramic phono pickup is cap-

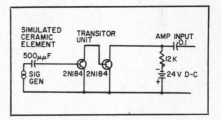

FIG. 1—Simple transistor amplifier can be built into phono pickup arm with power supply installed in amplifier

able of driving low load impedances.

The only change required in the existing equipment is the addition of a low-current d-c supply and two components to the preamplifier unit.

Chapter 7

RADIO AND TV BROADCAST EQUIPMENT

VIDEO PREAMPLIFIER FOR STUDIO MONITORS

By L. N. MERSON
Engineer, Philco Corporation, Philadelphia, Pa.

Compact transistor preamplifier unit boosts video line level to permit use of standard black-and-white or color receiver as studio line monitor. Unit may also be used as video line boost or distribution-line isolation amplifier

WHERE a television studio monitor or jeep is needed temporarily, a conventional tv receiver may be used by employing a jeep kit to boost the video line level to that required by the receiver. Figure 1 shows the necessary conversion circuitry. Fifty such kits and receivers were used at the 1956 presidential-nomination convention in Chicago.

Levels

The preamplifier provides a maximum output level of 3.5 v p-p from the normal video line level of 1.4 v p-p of video and sync from a 75-ohm line. This output level is sufficient to operate the receiver through its full dynamic range.

Bandwidth is flat to almost 8 mc. The small amount of degeneration introduced in the emitter cir-

cuit of Q_2 modifies the inherent transistor clamping action and prevents clipping on sudden surges of video. Capacitor C_1 provides d-c isolation enabling the input transistor to be properly biased, the 10,000-ohm input resistor prevents

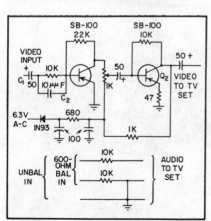

Top and bottom views of jeep kit illustrate compactness of unit

loading of the video line and C_2 provides high-frequency peaking.

Audio line levels are normally 0 to 4 vu across 600 ohms. These levels are usually adequate for the receiver audio circuits.

Receiver Modifications

A typical receiver modification is shown in Fig. 2. The video-detector load is opened at A and the jeep-kit output fed in. The 4.5-mc trap is shorted out and a 15-$\mu\mu$f shunt capacitor is added to flatten the video bandwidth.

The audio circuit is broken at B and the audio fed in.

These jeep kits can also be used as line boost amplifiers or as distribution-line isolation amplifiers. Their wide bandwidth permits them to be cascaded without bandwidth restriction.

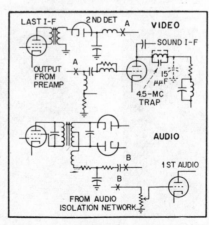

FIG. 1—Jeep kit incorporates video preamplier and audio isolation network

FIG. 2—Typical tv receiver circuits show where modifications are made

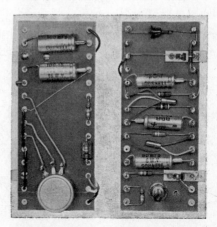

Jeep kit installation (lower right) on tv receiver

SYNC GENERATOR FOR PORTABLE TV CAMERA

By KOJIRO KINOSHITA, YASUSHI, FUJIMURA, YOSHIO KIHARA and NOBUO MII

TV Research Section, NHK Technical Research Laboratories, Tokyo, Japan

Japanese portable tv camera-transmitter uses sync generator comprising ten transistor flip-flop circuits with two feedback amplifiers to form a counter unit that divides twice the horizontal frequency by 525 to produce the field frequency. Unit uses modular design

SYNC GENERATORS for portable telecasting equipment must be light and compact. This article describes a transistorized sync generator now in use by NHK, the Broadcasting Corp. of Japan.

Housed in a 12 by 3.75 by 1.75-inch chassis, the unit has a total power consumption of about 500 miliwatts which is just under the power required to heat the filament of one 12AU7. Alloy-junction transistors are used throughout.

Generator

Figure 1 shows a block diagram of the complete sync generator. All 26 transistors are used in the common-emitter configuration. The unit is made up of 13 two-transistor modules. There are ten counter modules, a master oscillator and 2H output amplifier module, a 2H clipper and feedback amplifier module and a feedback and field frequency amplifier module.

The horizontal frequency is internally derived from a crystal controlled oscillator. A lockup signal picked up from the base station also may be used to provide the 2H frequency.

Circuits

The master oscillator output is clipped and fed to the counter chain. This is the main portion of the sync generator and consists of a total of ten counters employing two separate feedback loops. One counter chain and feedback loop divides the master frequency by 25 while the other divides by 21 resulting in a total division by 525 to give the field frequency. The 60-cps field frequency obtained from the tenth counter stage is amplified and fed to the camera-transmitter system.

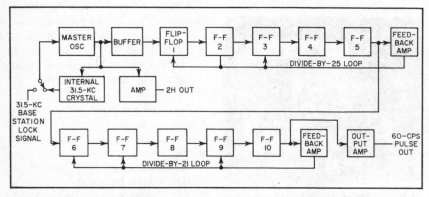

FIG. 1—Block of system shows feedback technique that produces count down from 31.5 kc, or twice the line frequency, to a field frequency of 60 cps

Portable tv camera-transmitter in use for over 4,000 hours (left) uses compact modular constructed sync generator (right)

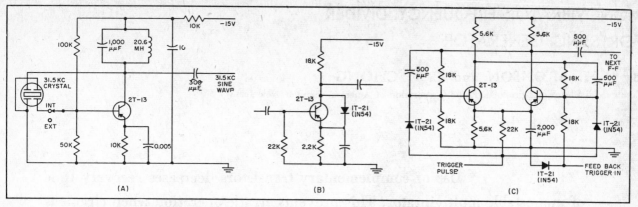

FIG. 2—Basic circuits. Output of crystal-controlled master oscillator (A) is clipped at (B) and triggers flip-flop counters as in (C)

The master oscillator shown in Fig. 2A uses a special miniature-type 31.5-kc crystal, utilizing the flexure mode of oscillation. It acts as the frequency controlling element in the feedback path between the transistor collector and base when S_1 is in the internal position. When S_1 is on external, the circuit works as a tuned amplifier.

The circuit shown in Fig. 2B uses a diode between the emitter and collector to minimize hole-storage effects and to clip the master oscillator signal to provide a trigger for the succeeding counter stages.

Each counter consists of a two-transistor flip-flop stage as shown in Fig. 2C. A feedback pulse is fed to selected counters which have the base return resistor of one transistor in series with a germanium diode. The trigger is inserted at the junction of the other diode-resistor combination which also acts as a buffer.

The pulse rise time at the col-

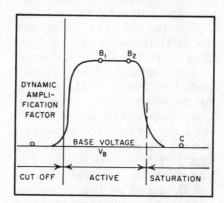

FIG. 3—Typical dynamic characteristic curve for junction transistor swiching circuit showing possible operating prints

lector is about 3 μsec while the delay time per stage is approximately 0.5 μsec. The ratio of minimum

trigger pulse amplitude to output trigger pulse amplitude is approximately two.

Major effort in the design of this sync generator was expended on the flip-flop circuits. Selection of the proper operating point on the dynamic characteristic curve associated with junction transistor switching circuits was of prime importance here. Such a curve is shown in Fig. 3.

If the operating point is chosen at C, minority storage effects will cause appreciable delay and insensitivity. To avoid this, the circuit elements were selected to provide an operating point in the active region. Choice here is based on the relation between circuit time constants and input pulse.

If an operating point is set at B_1, an appreciable overshoot occurs at the leading edge of the collector waveshape as shown in Fig. 4A; this overshoot will increase at each successive stage until a point is reached where one counter will be switched off and on again destroying the count down process.

By adjusting the circuit elements for operation at point B_2, the overshoot is minimzed as shown in Fig. 4B and good count down results.

The authors thank T. Nomura, and T. Shiromi for their advice and assistance and K. Iwama for his helpful discussions.

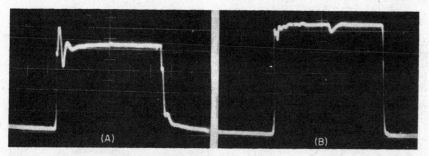

FIG. 4—Collector waveshapes show overshoot (A) caused by operation at B_1 in Fig. 3 and flat top (B) caused by operating at B_2 on the same dynamic characteristic curve

MONOVIBRATOR FREQUENCY DIVIDER FOR SYNC GENERATOR

By A. I. ARONSON and C. F. CHONG

Commercial Electronics Products, Radio Corporation of America, Camden, New Jersey

Use of complementary transistors decreases recovery time of monostable multivibrator. This prevents erratic operation when circuit is used in a television sync generator. Since both transistors are off during timing cycle, circuit is relatively insensitive to transistor variations and operates reliably from —50 to +70 C for input frequencies from 250 cps to one mc

GENERATION of accurate television line and frame frequency rates, a function common to all sync generators, is generally done by starting with a stable high-frequency crystal oscillator and counting down in proper multiples to the desired frequencies. Frequency division is often accomplished with multivibrator circuits.

Monostable Multivibrator

In one such circuit a monostable multivibrator uses the complementary symmetry property of transistors and is shown in Fig. 1. Transistors Q_1 and Q_2 are biased on. With an applied negative trigger, Q_1 is turned off, causing the voltage at A to go to slightly less than V_1. Since the voltage across C_1 cannot change instantaneously, it appears also at point B, the base of Q_2, turn-

ing Q_2 off. Capacitor C_1 discharges towards V_2 through R_2 and R_5.

When the voltage at B goes negative, Q_2 starts conducting and turns Q_1 on. Capacitor C_1 now discharges through a path comprising the saturation resistance of Q_1 and the input resistance of Q_2. After C_1 is fully discharged, Q_1 and Q_2 are on and ready for another input trigger to start the circuit recycling. Frequency division is accomplished by controlling the length of time Q_1 and Q_2 are off so that a specified number of input triggers can be accepted without causing circuit action.

Advantages

Use of complementary transistors in place of similar conductivity transistors provides a higher charge-to-discharge ratio of the

timing capacitor. In addition both transistors are off during the timing cycle thereby reducing the effects of transistor variations on timing accuracy.

Charge-Discharge Ratio

High charge-to-discharge ratio is desirable because unreliable operation can occur if an input trigger is coincident with the discharge of C_1.

In a conventional monostable circuit[1,2] the timing capacitor discharges through a collector load resistor of one transistor and the input resistance of the other while in the present circuit the capacitor discharges through the saturation resistance of one transistor and an input resistance of the other. Since

*Now with Remington Rand Univac, Phila., Pa.

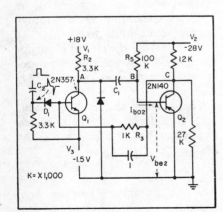

FIG. 1—Monostable multivibrator circuit uses complementary-symmetry facility of transistors to decrease recovery time

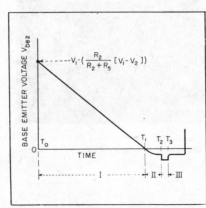

FIG. 2—Base-emitter recovery voltage of Q_2 sketched for three intervals of timing cycles. Intervals II and III are expanded

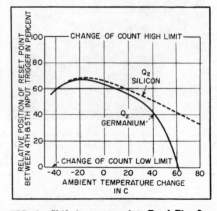

FIG. 3—Shift in reset point, T_2 of Fig. 2. A shift to 0 or 100 percent indicates a miscount of 4 or 6 respectively

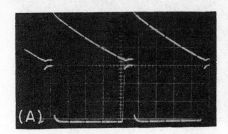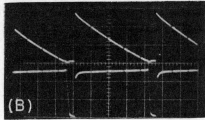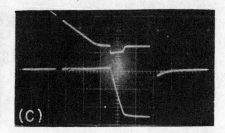

FIG. 4—Oscillograms of a 1-kc input signal for frequency-divider circuit. Waveforms (A) show base-emitter voltage V_{be2} (top) and collector voltage (bottom) of Q_2 with scale v=5v/divider and h=1 ms/divider, while (B) shows base-emitter voltage, V_{be2} (top) and collector voltage (bottom) of Q_1 with scale v=5v/divider and h=1 ms/divider and (C) shows waveform (B) on expanded scale

the sum of the latter resistances is low, an increase of five or more in charge-to-discharge ratio can be obtained with this circuit.

Figure 2 is a diagram of the recovery voltage V_{be2} on the base of Q_2 for three intervals of the timing cycle. When $T > T_1$ and V_{be2} crosses zero, Q_2 begins to conduct. Charging of C_1 is controlled by the base-emitter diode of Q_2. Interval III is the discharge time of C_1. Interval II and III each represent about 2 percent of the total recovery time.

The duration from the starting trigger to the time V_{be2} equals zero corresponding to interval I in Fig. 2 is given by

$$T_1 = (R_2 + R_5) C_1$$
$$ln\left\{ \frac{V_2 - I_{bo2} R_5 - V_1 + A}{V_2 - I_{bo2} R_5} \right\}$$

Where $A = [R_2/(R_2+R_5)] (V_1 - V_2)$

Values of C_1 and C_2 are a function of the operating frequency and the desired time constants.

Circuit Reliability

When both transistors are off, the timing is controlled at room temperature, primarily by passive elements. At elevated temperatures, however, the reverse saturation current I_{bo2} of the *pnp* unit approximately doubles for each 7 C of temperature rise.

The voltage drop across R_5 caused by current I_{bo2} is high enough at high temperatures to influence the charge time of C_1. Reducing the effects of I_{bo2} by decreasing R_5 is

undesirable because this also decreases the charge-to-discharge ratio of the circuit.

Use of a silicon transistor for Q_2 gives a 100-to-1 reduction in I_{bo2} and is the best means of reducing temperature effects on circuit recovery time. Improvement in circuit reliability using a silicon transistor for Q_2 is illustrated in Fig. 3. This curve shows the percentage

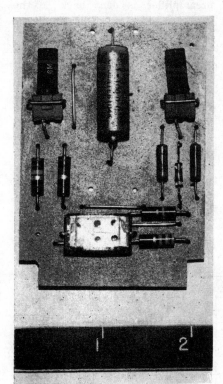

Monostable circuit constructed on a printed wiring plug-in board

shift in the reset point, T_2 of Fig. 2, of the circuit used as a divide-by-five counter as a function of temperature. A shift to either zero or 100 percent indicates a miscount to four or six respectively. For moderate temperature ranges a germanium transistor for Q_2 will be satisfactory.

Resistor R_3 helps reduce the input trigger power required to turn Q_1 off by limiting the base current in Q_1 The collector diode prevents Q_1 from going into saturation reducing further the trigger power required.

Application

A sync generator can be built using this circuit which occupies 54 cu in. and weighs one pound. It requires 0.7 w of power.

The frequency divider circuit operates reliably with input trigger from 250 cps to one mc. Various waveforms for a 1-kc input signal are shown in Fig. 4A and B. The voltage at the base of Q_2 is enlarged in Fig. 4C to indicate the change in the recovery-voltage waveform due to the input diode of Q_2.

The circuit operates reliably with input triggers from 0.2 to 20 v.

Credit is due Dr. H. J. Woll for the circuit proposal and for his encouragement during the development.

REFERENCES
(1) L. P. Hunter, "Handbook of Semiconductor Electronics", p 15-11, McGraw-Hill Book Co. Inc., New York, 1956.
(2) J. Millman and H. Taub, "Pulse and Digital Circuits", p 174, p 599, McGraw-Hill Book Co. Inc., New York, 1957.

TELEVISION STATION FOR REMOTE PICKUPS

By L. E. FLORY, G. W. GRAY, J. M. MORGAN
and W. S. PIKE

RCA Laboratories, Princeton, New Jersey

Designed to be carried on a man's back, nineteen-pound television transmitting station includes camera, monitor, sync generator, 2,000-mc transmitter and power supply. Total transistor complement of 72 is employed in addition to vidicon, kinescope and transmitter tubes. Camera delivers 1.5-v p-p composite video signal to line or transmitter

BROADCASTERS have need *for* lightweight portable television pickup equipment for spot news coverage and other events. An experimental portable television transmitting station has been developed that generates a picture of adequate quality for remote-broadcast purposes and transmits it for over a distance of approximately one mile. This equipment was used at recent political conventions to make possible remote pickups that would not have been possible by any other method.

System Description

Employing 72 transistors, the equipment includes, in addition to the camera, a sync generator, picture monitor and transmitter, as illustrated in the block diagram of Fig. 1.

The camera contains a video preamplifier, horizontal and vertical deflection circuits and a blanking amplifier in addition to the vidicon with its yoke and focus magnet. Horizontal drive pulses, vertical drive pulses and mixed blanking pulses, as well as d-c voltages are supplied to the camera over a four-foot cable from the backpack. The video signal from the camera video preamplifier is sent back to the backpack on a coaxial conductor braided into the same cable.

As in the miniature transistorized television camera described in the Jan. 1, 1957 ELECTRONICS, use is made of a new miniature vidicon only one-half inch in diameter.

The monitor contains horizontal and vertical deflection circuits and a video amplifier.

A switch on the backpack permits the operator to monitor either the video signal at a suitable point in the modulator or a rectified and amplified sample of the transmitter output. The latter gives the operator an instantaneous indication of transmitter overmodulation. The electronic viewfinder also permits the cameraman to check the vidicon electrical focus, which would be impossible with an orinary optical viewfinder.

The backpack contains a rechargeable battery, a transistor power converter to supply high voltage for the camera, monitor and transmitter tubes, a crystal-controlled sync generator, a 2,000-mc transmitter, a modulator amplifier in which d-c level setting, sync addition and set-up addition are performed, and a small auxiliary video amplifier and rectifier which samples the transmitter output. The transmitting antenna projects from the top of the backpack.

Compared to similar vacuum tube equipment built by the authors several years ago[1], the camera with its viewfinder weighs four pounds while the earlier unit, which used a one-inch vidicon and a similar monitor tube, weighed eight pounds. The backpack unit weighs 15 pounds and is 3 by 12 by 13 inches in size. The present equipment consumes 30 watts and operates for nearly five hours on a smaller size battery.

Camera

The circuit of the camera video preamplifier is shown in Fig 2, along with the blanking amplifier and the vidicon beam, focus and target voltage controls.

Common-emitter amplifier Q_4 is directly connected to the vidicon cathode. Negative mixed-blanking

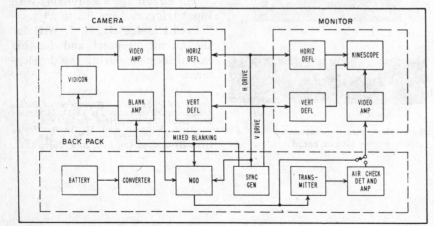

FIG. 1—Diagram of principal units of system—camera, monitor and backpack

signals from the sync generator in the backpack are applied to its base. Twenty volts peak-to-peak of positive mixed blanking are available at the collector of Q_4. This is adequate to prevent landing of the vidicon beam under any conditions during the horizontal and vertical retrace periods.

It has been established that in video amplifiers the best signal-to-noise condition is obtained if the vidicon is operated into a high impedance even though the input circuit has high capacity. The effect of this high capacity is compensated by a high-peaker circuit.

The input amplifier employs transistors, Q_1, Q_2 and Q_3 and exemplifies in Q_1 and Q_2 a type of feedback pair which has been freely used in later stages, particularly in the modulator. The pair consists of transistors of opposite conductivities, with feedback from the collector of the second to the emitter of the first.

This arrangement is d-c stable, has high input impedance and has low output impedance from which either output polarity may be derived. In this case, the collector of the second transistor is d-c connected to the base of the third, Q_3.

The response of this circuit is maintained flat to six mc with peaking inductance L_1 and by shunting the emitter resistor of Q_3 with C_1.

Bias control R_1 sets the operating point for this circuit.

Stage Q_5 is an ordinary common-emitter stage which drives the high peaker. High-frequency response is maintained by C_2 and overall low-frequency compensation is provided by C_3.

Stages, Q_6 through Q_9 are ordinary common-emitter amplifiers. They are, however, overcompensated as far as high-frequency response is concerned; the cumulative effect of overcompensation supplies aperture correction.

Video gain control R_2, compensated by C_4, is inserted between Q_8 and Q_9. Transistors Q_{10} and Q_{11} comprise another feedback pair which drives the interconnecting cable. The latter feedback pair delivers about 0.5 v of six-mc video signal to the 75-ohm line with good linearity. In this case, output is taken from the emitter of the sec-

Backpack houses sync generator (top left), power converter (lower left), transmitter (center) and rechargeable cells (right)

ond transistor of the pair.

Camera Deflection

The camera deflection circuits are shown in Fig. 3. The sync generator in the backpack supplies negative horizontal and positive vertical drive pulses to the camera. The negative horizontal pulses turn on Q_{31}. Amplified positive horizontal pulses appear at its collector.

Output transistor Q_{32} is driven by a step-down transformer. It acts as a switch to interrupt the build-up current in the coil. Ringing is prevented by the damper diode.

The vertical deflection circuit is driven by the positive vertical drive

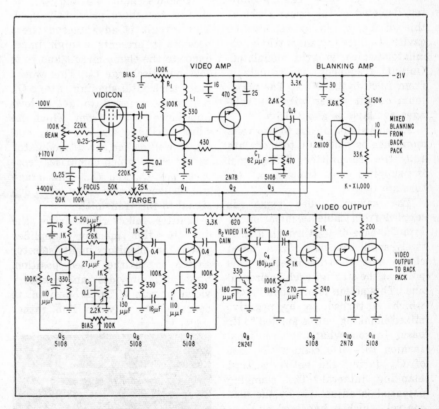

FIG. 2—Camera preamplifier employs 11 transistors and miniature vidicon pickup tube

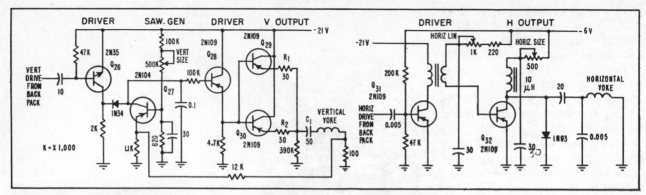

FIG. 3—Camera vertical (left) and horizontal (right) deflection circuits use low-power transistors because of miniature vidicon's modest deflector requirements. Vertical and horizontal drive are supplied via cable from backpack

pulses. These are inverted by Q_{26}, which drives a sawtooth generating circuit comprising Q_{27} and its associated components.

The vertical drive to the yoke is provided by two paralleled emitter followers, Q_{29} and Q_{30}. Resistors R_1 and R_2 equalize the currents in the transistors.

Negative feedback assists in linearizing the yoke current and circumventing the shortcomings of the necessarily small coupling capacitor, C_1.

Modulator and Transmitter

The modulator and transmitter are shown in Fig. 4. The transmitter, depicted at the upper right of this diagram, is a self-excited cavity-stabilized planar triode oscillator which is grid modulated. Output frequency is adjustable from 1,900 to 2,100 mc. The maximum output power with the voltages available is about 0.5 w.

A video signal of one to two volts amplitude is sufficient to modulate fully the transmitter. This signal is supplied by transistors Q_{12} through Q_{16}.

The camera video signal is coupled from cable termination R_1 into the base of Q_{12} through a relatively small capacitor, C_1. A keyed clamp sets the d-c level at this point at the start of each horizontal line. This action is provided by Q_{17}, which is turned on by negative mixed-blanking pulses applied to its base. In somewhat unconventional fashion this also clamps the base of Q_{12} during the entire vertical blanking interval. The clamping potential is adjustable by the bias control, which is bypassed by C_2.

Since d-c information is rein-

serted into the signal, the following stages must be d-c coupled for this information to appear in the output signal. By employing alternate *npn* and *pnp* transistors, a five-stage amplifier of adequate stability has been constructed. The stage gains of this amplifier are not high; the large number of stages is necessitated by the many other functions which must be performed. Mixed sync, for example, from sync amplifier Q_{21} is inserted on the emitter of Q_{12}. Adjustable set-up is inserted on the base of Q_{14}, by feeding in a small amount of mixed blanking of the correct polarity at this point.

Feedback pair Q_{12} and Q_{13}, is similar to others previously used. This circuit is advantageous here because it presents a high impedance to the clamp circuit and thus does not discharge C_1. This would cause horizontal shading. Stage Q_{14} has been included to get a low-impedance point from which to drive auxiliary amplifiers Q_{18} and Q_{19}, as well as the actual modulator output stage which comprises feedback pair Q_{15} and Q_{16}. This stage will deliver about 1.5 v of video signal to the transmitter.

Emitter follower Q_{18} drives the line to the monitor amplifier when the monitor input selector switch is in the video position. Stage Q_{19} provides an auxiliary video output if it is desired to operate into a video line rather

than use the transmitter.

Transistors Q_{22} and Q_{23} amplify the output of diode D_1, which rectifies a sample of the transmitter output. When the monitor input selector switch is in the air-check position, the output of this amplifier drives the video line to the monitor.

Voltage Regulator

Regulator Q_{20} assists in maintaining the correct operating point of the transmitter as the batteries discharge during operation. It was found that as the six-volt portion of the battery discharged, the slight change in the voltage of the transmitter heater was sufficient to upset its operating point.

As regulation of the 0.9-ampere current of this heater would be a difficult task, it was found that a slight readjustment of the transmitter bias would accomplish the necessary correction. However, the sense of the requisite bias change is opposite to the sense of the change of heater voltage causing the trouble; Q_{20} accomplishes the necessary sense reversal. Its base is connected to the 6-v battery, through the network comprising R_2, R_3, R_4 and D_2, which sets the base potential at the correct d-c operating point.

The collector of Q_{20} is connected to the potentiometer which sets the clamp potential, thus inserting the required correction at this point. Direct-current degeneration in this compensating circuit is provided by R_5, with the loop gain adjusted to maintain nearly perfect compensation during the useful operating life of the battery.

For stability reasons, it is also

Composite video output from camera

necessary to regulate the supply voltages of all these amplifiers. Regulated buses of six and 12 volts are supplied. The 12-v regulator comprises emitter follower Q_{25}, with the base potential set by zener diode D_3. The six volt regulator is Q_{24}, with base potential set by a potentiometer across the regulated 12-v supply.

Monitor

The monitor is shown in Fig. 5. As no suitable commercially available cathode-ray tube could be found for this unit, a special 1½-inch tube was made in the laboratory. The tube is magnetically deflected and electrostatically focussed. With an ultor voltage of 2,000 v, its resolution exceeds 300 lines. It fits a standard one-inch vidicon yoke, which requires about 400-ma peak-to-peak horizontal deflection current.

Complementary symmetry and negative feedback are used in the vertical circuit. Here Q_{37} and Q_{38} are power transistors with rated dissipations of one watt each, without heat sinks.

The video signal from the backpack is amplified by Q_{42} and Q_{43}. This circuit is a transistor version of the familiar electron tube long-tailed pair, providing a push-pull output with the cathode and grid of the monitor simultaneously driven in antiphase. By this means, the effective video drive on the monitor is made about 35 v peak-to-peak.

Driven clamp Q_{41}, which sets the d-c level of the monitor amplifier, is identical to that in the modulator except that only horizontal pulses from the monitor deflection circuit are used to drive it.

Sync Generator

The synchronizing generator, located in the backpack has two sections. The master oscillator and divider chain, comprising the first, is shown in Fig. 6 and the waveshaping circuits and line drivers comprising the second, is shown in Fig. 7.

Operating at twice the 15.75-kc horizontal scanning frequency, master oscillator, Q_{44} in Fig. 6, is a negative-resistance oscillator. Use is made of the fact that a parallel

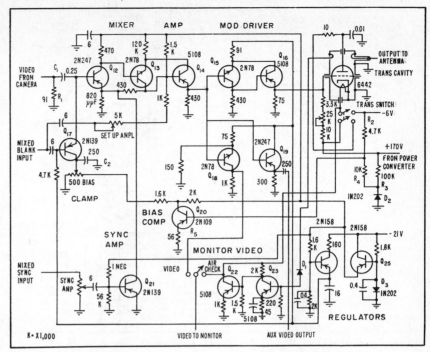

FIG. 4—Two-thousand-mc transmitter, video modulator and associated circuitry

L-C circuit in the emitter of a grounded-collector transistor amplifier will cause a negative resistance to appear at the base terminal of the transistor at frequencies for which the emitter load is capacitive. With a quartz crystal placed in the base circuit, a large negative resistance appears in series with it, if the resonant frequency of the emitter circuit is less than the operating frequency of the crystal, and oscillation occurs.

Output is taken across R_1 in the collector circuit to drive isolation amplifiers Q_{52} and Q_{45}.

Horizontal Divider

The horizontal divider, which divides by two, is blocking oscillator Q_{53} using an autotransformer

wound on a small toroidal ferrite core. A positive pulse of a few volts amplitude and about 1.5-μsec duration appears at the collector of Q_{53} each time the blocking oscillator fires. This is called the positive horizontal trigger and is used, after some processing, to initiate the leading edge of the horizontal blanking and drive pulses.

A delayed trigger pulse is derived from Q_{54}. An auxiliary winding on T_1 drives this transistor. The waveform on this winding is a considerably distorted single cycle of a sine wave; the width of each half-cycle is about 1.5 μsec. The winding polarity is such that the positive going half-cycle occurs first; Q_{54} does not conduct during

Side view of camera (bottom) and monitor (top) shows vidicon and kinescope yokes

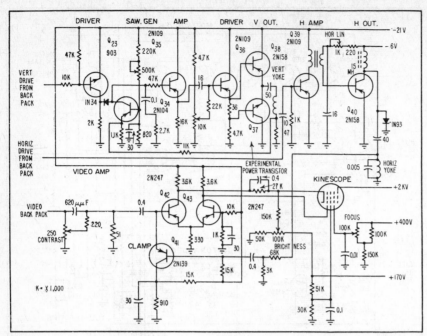

FIG. 5—Complete monitor circuitry for camera; kinescope is 1½ in. in diameter

this portion of the cycle.

On the following negative half-cycle, Q_{54} conducts and a positive pulse is generated at its collector. This pulse is delayed, with respect to the pulse from the collector of Q_{53} by about 1.5 μsec and is used as the delayed horizontal trigger pulse. The leading edges of the horizontal sync pulses and the serrations in the vertical sync pulses are derived from this delayed trigger.

This scheme sets the width of the front porch of the composite video output at 1.5 μsec, or the amount by which the delayed horizontal trigger lags the undelayed trigger.

Vertical Divider

The vertical divider chain comprises Q_{46} through Q_{49}. Each is a blocking oscillator with feedback from collector to base and the R-C circuit controling the repetition frequency in the emitter.

Temperature stability of this circuit is superior to that of other circuits in which the frequency-determining elements are in the base circuit because it isolates these elements from the effects of the temperature-dependent base-to-collector leakage current I_{co}.

As space is limited in the sync generator, the frequency determining elements of each divider are fixed and the amount of sync injection has been made adjustable to control the division ratio. Capacitors C_1 through C_4 are used for this purpose and require appreciably less volume than adjustable potentiometers. Voltage divider R_2 R_3 biases all the divider stages.

The vertical frequency pulses at the output of the last divider, Q_{49}, are too narrow to be used directly. Vertical pulses of the desired width

are generated by the flip-flop comprising Q_{50} and Q_{51}. Each time Q_{49} fires, a pulse from its collector cuts off Q_{51}. The usual regenerative action in the flip-flop assists this and turns on Q_{50}. This action is reversed on the 15th, 21st or 35th cycle of the master oscillator after the firing of Q_{49} by a pulse taken from the collector of Q_{47} and fed to Q_{50}. This scheme causes the flip-flop to generate clean positive and negative vertical pulses, set at widths of 7½, 10½ or 17½ horizontal lines by adjustment of the order in which the usual divisors of three, five, five and seven are assigned to the various dividers.

Pulse Forming

The output pulses from the divider chain consists of four kinds of pulses: positive horizontal trigger pulses, positive delayed horizontal trigger pulses, negative vertical pulses and positive vertical pulses. The waveshaper, shown in Fig. 7, synthesizes the following signals from these trigger pulses: positive vertical drive pulses; negative horizontal drive pulses; positive and negative mixed blanking pulses and negative composite sync pulses.

Vertical drive pulses are formed from the positive vertical pulses at the divider output by the amplifier and line driver comprising Q_{69} and Q_{70}. These are connected in the same feedback pair configuration used in the monitor. This amplifier makes positive vertical pulses of about five-volts amplitude available at a relatively low impedance.

Horizontal drive pulses are formed by an identical amplifier, Q_{67} and Q_{68}, driven from the undelayed horizontal trigger pulses at the divider output.

Sync generator consists of pulse shaper (left) with master oscillator chassis attached at rear (right) by hinge

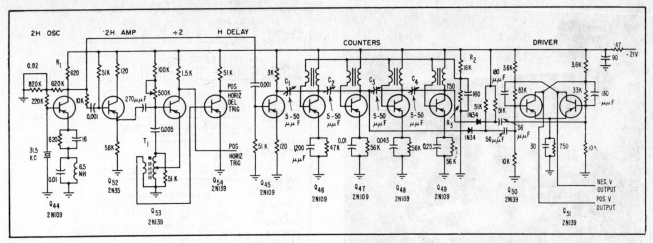

FIG. 6—Timing portion of sync generator includes 2H crystal-controlled master oscillator and blocking-oscillator count-down chain

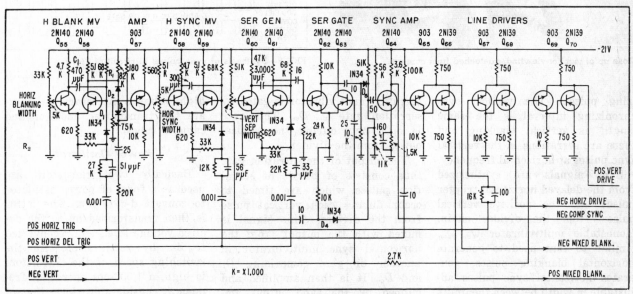

FIG. 7—Pulse shaping and gating section of sync generator turns out mixed blanking, composite sync and drive signals

Mixed blanking signals are synthesized from the undelayed horizontal trigger pulses and the negative vertical pulses from the divider by Q_{55}, Q_{56}, Q_{57}, Q_{65} and Q_{66}.

Monostable emitter-coupled multivibrator Q_{55} Q_{56} determines the horizontal blanking width. Normally Q_{55} is off and Q_{56} on. Horizontal trigger pulses are applied to the base of Q_{56} via diode D_1; each pulse turns off Q_{56} and turns on Q_{55}. The circuit remains in this state for a period of time determined by time constant of C_1 R_1 and the voltage at the slider of R_2. It then flips back to its original state. Adjustment of R_2 varies the

horizontal blanking pulse width between the limits of five and 15 μsec.

Negative horizontal blanking pulses from the collector of Q_{56} are mixed with the negative vertical pulses from the divider in an or gate comprising D_2 and D_3. The gate output is applied to Q_{56} and then to line driver Q_{65} Q_{66} from which both polarities of mixed blanking are available.

The three outputs from the sync generator just discussed must be transmitted via cable to the camera and monitor. Line driving amplifiers are provided for this reason.

The remaining transistors, Q_{58} through Q_{64}, are employed in generating composite sync. This signal is nonstandard, as illustrated in Fig. 8, in that there are no equal-

Rear view of sync generator shows master oscillator and counter chassis

Close up of monitor-viewfinder detached from camera

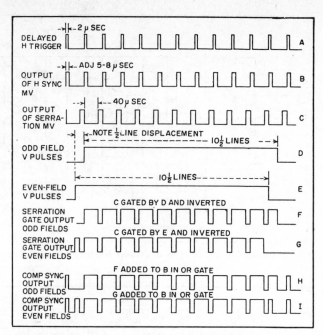

FIG. 8—Synthesis of sync pulses in circuit of Fig. 7

izing pulses. The vertical synchronizing interval is the same length as vertical blanking and there are serrations in the vertical sync pulses at horizontal frequency.

Sync signals are synthesized from the delayed horizontal trigger pulses and the positive vertical pulses from the divider chain. Monostable multivibrator Q_{58} Q_{59}, similar to that used to generate horizontal blanking pulses, generates horizontal sync pulses adjustable in width between the limits of five and eight μsec.

A similar multivibrator, Q_{60} Q_{61}, generates serration pulses of about 40 μsec width. The output of the

serration multivibrator is gated by serration gate Q_{62} Q_{63}. This gate is open only during the vertical pulses from the divider.

The output of the serration gate thus consists of bursts of serration pulses which are timed to occur during each vertical pulse from the divider. This signal is mixed with the output from the horizontal sync multivibrator in another or gate comprising D_4 and D_5. It is then amplified and clipped by the sync output amplifier Q_{64}.

The power converter, shown in Fig. 9, steps up the low voltage available from the battery to the

high voltages necessary to operate the transmitter, monitor and camera tubes.

Power Oscillator

Basically two transistors are used as a form of power oscillator to convert d-c to a-c. The latter is then transformed up to the desired voltage levels and reconverted to d-c by rectification. As the switching speed of the transistors is high, a 1,000-cps operating frequency, considerably above that of commercial vibrators, is employed to reduce the weight and size of the iron-cored components.

The power converter oscillator is similar to that described by Uchrin and Taylor. However it was found advantageous to slightly modify their circuit. Separate transformers are used to control the frequency of oscillation and to effect the necessary voltage transformation.

Transformer T_1 in Fig. 9 is wound on a small toroidal core and is designed to handle only the driving power for the base circuits of the transistors. A 1,000-cps square wave of about 90 v peak-to-peak appears from collector to collector of Q_{71} and Q_{72}. This signal is applied to the power transformer T_2 which steps it up to several different voltages.

Most of the power from the converter is delivered to the plus 170-v

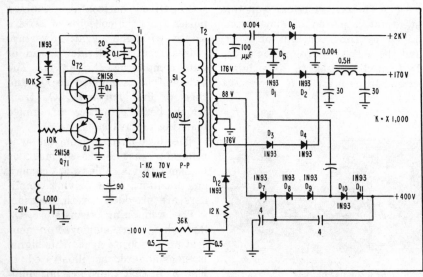

FIG. 9—Power converter uses one-kc square-wave generator for high-efficiency

bus. The principal load, about 50 ma, is the transmitter. This is supplied by the full-wave rectifier comprising D_1 through D_4. The 2,000-v ultor voltage for the monitor tube is supplied by the doubler comprising D_5 and D_6 driven by an additional winding.

The 400-v bus which supplies accelerating and focus potentials for the vidicon and monitor is served by the tripler comprising D_7 through D_{11}. The minus 100-v bus, which is used only for bias on the vidicon, is fed by half-wave rectifier D_{12}.

Overall efficiency of the power converter is about 90 percent.

Primary power for the entire unit is supplied by a bank of Yardney Silvercell batteries. Four LR-10 cells in series supply the six-volt bus which powers the heaters of the tubes and portions of the horizontal deflection circuits. Nine LR-5 cells in series with the six-volt portion of the battery power the 21-v bus. The flat voltage characteristic of these cells during discharge helps maintain stable operation of the more critical circuits, such as the sync generator dividers. This battery complement will operate the equipment for about five hours, before recharging.

Pickup Tube

The experimental miniature pickup tube, which was developed in an independent program of the Electronic Research Laboratory by A. D. Cope, has features which make it adaptable to transistor circuitry. The half-inch by three-inch vidicon operates in an axial

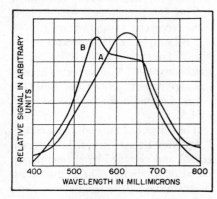

FIG. 10—Spectral response of miniature vidicon for equal values of radiant flux at all wavelengths

electromagnetic or permanent magnet field of about 80 gausses.

Power required to deflect the tube electromagnetically is about 20 ampere-turns or approximately $\frac{1}{4}$ that for the larger vidicon. This modest power requirement falls within the capabilities of low-power transistors of the audio amplifier variety.

The gun dissipates approximately 0.8 watt, a fourth that of the larger tube. Beam spot has been made smaller by reduction of the limiting gun aperture and path length so that the resolution is not greatly sacrificed over larger tubes, even though the scanning raster is only 0.24 by 0.18 in.

Photosurface

With such a small photolayer area, the use of a photoconductor having the sensitivity of that used in conventional vidicons would require an excessively fast lens or an increase in scene illumination. A new photoconductor surface has been developed which has sufficient operating sensitivity to more than offset the reduction in light-collecting area.

The photoconductive layer used has a sensitivity in the range from 1,000 to more than 2,000 μa per lumen when operating at signal levels of the order of 0.02 to 0.05 μa. The spectral response for the higher-sensitivity layers peaks at 650 millimicrons, as shown by curve A in Fig. 10. A more panchromatic layer with a peak response at 550 millimicrons, as shown by curve B, can also be made.

The gamma of the photosensitive surface depends upon the current density. At low illumination levels the signal varies linearly with light. At higher levels, for an output signal of 0.05 μa or greater, the gamma drops to 0.5.

The signal decay as a function of time for the new photosurface when compared with the photosurface used in earlier vidicons is somewhat different. The initial decay is not as rapid, but there is a smaller long term component of lag. This makes exact comparison difficult by subjective tests.

Tests were conducted with an experimental one-half inch vidicon and a typical one-inch tube operat-

Experimental pickup tube has half-inch diameter and is three inches long

ing in the same type of equipment and producing comparable images of the same scene. With 10 to 20 footcandles falling on the scene, equivalent output signals with comparable lag were observed in each picture when the small camera was operating with the lens at $f/4$ and the lens of the large camera at $f/2$. The combined difference of two lens stops and the four-to-one area difference indicates an operating sensitivity for the small tube 16 times greater than for the larger tube.

The authors wish to acknowledge the direction of V. K. Zworykin in this project and the cooperation of A. D. Cope who supplied the miniature vidicon. The assistance of Lawrence A. Boyer in design and construction contributed materially to the success of the work.

REFERENCES

(1) L. E. Flory, J. E. Dilley, J. M. Morgan and W. S. Pike, A Developmental Portable Television Pickup Station, *RCA Rev*, p 58, Mar. 1952.

(2) G. C. Uchrin and W. O. Taylor, A New Self-Excited Square Wave Transistor Power Oscillator, *Proc IRE*, p 99, Jan. 1955.

(3) A. D. Cope, A Miniature Vidicon of High Sensitivity, *RCA Rev*, Dec. 1956.

WIRELESS MICROPHONE

By G. FRANKLIN MONTGOMERY

National Bureau of Standards, Washington, D. C.

Transistorized wireless microphone operating at 460 kc, establishes induction field around transmitter within usable area without exceeding FCC radiation field limitation. Normal speaking voice produces peak f-m deviation of about 10 kc when used in lecture halls and auditoriums. Fixed f-m superheterodyne receiver recovers audio signal and feeds public-address or speech amplifier

WIRELESS MICROPHONES are useful in studio or lecture hall where the announcer or speaker must be able to move without being restricted by microphone cabling. The unit described here consists of a dynamic microphone with a self-contained wireless transmitter.

Applications

A fixed receiver, whose electrical output substitutes for the direct microphone output, is used to feed the public-address or speech preamplifier. Both the wireless microphone transmitter and the fixed receiver were originally developed for the main lecture auditorium at the National Bureau of Standards.

Wireless microphones commonly use vacuum-tube transmitters and usually operate on a frequency within the vhf band. Vacuum-tube power drain often limits transmitter battery life to several hours. In addition, a troublesome bar to vhf operation is the FCC requirement that the radiation from an unlicensed transmitter be limited to 15 microvolts/meter at a range of $\lambda/2\pi$, where λ represents wavelength.

Specifications

For a moderate distance between transmitter and receiver, a signal with an adequate signal-to-noise ratio may be impossible to transmit without exceeding the FCC specification. The transmitter described in this article uses transistors to permit relatively long battery life and operates at 460 kc, a frequency at which an induction field can be established around the transmitter within a usable area without exceeding the radiation field limitation.

Transmitter

The transmitter circuit diagram is shown in Fig. 1. Frequency modulation was chosen principally because of its inherent automatic-volume-control action. The transmitter radiates a signal directly from the tank circuit which consists of coil L_1, wound on a ferrite rod.

The radiated power, determined by the transmitter coil current and its radiation resistance, is about

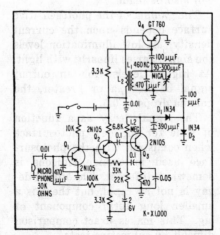

FIG. 1—Circuit for wireless microphone transmitter. Operating at about 500 kc, transmitter radiates about 2.2 × 10⁻¹³ watt directly from tank circuit

2.2×10^{-13} watt. Power is delivered to the tank by Q_4 operating as an oscillator at 460 kc.

Modulation

Diodes D_1 and D_2 control the r-f current through a 100-$\mu\mu$f capacitor shunted across the tank. The direct diode current is controlled by the audio-frequency current delivered to Q_3. A change in this direct current produces an approximately linear change in tank-circuit susceptance. The oscillator is therefore frequency modulated by the audio signal. Transistors Q_1 and Q_2 are audio-frequency amplifiers for the microphone output.

The coupling components were chosen to attenuate voice frequencies below about 700 cps. With the circuit shown, a normal speaking voice produces a peak deviation of about 10 kc. The total battery drain is about 15 milliamperes, so that with the single mercury cell the battery life is about 150 hours.

The vacuum-tube superheterodyne receiver is shown in Fig. 2. An r-f amplifier, V_1, operates at the signal frequency of 460 kc. Pentagrid converter, V_2, converts the signal frequency to 50 kc and the signal is amplified and limited by V_3 and V_4.

The output of stage V_4, a variable-frequency square wave, is differentiated and rectified by the counting-detector diode D_2. The audio signal is recovered by passing the diode output through a low-pass filter and amplified by V_5. Peak audio output is about 0.5 volt.

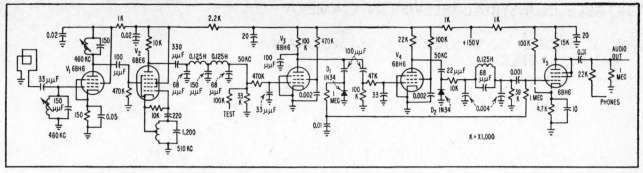

FIG. 2—Circuit for fixed receiver. Operating frequency of transmitter is converted to 50 kc by V_2 and is amplified and limited by V_3 and V_4. Audio signal is recovered after passing through low-pass filter. Peak audio output of receiver is about 0.5 volt

Receiver Unblocking

Squelch action is provided by diode D_1 which rectifies the received signal and unblocks the audio output stage during operation. This feature is useful in urban locations where the noise level is usually high.

The f-m superheterodyne receiver consumes little power and uses low plate and screen voltages consequently the operating temperature is low and the components should have long life.

Audio Response

The overall audio-frequency response of the system from microphone input to the receiver output is shown graphically in Fig. 3. No audio-gain control has been included in the transmitter but a control may be desirable to keep the deviation at maximum to compensate for differences in speakers.

The maximum size of the receiving loop depends on local noise field. A loop 5 meters square was used for this equipment. The loop could be made larger by using more transmitter power or by reducing

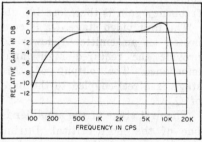

FIG. 3—Overall audio frequency response of transmitter-receiver system

local noise such as that generated by electrical machinery.

Shielding

In noisy locations, it is helpful to use a shielded receiving loop made of small coaxial cable. One end of the inner conductor is connected to the ungrounded receiver antenna terminal, the other end of the inner conductor and the shield braid are connected to the ground terminal. At a point halfway around the loop, the outer braid is cut and stripped for a length of an inch or two. Shielding is effective in reducing electrostatic pickup from nearby noise sources.

TWO-CHANNEL MIXER FOR TAPE RECORDING

By R. C. FERRARA
General Electric Co., Syracuse, N. Y.

A TRANSISTORIZED two-channel audio mixer described here can be used as an input control device for tape recorders or audio systems.

▶ Circuit — The mixer circuit shown uses two 2N107 *pnp* transistors. Each channel contains an input jack, gain control, transistor and filter network. The mixer input is fed to the base of the transistor through a 250,000-ohm gain control. Emitter voltage is filtered by a network consisting of a 2,000-ohm resistor and a 40-μf capacitor.

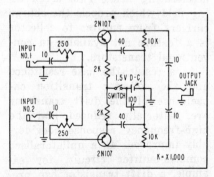

Circuit for the two-channel audio mixer. Microphones, phone pickups and speaker voice coils in any combination can be used for mixing. The number of audio channels can be expanded to include more than two

The collector current feeds into a 10,000-ohm load resistor. The output of each channel, taken across the load resistors through 10-μf blocking capacitor, is combined and fed to the output jack. A 100-μf capacitor is used across the battery as a filter and as an aid in maintaining battery life. The power required for the mixer circuit is derived from a 1.5-v battery, with little current drain. In general, the circuit is well filtered against noise.

PORTABLE INDUSTRIAL-TELEVISION CAMERA

By L. E. FLORY, G. W. GRAY, J. M. MORGAN
and W. S. PIKE
RCA Laboratories, Princeton, New Jersey

Twelve-transistor industrial-television camera puts out video-modulated r-f signal on either of two vhf channels. Video amplifier stages use drift transistors to achieve 4-mc bandwidth. Pickup tube is only one-half inch in diameter and three inches long. Total camera power consumption is 5.2 watts on a-c line operation

DEVELOPMENT of half-inch-diameter vidicons has made possible the design of transistorized television cameras of small size, low weight, low power consumption and high sensitivity.

The experimental camera to be described is a simple unit which, in connection with any standard television receiver, forms a complete closed-circuit television system. Video information is modulated on a carrier frequency which may be adjusted to channel 2 or channel 3 of the vhf band.

The block and schematic diagrams of the camera are shown in Fig. 1 and 2.

Video Amplifier

For best signal-to-noise ratio, the input impedance is made high, about 50,000 ohms. The vidicon signal current develops about 5 mv across this load at low frequencies.

Since the modulator requires about 200 mv at about 200-ohms impedance, a power gain of approximately 57 db is required. The high-peaker stage requires about 26 db more gain to equalize the falling high-frequency response of the input network; it is advantageous to provide an additional 22 db of high-frequency gain to compensate for the apparent spot size of the camera tube (aperture correction). Total required high-frequency gain is therefore 105 db.

The first video stage, Q_1, is connected as an emitter follower to provide the required high input impedance. The output of this stage is direct-coupled to the base of Q_2.

The first stage of the amplifier is one of the most critical. During development of the camera, best results were obtained in this stage with experimental RCA drift transistors.

Drift Transistor

The drift transistor utilizes the alloy junction technique but features a base region in which the impurity distribution is carefully controlled to produce a built-in accelerating field. This accelerating field which propels the charge carriers from emitter to collector is a feature not available in conventional transistors.

As a result, the base resistance and the collector transition capacitance of the drift transistor are materially reduced and the high-frequency response considerably improved. In a unilateralized common-emitter circuit, for example, a drift transistor can provide a power gain as high as 45 db at 1.5 mc or 24 db at 10.7 mc.

These units are now commercially available as type 2N247.

With present manufacturing tolerances, it may be necessary to select transistors for Q_1 and Q_5 to obtain satisfactorily high input impedances in these stages.

The second and third stages of the amplifier must be considered together. Neglecting their emitter circuits, they are conventional a-c coupled common-emitter amplifiers. A shunt peaking inductance in the collector circuit of Q_2 maintains the high-frequency response of the stages to 4 megacycles.

In each emitter circuit, a 470-ohm resistor provides a d-c path of sufficient resistance to obtain adequate d-c stability. Such a high resistance would reduce the stage gain to a low value. Therefore R-C networks comprising R_1C_1 and R_2C_2 are shunted across these emitters. This does not remove all the a-c degeneration. High-frequency positive feedback from the emitter of Q_3 to the emitter of Q_2, via a capacitive divider, sufficiently increases the high-frequency gain of the circuit to provide the necessary aperture correction.

Transistors Q_1 and Q_3 are biased by common bleeder network R_3R_4, while Q_2 is biased by its direct connection to Q_1.

The adjustable high-peaker, located between the collector of Q_3 and the base of Q_4, is a simple adjustable high-frequency equal-

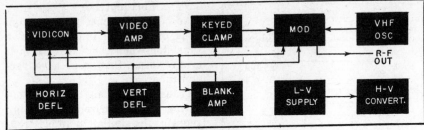

FIG. 1—System block diagram of transistor industrial-television camera

izer. The amount of high-frequency equalization is fixed by the circuit resistances.

Turnover frequency is adjusted by C_5 to match the loss characteristic of the input network. This adjustment is most readily carried out by observing a test pattern and visually adjusting the high-peaker to minimize trailing.

Transistor Q_4 is a conventional common-emitter stage. Its emitter is bypassed and a peaking inductance in its collector maintains its high-frequency gain. Bias is applied via the high-peaker resistors to its base.

Low-Frequency Response

Transmission of a good televi-sion picture imposes special requirements on the low-frequency response of the system amplifiers. Making the response good to d-c is usually impractical. A better technique is to adjust the amplifier low-frequency gain and phase response to adequately reproduce a square wave at the horizontal scanning frequency and then employ clamping or d-c restoration to reinsert the d-c component. This technique results in smaller coupling and bypass capacitors.

In this equipment, overall low-frequency correction is made by placing capacitor C_6 across R_5, the resistor being required to bias Q_4 properly. This one capacitor adequately compensates the low-fre-quency response of the amplifier through Q_4.

The signal from the collector of Q_4, is a-c coupled to the base of Q_5 via a relatively small capacitor, C_7. A driven clamp comprising diodes D_1 and D_2 sets the base potential of Q_5 at the start of each horizontal line. Negative drive pulses for the clamp are derived from the horizontal deflection circuit via R_6, R_7 and C_8. On each pulse the diodes conduct, clamping the base of Q_5 to a potential which is adjusted for optimum operation of the modulator by R_8; d-c information is thus reinserted at this point in the video amplifier chain.

Modulator D_3 operates as a variable impedance in series with the carrier oscillator output. This impedance is caused to vary in accordance with the video information.

Carrier Oscillator

Transistor oscillator Q_6 is tunable from 54 to 66 mc. Oscillation is maintained by collector-to-base feedback via the tuned transformer. Energy from the oscillator

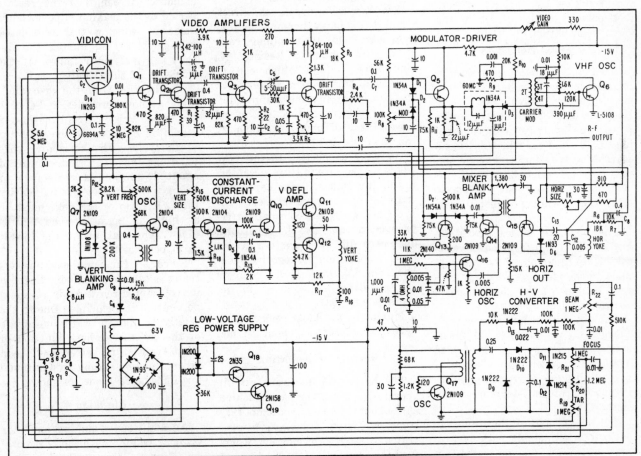

FIG. 2—Camera uses 12 transistors, 20 diodes and one photodiode in addition to vidicon pickup tube

is applied to the anode of D_3 via a two-turn coupling coil on the oscillator transformer. Bias is applied at the same point from a divider comprising R_9 and R_{10}.

The cathode of the modulator diode is connected to the emitter of Q_5 through a trap broadly tuned to the carrier frequency. The modulator diode is adjusted to be conducting with R_8. The r-f from oscillator Q_6 can then pass through the modulator diode to the output terminal of the camera via capacitor C_7. Video signals at the emitter of Q_5 vary the diode impedance, thus modulating the video signal on the carrier.

Synchronizing pulses are also added to the outgoing signal in the modulator. Negative horizontal sync pulses from the deflection circuit are applied via R_{11}, which imposes a negligible load on the video signal.

Positive vertical-sync pulses from vertical blanking amplifier Q_7 are applied to the other electrode of the modulator diode by R_{12}. The resulting composite output signal has both horizontal and vertical synchronizing components of correct polarity with respect to the video information.

Vertical Deflection

Blocking oscillator Q_8 generates negative pulses across R_{13} at vertical frequency. During a-c operation this oscillator is locked to the power line by the circuit comprising D_4, R_{14} and C_9, which injects a suitably shaped line-frequency signal into the base circuit of the oscillator. During battery operation, the blocking oscillator runs free.

The negative pulses across R_{13} are amplified and inverted in vertical blanking amplifier Q_7 and applied to the vidicon cathode to prevent landing of the electron beam during vertical flyback.

Each negative pulse across R_{13} charges C_{10} via diode D_5. A linear sawtooth voltage is generated across this capacitor by allowing it to discharge between pulses through constant-current transistor Q_9.

The magnitude of the discharge current is adjustable with vertical size control R_{15}. The generated sawtooth voltage waveform has a

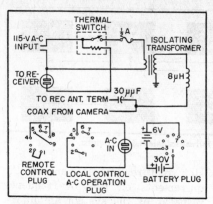

FIG. 3—Adaptor box for remote control of camera (top) and plug wiring diagrams for various types of operation (bottom)

maximum amplitude of about 10 volts; this limitation being set by the magnitude of the pulse available across R_{13}.

The vertical yoke windings have a resistance of about 200 ohms and an inductance of about 57 mh. The latter is negligible. The sawtooth voltage appearing across C_{10} is coupled into these windings, causing an accurately linear sawtooth current to flow in them. Inpedance conversion is performed in the following stages as a first step in accomplishing this purpose.

Emitter follower Q_{10} is direct-coupled to a class B complementary symmetry output amplifier comprising Q_{11} and Q_{12}. Resistor R_{16} limits the voltage which can be applied to this system, swamps out the somewhat nonlinear input impedance characteristics of Q_{10} and raises the load impedance seen by the sawtooth generating circuit.

As a d-c connection to the yoke would cause prohibitive decentering of the vidicon beam, capacitive coupling to the yoke is used.

Capacitor C_{10} is too small to reproduce adequately the sawtooth, but space did not permit a larger unit. Application of negative feedback via R_{16}, R_{17} and R_{18} corrects the resulting nonlinearity in acceptable fashion.

Maximum available peak-to-peak deflection current is about 25 ma.

Horizontal Deflection

The horizontal deflection circuit operates in much the same fashion as its counterpart in contemporary home television receivers; it generates a sawtooth current waveform by periodic interruption of the current flowing in an induct-

ance. Fortunately, transistors make better switches than vacuum tubes in that the impedance of a transistor when conducting is materially lower than that of a vacuum tube.

Transistor Q_{16} an oscillator operating at the horizontal scanning frequency, is similar to an electron-tube Colpitts oscillator. Some temperature compensation is applied for stability by temperature-sensitive capacitor C_{11} which is mounted on Q_{16}.

Although the waveform across the oscillator tuned circuit is a good sine wave, the waveform in its collector circuit is a negative pulse. This keys on transistor Q_{14}. The large positive pulse on the collector of Q_{14} is coupled into the base of horizontal output transistor Q_{15} without inversion by the stepdown transformer. This cuts off Q_{15}, causing a large pulse to appear across the yoke windings during the horizontal retrace period.

Diode D_6 is the usual damper diode, as in a conventional receiver. Capacitor C_{12} adjusts the resonant frequency of the yoke system. It is chosen to lengthen the retrace period to the maximum permissible amount, reducing the peak pulse voltage occurring during the flyback and materially improving the efficiency of the circuit.

To block d-c from the yoke windings, a choke provides a d-c path for the collector current of Q_{15} and capacitor C_{13} couples the collector to the yoke. The horizontal yoke windings have an inductance of about 1 mh and a resistance of about 3 ohms. For good linearity the L/R ratio of the yoke must be as high as possible. The figures given represent about the lowest ratio permissible.

Since a transistor such as Q_{15} really contains a built-in diode in its collector circuit, diode D_6 could be omitted in principle. Due to the relatively high forward impedance of the collector-to-base junction of small transistors, better results can be obtained by adding an additional low-impedance diode as shown.

Blanking

An additional blanking amplifier Q_{13} is associated with the horizontal deflection system. The vidicon requires about 15 to 20 volts of positive blanking signal on its cathode to positively prevent beam landing during the retrace periods under all conditions of beam current and illumination. It is difficult to provide this in a camera in which the maximum supply voltage is 15 volts. For this reason, additional blanking is applied to the vidicon grid to reduce the beam current during retrace.

Diodes D_7 and D_8, nonlinearly mix positive horizontal pulses from Q_{14} and positive vertical pulses from Q_7 by forming an orgate. The resultant is applied to the base of Q_{13}. Amplified negative blanking pulses at the collector of Q_{13} are applied to the vidicon grid.

This arrangement, in conjunction with Q_7, results in the application of positive vertical blanking pulses to the vidicon cathode and negative mixed blanking pulses to the vidicon grid. Thus double blanking is supplied during the vertical blanking interval, which is the more critical as it encompasses a number of complete horizontal scans, during the forward scanning portions of which the beam is moving relatively slowly. On the other hand, during the horizontal retrace the beam is always in rapid motion and the resulting current density at the vidicon target is proportionately lower.

High-Voltage Supply

The vidicon requires a positive accelerating potential of about 300 volts, a positive focussing voltage adjustable from 250 to 300 volts and a negative source of bias adjustable from zero to about −100 volts for beam current control. All voltages are referred to the vidicon cathode, which is near ground potential, being returned to the collector of Q_7.

These voltages are supplied by converter Q_{17}, a square-wave oscillator operating from the −15-volt bus which powers all the transistor circuits of the camera.

As the converter operates at 15 kc, the size and weight of the required filter components is greatly reduced. The operating frequency is controlled by the associated transformer.

The waveform at the collector of Q_{17} is nearly a square wave whose amplitude approaches the supply

The voltage at the transistor collector is stepped up in the transformer and applied to two separate rectifier systems. The first, comprising D_9 and D_{10} is a doubler. Its output is regulated by a pair of silicon zener diodes, D_{11} and D_{12}. Two diodes in series are used to reduce the dissipated power.

The output voltage is adjusted to be from 280 to 300 volts by selection of the regulator diodes, which have fairly wide tolerances as received from the manufacturer.

The bleeder comprising R_{19}, R_{20} and R_{21} provides the wall, focus and target potentials for the vidicon. The maximum current available from this rectifier system is about 800 microamperes. Rectifier D_{13} supplies about 100 volts negative to beam current control R_{22}.

The camera may be operated either from the a-c line or from batteries. During a-c operation, power for the camera may either be taken from an a-c outlet or it may be sent to the camera from the remote viewing point over the same coaxial cable which carries the r-f output of the camera. For either the local or remote a-c mode of operation, the 15-volt bus which operates all the transistor circuits in the camera is energized by the low-voltage regulated supply comprising the power transformer, rectifiers and transistorized voltage regulator Q_{18} Q_{19}.

Using the remote method of supplying power, the plug shown in Fig. 3 is inserted into the socket to connect the primary of the power transformer across the r-f output connector through the 8-microhenry r-f choke. Power is applied to the end of the output coaxial cable remote from the camera by the adapter box shown in Fig. 3, which interconnects the cable and the remote receiver.

Because of the great frequency separation between the 60-mc camera output and the power-line frequency the simple chokes and capacitors shown adequately isolate the two signals.

The action of the transistor regulator is identical to an electron-tube series regulator In this case transistor Q_{19} is the series element and Q_{18} is the amplifier. A pair of zener diodes provide the necessary volt-

voltage. The transistor in this circuit is switched rapidly between its fully conducting state in which its current is high but the voltage across it is nearly zero, and its off state in which its current is zero and the voltage across it is considerable. Under these conditions the power dissipated in the transistor is low and overall efficiencies of the order of 70 to 90 percent may be achieved. age reference to the regulator.

The regulator delivers about 150 ma to the 15-volt bus and will hold the output voltage within 0.1 volt of the specified value for any input voltage between 100 and 130 volts. It also reduces the 2-volt ripple content at the output of the bridge rectifier by a factor of approximately 100.

The camera may be operated on batteries by inserting the proper plug, also shown in Fig. 3.

Automatic Sensitivity Control

A novel feature of the camera is an experimental arrangement for automatically varying the camera sensitivity as a function of the ambient illumination of the scene being viewed. A photo-diode is arranged at the proper distance behind a hole of suitable diameter in the front surface of the camera so that it sees a solid angle identical to that of the camera lens.

Increasing light increases the current through the photodiode. This is arranged to reduce the target voltage of the vidicon, thus reducing its sensitivity.

Zener diode D_{14} provides a d-c level shift in this circuit merely for convenience in selecting appropriate operating points for the vidicon and the photodiode. Compensation for ambient illumination changes of the order of 50 to 1 may be provided by this method. Such compensation is useful where the camera operator is unskilled or the camera itself is inaccessible.

The experimental miniature vidicon[1] has the advantage, for use with transistor circuits, of requiring only about 20 ampere turns of deflection field. This is about one-third the amount required by conventional vidicons. The tube may also be made somewhat more sensitive than its predecessor.

Chapter 8

R-F COMMUNICATIONS CIRCUITS

SUPERREGENERATIVE 52-MC TRANSCEIVER

By W. F. CHOW

General Electric Company, Syracuse, New York

Three-transistor transceiver uses tetrode as 52-mc oscillator for transmitting and as forced-quench *superregenerative* detector for receiving. Theory of operation is covered for both self-quenched and forced-quenched modes of regenerative oscillation

BASIC advantages of superregenerative detection, sensitivity and simplicity, are further enchanted by the merits of a transistor.

Because of the dependence of transistor parameters on the operating points, the loop gain of a transistor oscillator can be controlled by changing the emitter current and/or the collector voltage. This property makes possible the control of a transistor by an external quenching signal.

Figure 1 shows the a-c circuit of a junction transistor oscillator. Inductance L_1 and C_1 form a tank circuit and a feedback path is provided by C_2. Choke L_2 prevents the feedback signal from being shunted to ground.

Self-Quenched Oscillator

If the oscillator circuit is so designed that the resistance of the circuit between emitter and the base is small, but there is a finite d-c resistance in the collector circuit as shown in Fig. 2 and if C_3 is large enough, it will function as a self-quenched oscillator.

Initially, the emitter d-c bias current is very small. It is obtained from the d-c voltage drop across the base. Since the collector current is also small, the d-c voltage drop across the r in Fig. 3 is negligible and collector bias voltage V_c is practically equal to the battery supply voltage.

If under this initial d-c bias condition the loop gain of the oscillator circuit is adequate, the oscillation starts and its amplitude builds up. At the same time the instantaneous emitter current increases according to the envelope of oscillation by the rectification action of the emitter diode. At first, the effect of increased emitter current overcomes the effect of slightly increased collector bias. Thus, the loop gain is improved.

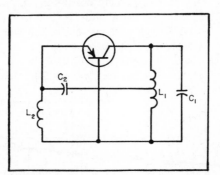

FIG. 1—Basic a-c circuit of junction transistor oscillator

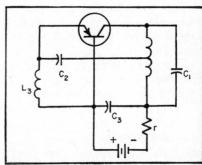

FIG. 2—Self-quenced oscillator for superregenerative detection

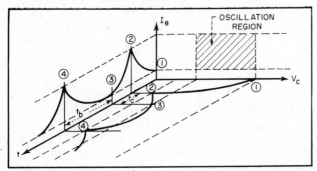

FIG. 3—Bias current and voltage of self-quenched oscillator

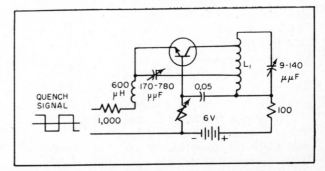

FIG. 4—Experimental circuit of forced-quench transistor oscillator

The oscillation builds up at a high rate and the emitter and collector current increase correspondingly. Due to the presence of r in the collector circuit, the collector voltage decreases.

The amplitude of oscillation soon reaches a point at which the emitter current is large and the collector bias voltage is small. Then the loop gain drops below unity and the oscillation dies rapidly, the emitter and collector currents decreasing at the same time. Now V_c recovers, but since C_3 and C_2 are connected across the collector and the base of the transistor as far as the quench frequency is concerned, the instantaneous V_c will be the same as the voltage across the capacitance.

A finite time is required to build up the voltage across these two capacitances through r. If the time required to charge these two capacitances is longer than the time required to let the oscillation stop, or the amplitude of the oscillation decreases to a very small magnitude, the oscillator appears as a self-quench oscillator.

Bias Variation

Figure 3 shows qualitatively the variation of the bias current and voltage of a self-quenching oscillator with time. The shaded area indicates the oscillation region in which the values of I_e and V_c provide a proper bias point for the transistor.

At time $t = 0$, the I_e and V_c start from the values marked 1. As the amplitude of oscillation builds up, I_e and V_c reach the values marked 2. At this bias point, the gain of the transistor is not enough to maintain oscillation. The amplitude of oscillation comes down and V_c builds up.

When V_c reaches point 3, the oscillation begins to build up again. The time t_c required to charge the capacitor plus the time t_b required to build up the oscillation to a peak point determines the quench frequency. When there is an r-f or a noise signal, t_c will be shortened.

The quench frequency is modulated by the envelope of the r-f signal. Since the amplitude of oscillation is the same for all quench frequencies, an increase of frequency indicates an increase of the area under the oscillation envelope. Consequently, a gain of I_e is obtained, and it follows the envelope of the r-f signal.

Forced Quenching

When the oscillator oscillates strongly without any self-quenching action, it can be quenched by applying an auxiliary signal. This auxiliary quench signal is used to move the bias point of the transistor to a value such that the loop gain of the oscillator is too small to maintain the oscillation.

If the quench signal is a periodic wave, oscillation will start and stop for each cycle. The length of the period during which oscillation stops depends entirely upon the properties of the oscillator and the magnitude and frequency of the quenching signal.

If the circuit design of the oscillator is such that the oscillator is not self-starting, an outside quenching signal will be able to start the oscillator and maintain the oscillation for a period of time.

The principle of applying a quench signal to start and stop an oscillation is similar to the externally quenched superregenerative action of a tube circuit.

Figure 4 shows a test circuit of the forced-quench oscillator. Coil L_1 has two taps, one for the feedback circuit and the other for the collector. When the collector is connected across a tap point instead of across the whole inductance, the loading effect of the transistor on the Q of the tank is reduced. The Q of the coil at one mc is 120.

The variable resistance in the base circuit adjusts the bias point for no oscillation when the quench signal is zero or of the wrong polarity. The resistance in the collector circuit is small, so V_c does not change too much. Consequently, the bias point of the transistor is controlled predominantly by the emitter current.

When a 10-kc square-wave signal is applied, the duration of each half-cycle of the quench signal is long enough to allow the oscillation to build up and to die down to such a small value that the envelope of the oscillation is controlled by the outside r-f signal. Since the duration of the half-cycle for build up of the oscillation is long enough to let the oscillation reaching the saturation region, the operation is in the logarithmic mode.

If the duration of the building up of oscillation is cut down, the oscillation would be in the linear mode. However, there is a minimum required length for the duration of the quench period. When that period is less than this minimum value, the oscillation is unable to die down to less than the level of noise and outside r-f signals. Consequently, the oscillator oscillates continuously and cannot be controlled.

When a 25-kc quench signal is fed into the emitter, the duration

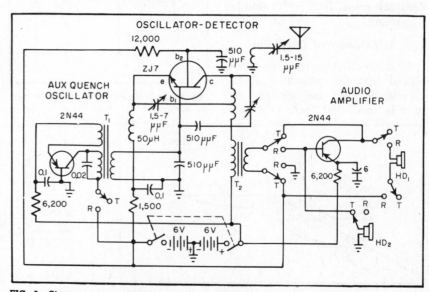

FIG. 5—Six-meter transceiver uses high-impedance headphone as dynamic microphone

of quenching period is not long enough to quench the oscillation. Using the 25-kc quench signal, a quench oscillation can be obtained by increasing the rate of decay of oscillation. This can be done by increasing the loss in the tank.

Self-Quench Test

The circuit of Fig. 2 was used to test self-quenched oscillation. The inductance of the tank circuit was the same as that used in the forced-quench circuit. Capacitance C_3 and resistance r were varied to study the self-generated quench frequency.

When the transistor was inserted in the circuit with the feedback path open, the bias point was $I_e = 90\ \mu a$; $V_e = 6$ v. After the feedback path was closed, oscillation started. The emitter bias current increased by the rectification of the a-c voltage feedback to the emitter.

If the values of C_3 and r are

changed, the self-generated quench frequency is changed.

Six-Meter Tranceiver

Figure 5 shows the circuit of a completely transistorized six-meter transceiver.

A ZJ7 tetrode is used as the 52-mc oscillator in the transmitting position and as the superregenerative detector in the receiving position. Two 2N44 junction transistors are used, one as the auxiliary quench oscillator and the other as the audio amplifier. Two 2,000-ohm head sets are used, one as the headphone and the other as the dynamic microphone.

In the receiving position, the 20-kc quench signal of the auxiliary quench oscillator is fed into base b_1 of the tetrode through T_1. Consequently, the oscillation of the tetrode is quenched periodically and superregenerative detection is obtained. The audio output of the

detector is fed into the base of the audio amplifier through T_2.

In the transmitting position, the collector bias supply of the auxiliary quench oscillator is removed. Hence, there is no quench signal and the tetrode oscillates strongly at 52 mc.

When a voice is spoken into the microphone, the induced audio is amplified and fed into the collector circuit of the tetrode through T_2. Consequently, the oscillator is amplitude modulated by the voice signal.

The modulated r-f is inductively coupled to the antenna.

Field tests of two transceivers indicate a usable distance of one-half mile. Battery drain of each unit is 48-mw for receiving and 60 mw for transmitting.

FORTY-METER TRANSMITTER

Powered by two flashlight cells, the crystal-controlled transmitter shown in the diagram has been used in the amateur 7-mc band to communicate with stations all over the United States, parts of Canada and the Hawaiian Islands.

The equipment requires 120 milliwatts input. Several precautions have been taken to insure stable operation. Grounded base is used in the amplifier to reduce feedback, obviating the need for neutralization. The small amount of capacitive feedback is regenerative but no instability has resulted.

▶ **Stabilization** — A minimum resistance in the emitter is essential to temperature stability and to lessen the danger of current runaway, but series dropping resistors are otherwise minimized. Capacitor C_1 is a phasing adjustment to be set at the point of best stable operation of the oscillator under keying. It also permits slight frequency change.

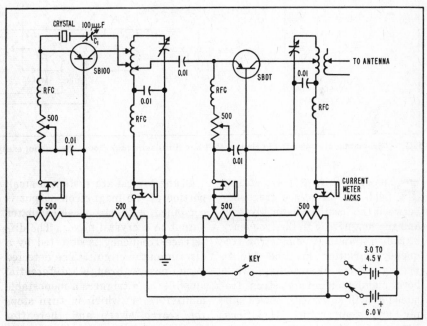

Circuit of the two-transistor 7-mc transmitter

AFC USING TRIANGULAR SEARCH SWEEP

By H. H. HOGE and D. L. SPOTTEN

Air Arm Division, Westinghouse Electric Corp., Baltimore, Md.

Miniaturized plug-in unit for radar uses transistor multivibrator with Zener diodes. Search sweep varies local oscillator klytron repeller voltage, holding 30-mc i-f to bandwidth of 200 kc. Vacuum-tube discriminator feeds three-stage video amplifier triggering sweep generator. Low-voltage, low-power components afford greater reliability at less than half the power consumption of conventional vacuum-tube sawtooth systems

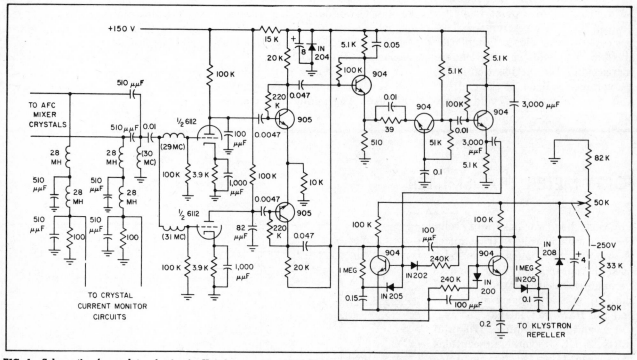

FIG. 1—Schematic of complete afc circuit. Unit has transistors and diodes throughout except for vacuum-tube discriminator stage

RADARS require afc systems to hold the difference frequency between the local oscillator klystron and the magnetron to the frequency of the i-f amplifier. Magnetron frequency variation may be due to pulling effects of scanning, while both magnetron and klystron frequencies will shift with power supply and temperature variations. The receiver bandwidth must pass the signal even if the magnetron frequency shifts. But, since the receiver performance is lowered as the bandwidth increases, the afc must hold the local oscillator to close limits, usually about 200 kc for a 30-mc i-f.

In the typical afc system, a small portion of the magnetron output is combined with the local oscillator output in a crystal mixer. The difference frequency is detected by a discriminator circuit. The detected signal passes through a differential amplifier and triggers a monostable multivibrator, which in turn stops the search sweep and thereafter corrects the local oscillator frequency at the repetition rate. A conventional afc unit consumes 10 to 20 watts in these functions.

Afc With Transistors

The semi-transistorized afc of Fig. 1 employs vacuum-tube detec-

tion because the low alpha cutoff of present silicon transistors yields a detected signal too greatly attenuated for practical use. The outputs of the two detectors feed the base of an emitter-coupled differential amplifier using two silicon transistors.

Output is taken from one collector to invert the signal from one detector and produce an S-type discriminator curve. The signal is fed to a three-stage video amplifier, using silicon junction transistors, whose output triggers a sweep generator. A common-collector stage matches the high impedance of the differential amplifier collector cir-

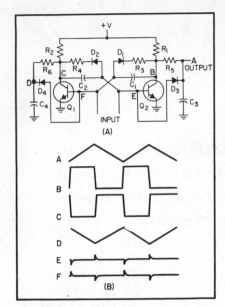

FIG. 2—Flip-flop circuit and waveforms

End-on mounting of some components is employed in plug-in unit

cuit to the low impedance of the second video stage which has a common-base connection.

The output of this stage couples to the sweep generator through a split-load phase inverter. Manual adjustment of the repeller voltage is by the dual potentiometer. A range of about 55 volts insures full coverage of the klystron power mode. The sweep circuit and divider network draw less than 2 ma.

Basic Sweep Circuit

The sweep circuit of Fig. 2A comprises two transistors and four silicon diodes operating in the Zener region. Diode D_1 has a lower Zener voltage than D_2 and the current through R_1, R_3 and D_1 saturates Q_1, making point C only a few volts above ground.

Transistor Q_2 remains cut off, while point B is at a high potential and C_3 charges through R_1 and R_5 until point A reaches the Zener voltage of D_3.

The current through D_3 then passes through the base of Q_2 to

produce an amplified negative signal at point B. This signal is fed through C_1 and amplified by Q_1. The positive signal which appears at point C is then coupled back into Q_2. The signal regenerates until the potential at C reaches the Zener rating of D_2, thus locking the circuit with Q_2 saturated, point B a few volts above ground and Q_1 cut off.

Point C is now at high potential and C_4 charges through R_2 and R_6 until the Zener rating of D_4 is exceeded. Capacitor C_3 is simultaneously discharging through R_5 and Q_2 which is held in saturation. When the Zener rating of D_4 is exceeded the multivibrator will exchange status with C_4, discharging while C_3 recharges.

The resulting triangular waveform is applied to the repeller of the local oscillator until a change in discriminator output is effected. This change in polarity is fed to the video amplifier and then to the correct base to stop the search sweep. As the charge on C_3 leaks off, the repeller voltage changes and

the resulting frequency shift is detected at the next pulse of the magnetron.

The sweep may be stopped at any point on the triangular search pattern. The range is set by the Zener diodes D_3 and D_4. The response of network R_5 and C_3 at the repetition rate is such that the voltage swing at C_3 represents about a 200-kc change of the klystron frequency. The search sweep is therefore disconnected and the klystron is shifted only enough to see-saw across the i-f. between repetition rate pulses. This results in a lock-on saw-tooth sweep of about 0.2 volt p-p, which is well within the bandwidth of the receiver.

The loop is degenerative and tight enough to prevent false triggering due to stray radiation. The entire system uses approximately six watts. Assembled unit is less than 6 in. square, 2.5 in. high. Greater reliability and subminiaturization are therefore achieved through the use of low-power components with semiconducting devices.

MOBILE 12.5-MC F-M RECEIVER

By A. M. BOOTH

Chief Design Engineer, Electronic Products Division,
Gruen Industries, Inc., Cincinnati, Ohio

Mobile f-m receiver with nineteen transistors is designed for mass production using printed circuits and available transistors. Circuit operates over temperature range from −67 F to +149 F with simultaneous supply voltage variation from 22 v to 30 v without degradation

SOME MILITARY vehicular communications require a reliable transistorized 12.5-mc f-m receiver, capable of being mass produced without special selection of components. The receiver described here is coupled to an electron-tube variable-frequency tuner to provide a receiver covering the 20 to 70-mc band. The 12.5-mc i-f signal produced by the tuner is used as the r-f input to the fixed-tuned transistor receiver.

Auxiliary circuits provide a squelch relay, a beat-frequency oscillator, and a 1-mc crystal calibrator. All electrical components are designed for printed-circuit wiring and dip-soldering techniques, except the 1.3-mc selectivity filter.

Sensitivity is 12 microvolts into 50 ohms for a signal plus noise-to-noise ratio of 10 db. A 100-μv signal into 50 ohms is the minimum required to drive the receiver to a rated 500-mw output with a 600-ohm resistive load. A 12.5-mc signal modulated at 1,000 cps with 15-kc deviation is considered standard. Estimated average power consumption is 1.2 watts under normal voice-operating conditions.

Standby current drain is 37 ma with a 26-volt battery supply. Full output is obtained at approximately 100-ma current drain at 26 volts.

The unit is capable of operating without degradation in performance over an ambient temperature range of −67 F to +149F with a simultaneous battery voltage variation of 22 to 30 volts.

The mixer, i-f and discriminator circuits are shown in Fig. 1. A 12.5-mc r-f selectivity is achieved by a 6-section L-C band-pass filter of a modified constant-K type. A bandwidth of 250 kc to 1,000 kc for the 6-db/60-db points provides 85-db attenuation for spurious frequency responses outside the ±200-kc points of the 12.5-mc r-f reference frequency.

The local oscillator employs a 11.2-mc series-resonant crystal as a feedback network between the tap on the collector tank coil and the emitter. This unit has a frequency-stability of ±0.007 percent for a simultaneous voltage variation of 22 to 30 volts and temperature change of −67 F to + 149 F.

A 6-volt Zener diode is used for supply voltage and mixer bias regulation. Voltage change across the diode is 0.1v with a supply variation of 22 to 30 volts.

Mixer

The mixer circuit is of the common-emitter type producing approximately 12-db of conversion gain. Single-tuned circuits are used for ease of production alignment and field maintenance. To

Three-section receiver consists of an r-f chassis (top) with 12.5-mc filter in metal can, an i-f chassis (center) and audio chassis (bottom) with three power transistors at upper left

provide the required bandwidth it is necessary to sacrifice conversion gain by impedance mismatch in the collector circuit. For narrower bandwidth requirements, conversion gains of approximately 20 db would be realizable.

The first 1.3-mc i-f amplifier is located on the 12.5-mc r-f chassis as an aid in testing for normal r-f unit conditions with a standard vtvm. It has a bandwidth of approximately 600 kc to avoid attenuation of the response curve of the 1.3-mc selectivity filter.

The selectivity filter serves as the output unit for the r-f chassis and the input unit for the i-f chassis. A 12-section L-C bandpass filter of modified constant-K type is used. It establishes the entire 1.3-mc i-f system selectivity and has a bandwidth of 100 kc to 200 kc for the 6-db/60-db points.

Each of the parallel-tuned circuits is set on the high-side 60-db cut-off point. These are coupled with a capacitor which series resonates with the parallel-tuned tanks to the low-side 60-db cut-off point.

The input and output load resistances were chosen to be 5,600 ohms so that the values of L-C would be of practical size.

I-F Amplifiers

Three stages of 1.3-mc amplification employ fixed single-tuned transformers eliminating production tuning and simplifying field maintenance. These are designed for a collector load to input impedance ratio of 5,000 to 1,000 respectively. A deliberate mismatch was required to realize an overall 320-kc bandwidth at the 6-db points for the three stages. This bandwidth prevents excessive attenuation of the 1.3-mc selectivity filter characteristic.

A 5,000-ohm collector-load impedance was chosen so that a fixed tuning capacitance of sufficient value could be used to minimize the detuning effects of the variation of C_o of the transistor and wiring capacitance. The surface barrier SB-100 transistor is used because of its 2 to 3-$\mu\mu f$ value of C_o and the small change in value with voltage variation.

The three interstage transformers are encapsulated so that they can be inserted directly into the printed-circuit board. The windings are coated with a cement prior to resin encapsulation. This reduces the capacitance effect of the encapsulating material. Changes in the type of encapsulating material

also require a change in the fixed-tuning capacitor due to variations in added capacitance effects.

Limiter

Limiter action depends primarily on collector-to-base voltage saturation. The emitter bias is 300 μa and collector-to-base d-c voltage is 1 volt. A 12-volt Zener diode stabilizes bias conditions for both the limiter and discriminator driver. The gain of this stage is approximately 8 db at initial limiting conditions and is restricted by the low value of bias. The limiter characteristic curve is flat and well within the specified 1-db requirements for a standard signal input variation of 100 μv to 100 mv.

Discriminator

The discriminator driver power gain is 20 db. The power level is raised at this point to obtain better rectification efficiency of the detector diodes. A conventional Foster-Seeley type is used. Peak-to-peak spacing is 130 kc with good linearity over a 90-kc range.

Audio Amplifier

In the audio amplifier section, shown in Fig. 2, a low-pass noise filter cuts off noise energy above

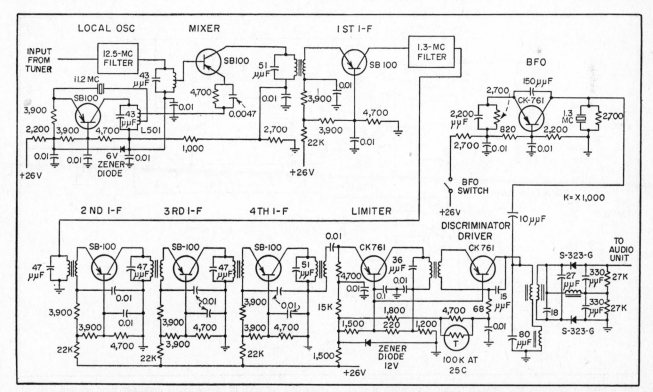

FIG. 1—Mixer, i-f and discriminator sections of f-m mobile receiver.   Input to this section is from electron-tube variable tuner

* Work done while with Crosley Division, Avco Mfg. Co.

3,200 cps for 6-db signal-to-noise ratio improvement. Its attenuation characteristic is 12 db per octave above 3,200 cps. Insertion loss in the passband is negligible.

Attenuation of the highs due to preemphasis in the transmitter is produced by the network R_1 and C_1. The 30-db power gain of the de-emphasis amplifier compensates for the loss of the network.

A high input impedance is required to avoid disturbing the frequency characteristics of this network and to allow a reasonable value for C_1. The high impedance is obtained by emitter degeneration. High resistance base-biasing resistors are employed to avoid loss; and a low I_{eo} transistor is required to maintain stability.

As this stage is the first in the high-gain audio amplifying system, a transistor with low inherent noise is also required.

The audio amplifier section delivers 500 milliwatts output to a 600-ohm resistive load under standard r-f input signal conditions. Total harmonic distortion does not exceed 5 percent for audio signals at 250 to 3,200 cps at 500-mw output. The audio output is flat within ±2 db at a 200-milliwatt reference level between 250 to 3,200 cps.

Four direct-coupled stages are used with the final two transistors connected in a complementary symmetry circuit. Since this circuit utilizes no interstage transformer it occupies a minimum of space.

Direct coupling permits the use of d-c overall feedback for temperature stabilization. In addition, a-c feedback is employed to stabilize power gain with respect to both temperature and supply voltage changes and also to reduce distortion. The 0.1 μa value of I_{eo} for the TI904 silicon transistor was used as a criterion for selecting it as the first audio amplifier transistor. This contributed to overall power output stability at high ambient temperature.

The second audio amplifier transistor is a GT-691 chosen for its power-dissipation rating, relatively low I_{eo}, and convenience of direct coupling a pnp type to the preceding npn type transistor. As used, it supplies the proper impedance match as well as gain.

The 6-volt Zener diode is used as a direct-coupling device to provide an optimum operating bias for the second audio amplifier. This increases the power gain and temperature stability.

The driver and final stages use similar types GT-731 and GT-732 with ratings of 1 to 2.5 watts at

25 C depending upon the size of the heat sink employed. These power transistors were the only available pnp and npn types having the power ratings needed.

The driver amplifier is common-emitter connected and operates class A. It provides power gain and is directly coupled into the final stage.

The final class-B power-amplifier stage consists of a complementary symmetry pair. A common collector connection is employed primarily for its inherently good temperature stability but at a sacrifice of power gain in comparison to common-base and common-emitter connections.

The use of overall d-c feedback for temperature stabilization is provided by a 2,200-ohm resistor in the output feeding back to the first amplifier stage. A change in output current at the emitter junction of the final stage is used to develop the proper corrective bias in the input circuit.

Stabilization of power gain with respect to temperature and d-c supply voltage variations and reduction of harmonic distortion is accomplished by the a-c feedback network of resistors R_2, R_3 and capacitor C_2. Capacitor C_3 eliminates a tendency to regeneration at

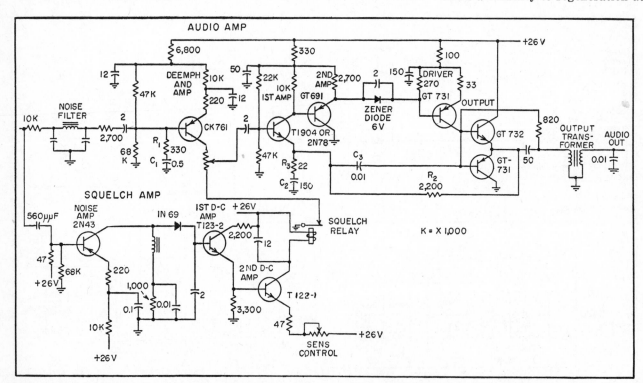

FIG. 2—Audio amplifier with squelch circuit. Operation of squelch relay lifts collector of deemphasis amplifier from ground

higher audio frequencies.

Measurements at 500-mw output pass all specification requirements. Values of 1-percent distortion at nominal 26-v supply are common and reproducible. Less than 5-percent distortion is maintained at 75 C operating temperatures.

Replacing and interchanging transistors produce no noticeable difference in performance. Transistors capable of operating at higher temperatures and having lower values of I_{co} would aid greatly in improving this circuit for use over a wider temperature range.

The squelch circuit is operated by the noise level present at the discriminator output. The first amplifying stage is tuned to accept a noise spectrum centered at approximately 25 kc and to reject audio-frequency signals. The noise is rectified and amplified to energize the squelch relay with a pull-in current of 7.5 ma.

At high noise levels, the relay contacts open, lifting the volume control and collector lead off ground in the deemphasis audio amplifier.

When a signal is received the noise level will be reduced and the relay will become deenergized. This closes the contacts and the receiver is returned to normal operation.

Overall sensitivity is such that an equivalent noise input of 0.15-v rms at the discriminator output will produce silencing. A 2½-db reduction of the 0.15-v rms input will unsilence the receiver.

Beat Frequency Oscillator

A Clapp oscillator circuit which does not require a tank circuit is used for the bfo. In addition, use of a Zener diode for voltage stabilizations was found unnecessary. A series-resonant 1.3-mc crystal, is employed.

An inherent frequency stability of ±2 cps is obtained for ±20 percent supply voltage variation. Tem-

Complete receiver with electron-tube tuner. Transistor r-f, i-f and audio sections are in vertical chassis mounted at rear of vacuum-tube unit

perature changes from −55 C to +85 C produce a ±60-cps frequency change. This easily meets the stability specification requirement of ±0.02 percent (±250 cps) for the bfo.

Crystal Calibrator

A 1-mc crystal oscillator shown in Fig. 3A is employed for calibrating 1-mc interval points on the 20 to 70-mc tuner dial.

An output sufficiently rich in harmonics is generated by overdriving the junction-type CK-761 transistor to produce clipping of the output wave as shown in Fig. 3B.

The author acknowledges the assistance of G. Bruck, D. E. Kammer, F. M. Brauer, I. M. Wilbur, H. H. Lenk and W. Worth. The work described here was sponsored by the short-range communications section of the Signal Corps Engineering Laboratory.

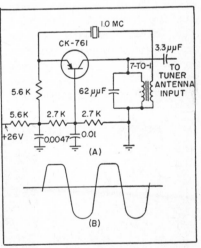

FIG. 3—Crystal calibrator (A) provides 1-mc output (B) rich in harmonics

RURAL CARRIER SYSTEM WITH SELECTIVE CALLING

By B. R. STACHIEWICZ

Telecommunication Division, Stromberg-Carlson Co., A Division
of General Dynamics Corp., Rochester, New York

Selective calling of up to ten telephone subscribers per wire
pair is possible with transistorized transmitter and receiver using double-
sideband transmission and operating over line losses of 35 db. Subscriber
receiver response is flat within 0.5 db as a result of circuit stabilization

APPLICATION OF CARRIER transmission to rural telephone lines has been rather slow for many reasons, mainly economical. Systems could not, until recently, be proven-in over such short distances due to high cost per channel. Now that the cost of the equipment and maintenance has become competitive with that of constructing new wire lines, subscriber carrier operation becomes feasible.

Numerous requirements must be met. Overall quality of transmission should not be substantially inferior to that in other types of carrier equipment, which means that the transmitted bandwidth response should be flat between 250 to 2,500 cps. Crosstalk and noise must be kept low and reliable operation under extreme climatic conditions should be maintained.

Use of transistors takes of advantage of low power drain and small size, which is always attractive for pole-mounted units especially if standby power is required.

System

The system shown in Fig. 1 uses double sideband transmission and operates over line losses up to 35 db at carrier frequencies.

The five channels at the central office connect to the line through a line coil. The subscriber terminals drop off at convenient points along the line. Since carrier channels operate over voice-frequency physical circuits, it is necessary to isolate all subscriber telephones connected across the line. This isolation is accomplished by pole-mounted low-pass filters, which present high bridged impedance to the carrier frequencies.

Ringing and dialing are performed by interrupting the carrier while compandors are used, optionally, in cases where excessive line noise or crosstalk is encountered.

Fig. 2 and Fig. 3 show in block form the arrangement of circuit.

Voice signals from the central office line are routed through the hybrid and level adjusting pads to the modulator in the transmitter. When compandors are used, the compressor unit replaces the pads. The modulated wave is amplified to the line level and passed through transmitting filter to the line.

Subscriber Signals

The signals from the subscriber are selected by the receiving filter, adjusted to proper level, amplified, detected and delivered to the hybrid through the expandor, if used, or through the voice-frequency receiving pads.

The voice-frequency signal is then routed through the hybrid into the central office.

Transmission at the subscriber end is essentially the same. Dialing by subscriber is accomplished as

Laboratory setup for checking overall transmission quality of transistorized carrier system

Central office terminal shows how individual circuits are arranged on cards for easy insertion and removal

follows: when the subscriber lifts his receiver a d-c circuit through the dial contacts, voice-frequency loop and the winding of relay *SR* is completed. Direct current, supplied by a power pack located on the pole-mounted terminal, flows in this circuit. Upon dialing, this current is interrupted causing the *SR* relay to operate. The contacts of this relay are arranged to short

tral office relay winding, which reproduce the dialing pulses generated at the subscriber end.

To call the subscriber, the ringing voltage of any frequency up to 66⅔ cycles from the central office modulates the carrier wave transmitted to the subscriber. At the subscriber terminal those signals are amplified and detected in the receiver amplifier circuit. A special

relay are wired to an electronic gate which, when operated, disables the ringing amplifier by applying negative bias to the emitter of the first transistor.

The 1-kc tone generator located on the ringing card at the central office terminal identifies the ground on the ring wire. With no ground, the oscillator is shut-off while grounding turns the oscillator on

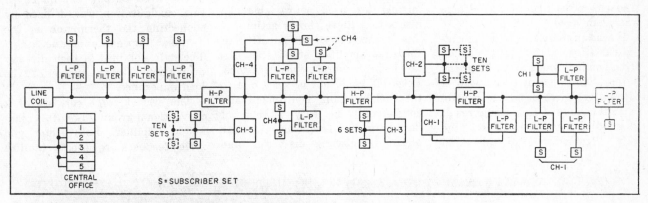

FIG. 1—Block diagram shows how rural subscribers are serviced by one wire pair

the output of the transmitter when the station is on-hook or the dial contacts are open. When the subscriber goes off-hook, the channel carrier is automatically transmitted. Thus dialing information is transmitted by switching the carrier frequency on and off.

At the central office the interrupted carrier is picked-off at the base of the detector stage, amplified and rectified by the signal a-c amplifier and used to drive relay contacts in series with the ring wire from the central office. The operation of this relay completes the d-c circuit through the hybrid winding and the central office relay. Pulses of d-c are thus obtained in the cen-

d-c amplifier located on the ringer card amplifies and rectifies the ringing signal and drives a mercury relay. The armature of the relay vibrates at the ringing frequency and the contacts are connected in the vibrator circuit. The output of the vibrator is applied to the subscribers voice-frequency extension at a magnitude that operate the bell in the subscriber telephone.

The tripping during the ringing period is accomplished by the contacts of relay *RT* which open the vibrator circuit as soon as the subscriber answers the phone.

Tripping during silent period is accomplished through the operation of relay *SR*. The contacts of this

and its frequency modulates the outgoing carrier. At the subscriber end the 1-kc tone is picked-off, amplified and operates a relay that applies ground to the outgoing ring wire. When the called subscriber answers, the tone amplifier is disconnected from the circuit and does not interfere with speech transmission.

Receiver

Figure 4 shows the receiver circuit. The received amplitude-modulated wave is passed through the input transformer and into a common-base amplifier-regulator stage. Partial regulation is achieved in this stage and signals are then

113

amplified in a high-gain, feedback-stabilized, common-emitter amplifier. The following stage acts as a detector, amplifier and regulator for the detected voice-frequency signal. Carrier leak is eliminated by a low-pass filter following the detector and final amplification is achieved in a feedback-stabilized grounded emitter voice amplifier.

Regulation occurs in the input stage and the detector. This approach eliminates the necessity of using a d-c amplifier in the control loop. Transistor d-c amplifiers are difficult to stabilize against variations of operating point resulting from the effects of I_{co}, especially if ambient temperature varies considerably. This decision was also supported by the fact that the wide range of ambient temperatures over which the equipment should operate would make it difficult, in the case of a d-c amplifier, to stay within power ratings of germanium transistors. Since economy considerations did not permit use of silicon transistors the idea of a d-c stage was abandoned.

It was proposed to regulate signal variations of +5 db to −10 db from nominal input to ±1 db at the output. This meant that with a conventional regulator, a control current change of 1 db should result in a 10 db change in input stage gain. Performance of this kind

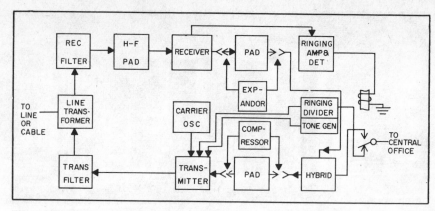

FIG. 2—Central office terminal equipment block diagram shows how subscribers' dialing to central office is fed from receiver to ringing amplifier and detector to operate relay connecting hybrid into circuit

may be achieved by multiplication of control current or the use of bucking current to increase the ratio of current variation. However, the solution adopted consists in regulating the input stage only as much as would be required to prevent circuit overload under the condition of maximum input signal and to achieve the remaining regulation in the output stage. Thermistors are used as variable elements to achieve slower action.

A common-base configuration of the first amplifier stage was dictated by several considerations. Feedback could not be used due to the variable gain requirement and consequently no easy method of flattening the frequency response and of combating distortion and

intermodulation effects was on hand. The frequency cut-off of the grounded-base transistor is extended by a factor of 1/1-α over the grounded emitter response. Better linearity of characteristics can also be obtained with the base grounded resulting in decreased intermodulation and distortion products.

The high output impedance of this configuration was utilized by connecting the thermistor as the a-c load impedance of the stage. The gain obtained is then directly proportional to the value of the load impedance.

The second stage consists of a conventional grounded-emitter transistor amplifier. Both shunt and series feedback are used to stabilize

Card technique is demonstrated in construction of filters (upper left) and receiver (lower left). Underside wiring of each card is shown adjacent to it

114

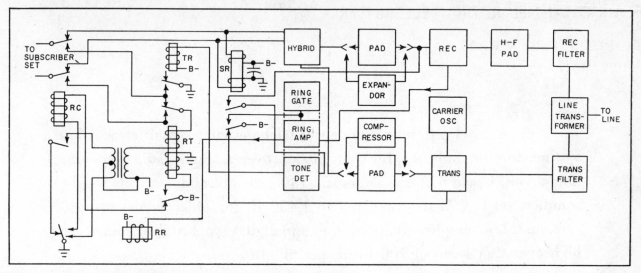

FIG. 3—Subscriber terminal equipment block diagram shows how ringing signal is fed from receiver to ringing amplifier to relay *RR* which vibrates at the ringing signal frequency and activates other relays to feed signal to subscriber set

the input and output impedances. Frequency response over the required range of 24 to 138 kc is held flat within ±0.5 db. Feedback is sufficient to stabilize the gain of the stage against manufacturing variations in transistors and against varying temperature and voltages.

Output

The output of this amplifier is transformer-coupled to the detector stage. The speech intelligence is recovered and the d-c current resulting from the rectified carrier is used as the control current in the regulator loop. A thermistor, the resistance of which varies with this current, is the load. Stage gain is thus proportional to the load impedance and additional regulation is obtained.

The low-pass filter following this stage eliminates higher frequencies present and passes the voice signals to the output amplifier. Here, shunt and series feedback are used to stabilize the gain and the impedances. An output of up to +7 dbm is obtained with distortion not exceeding 5 percent.

Indirect heating is provided for thermistors under low ambient temperatures to insure proper circuit operation.

Climatic tests performed on this circuitry indicated an output stability of ±1 db under all conditions with distortion not exceeding 4

percent at the output.

The author acknowledges the work of R. L. Layburn who developed most of the circuitry and D. F. Jamieson and H. V. Buck, the electrical and mechanical engineers. Filters and transformers were designed under W. L. Brune.

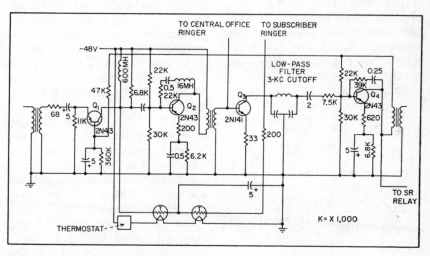

FIG. 4—Schematic of transistorized receiver (right) used in both terminal sets

FOUR-CHANNEL RADIO TELEPRINTER

By PHILLIP G. WRAY

Teletype Corp, Chicago, Ill.

Up to four channels of teletypewriter signals are combined by time-division multiplex for transmission over a single radio communication circuit. Use of 572 transistors and 739 germanium diodes cuts weight of complete set to 275 lb, compared to 1,450 lb for older model using 357 vacuum tubes. Simplified ring counter and digital synchronizer contribute to high operating speed of 100 words per minute

A SMALL, LIGHTWEIGHT multiplex telegraph set is under development for the Navy Bureau of Ships, to provide additional radio-teletypewriter channels for mobile and shipboard service. Use of transistor circuitry will allow an 80-percent reduction in volume and weight and a 95-percent reduction in power consumption over its electron-tube equivalent, which was developed earlier for fixed-station installations. Time-division multiplex equipment is used to combine two, three or four channels of teletypewriter information for transmission over a single radio communication circuit.

Although some of the circuits

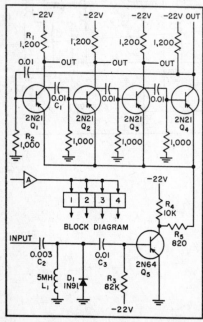

FIG. 1—Ring counter and drive amplifier

used in the transistorized version were derived from the rapidly growing field of transistor pulse techniques, several unique circuits were developed. This article describes two of these, a simplified ring counter and a digital synchronizer.

Simplified Ring Counter

A ring counter circuit was needed to perform the basic function of distribution (conversion between serial and parallel forms of information). Since the complete multiplex set requires twelve of these ring counter distributors, it is important for the counter to be of simple design, requiring a minimum of components.

The new ring counter achieves its simplicity through use of the inherent negative resistance characteristic of the point contact transistor. Only three components per stage, in addition to the transistors, are required.

The circuit of the simplified ring counter is given in Fig. 1. While only four stages are shown, the ring may be constructed of any number ranging from two upward. Rings of this type have been reliably operated with as many as 26 stages.

When power is applied, transistor Q_5 will conduct into saturation because of the negative bias supplied to its base through R_3. As a result, the collector of Q_5 will be essentially at ground potential. Simultaneously, the counter element with the highest combination

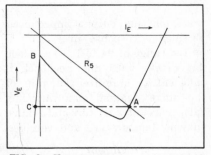

FIG. 2—Characteristic curve for typical ring counter element

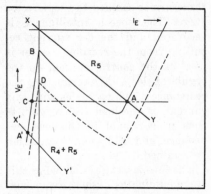

FIG. 3—Stepping action of counter

of reverse collector to base leakage current (I_{co}) and emitter current amplification (alpha) will conduct into saturation. Current flowing through the common emitter resistor R_5 develops a bias voltage which, when applied to the emitters of the remaining off elements, holds them sufficiently in the reverse direction to prevent their conduction.

Figure 2 shows the input negative resistance characteristic curve of emitter voltage and emitter current for a typical ring stage. With the element nonconducting, the voltage at point B is mainly the re-

New transistorized multiplex AN/UGC-1 on table has volume of only 10.9 cu ft. compared to 55 cu ft for older two-rack electron-tube version on floor at right

New set, in 15-drawer aluminum cabinet, makes extensive use of etched wiring

sult of I_{co} flowing through base resistor R_2. When conducting, the element will be stable at point A. Because the common emitter resistor produces a voltage which is common to both on and off elements, the emitters of the off elements must be located at point C. The off counter elements are held at this point, which corresponds to a slightly negative emitter current, by a bias voltage equal to BC.

Stepping Action

The counter is stepped from one conducting element to the next by means of negative stepping pulses. Each pulse applied to the common emitter connection causes the conducting element to switch to the nonconducting state.

When the collector voltage of the on element switches from slightly negative to highly negative, a differentiated negative waveform is passed through coupling capacitor C_1 and impressed on the base resistor of the following element, thus priming that element. When the stepping or triggering pulse is removed, the primed element conducts into saturation.

Referring to Fig. 3, the switch-ing of the on element to the nonconducting state is shown as the displacement of load line XY to $X'Y'$, with the resulting change in operating point from A to A'. As described, when the collector voltage of the on element switches from slightly negative to highly negative, a differentiated negative waveform is generated and applied to the base of the following element. This priming effect is shown as a displacement of the characteristic curve for the primed element to the position indicated by the dotted curve. The emitter of the primed element is now more positive with respect to its base by a voltage equal to BD than any of the other elements in the distributor ring.

The time constant of the distributor coupling circuit has been designed to hold the element in this primed condition for a period longer than the duration of the trigger pulse. Therefore, when the trigger pulse is removed, indicated by a shift in the load line from $X'Y'$ back to XY, the primed element will be the one that is switched into conduction.

Each ring counter requires a drive amplifier (Q_5 in Fig. 1) to produce negative triggering pulses of the required width and magnitude to step the counter. To control Q_5, a square-wave input signal is applied to series-resonant ringing circuit C_2-L_1. The ringing circuit has a resonant frequency of approximately 50 kc. The positive-going leading edge of the square wave shock-excites the resonant circuit into damped oscillation.

Clamp diode D_1 prevents the oscillation from continuing past the first half-cycle by dissipating the energy in the resonant circuit when the voltage swings negative. The result is a single positive output pulse approximately 10 microseconds in width. This positive pulse is coupled by C_3 to the base of normally conducting transistor Q_5, driving the transistor out of saturation and into cutoff for the duration of the pulse. The collector output of Q_5 is thus a 10-microsecond trigger pulse swinging from ground potential to approximately -20 volts.

Ring Counter Waveforms

A group of typical waveforms of the ring counter and its drive amplifier is shown in Fig. 4. Waveform *B* shows the positive output pulse that is obtained from the ringing circuit and applied to Q_5. The resulting 10-microsecond trigger pulse is shown in waveform *C*.

At the instant the trigger pulse is received, the base is driven to −6 volts by the differentiated negative pulse resulting from the preceding element turning off, as shown in base waveform *H*. The base voltage then begins to decay exponentially, reaching approximately −4 volts when the trigger pulse is removed. At this instant the voltage of the primed base is approximately 3 volts more negative than the base voltage of any of the other ring elements, and therefore the primed element conducts, as shown in waveforms *G* and *I*.

Satisfactory operating margins are achieved in the simplified ring

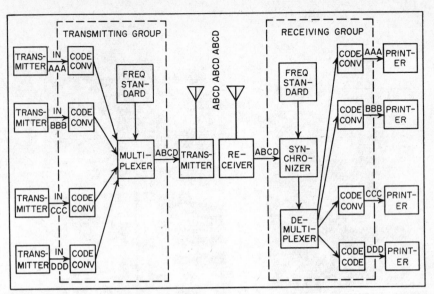

FIG. 5—Complete new multiplex system

counter without adjustment of the circuit constants. The basic ring counter element is biased to be stable only in the on condition and therefore is not dependent upon interception of the emitter load line (R_5 in Fig. 2) more than once with the emitter input characteristic curve. No stabilization circuits are required.

This circuit was designed to operate with transistors which have emitter to collector current gains as low as 1.8 and which allow reverse current flow at 20 volts and 60 C to be as high as 3 ma.

Use of Ring Counter

A block diagram of the complete multiplex system is shown in Fig. 5. It consists of the transmitting and receiving terminal groups and the associated telegraph and radio equipment. On the sending side, independent transmitters deliver randomly timed start-stop telegraph signals to the individual transmitting group code converters.

A ring counter distributor in each converter assists in transforming the information from serial to parallel form. The converter outputs are scanned, channel by channel, by another ring counter distributor within the multiplexer, picking up a complete character in turn from each. The signal elements within these characters are then transmitted in sequence, under the control of a third ring counter, to the external equip-

ment at the multiplex frequency rate.

On the receiving side, the operation is exactly reversed, with distributors performing the complementary functions. The information contained in the incoming multiplex signal is separated, channel by channel, and delivered to the proper code converters. The start-stop signals are then reconstructed and delivered to the telegraph receivers. Initial framing of the receiving channel ring distributors is necessary to insure that the information will be delivered to the proper channels.

The operating speed of each terminal is controlled by a frequency standard using a transistorized crystal oscillator. To obtain high stability, the oscillators operate at the relatively high frequency of 63 kc. Ring counters are used, because of their simplicity, to divide this down to the required scanning frequency. At the receiving terminal, a synchronizer operates in conjunction with the frequency dividers to maintain exact synchronism between the incoming signal and the demultiplexer.

Figure 6 shows a block diagram of the oscillator, frequency-dividing and multiplexer distributing circuits used in the transmitting terminal group. The oscillator output is applied directly to the drive amplifier transistor for the first ring counter. The square-wave output from the first element is used to drive the second ring counter. This

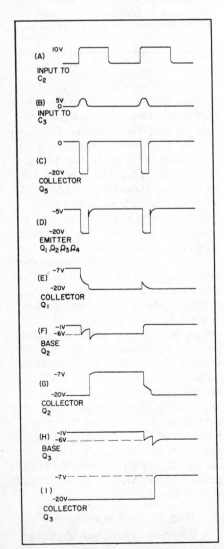

FIG. 4—Waveforms of ring counter

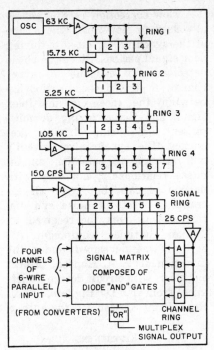

FIG. 6—Transmitting terminal frequency dividers and multiplexer

procedure is continued throughout the remainder of the dividing and distributing circuits. In this way, counter rings 1 through 4 divide the 63 kc exactly 420 times to generate the multiplexer signal ring driving frequency of 150 cps.

The signal ring, channel ring and signal matrix comprise the multiplexer portion of the transmitting terminal. Here the six-wire parallel input from each of the code converters is transformed into a sequential channel-by-channel multiplex output signal.

The signal ring contains six elements, corresponding to the six pulses contained in the signal code. The channel ring contains four elements, corresponding to the four channels of information handled. The signal matrix is composed of 24 diode AND gates. Each AND gate has three inputs: signal ring, channel ring and code converter output, all of which must be present simultaneously to produce an output signal. The output signals of the 24 AND gates are then combined, by means of a diode OR gate, into the sequential multiplex signal output.

Digital Synchronizer

To maintain the receiving terminal in exact synchronism with the incoming multiplex signals, it is necessary to compensate for the effects of oscillator drift and long-term signal distortion resulting from radio circuit multipath. This is accomplished by a unique digital synchronizer circuit designed to operate in conjunction with the receiving terminal frequency dividers. By noting the time location of on-off transitions in the incoming multiplex signal, the exact phase relationship between the incoming signal and the demultiplexer signal ring driving frequency is established.

Common Oscillator

Synchronizer action is accomplished by adding or subtracting drive pulses to the second frequency divider stage until the proper time relationship is achieved. This system of synchronizing permits the use of a single crystal oscillator to drive both the transmitting and receiving groups in a given location, since the frequency of the crystal oscillator remains fixed.

Figure 7 shows a block diagram of the oscillator and frequency-dividing circuits used in the receiving terminal group. These circuits, and the demultiplexer not shown, are essentially the same as those of the transmitting group. The synchronizing circuits (Fig. 8) have been added to the block diagram of Fig. 7.

A normally open SUBTRACT gate has been placed in series with the normal drive input to dividing ring 2 in Fig. 7. This provides a means of subtracting drive pulses from the normal synchronous drive to ring 2. In addition, an output from the third element of ring 1 has been connected through a normally closed ADD gate to the normal drive input to ring 2. This arrangement provides a means of adding extra drive pulses to the normal drive.

The synchronizer circuits also include a group of phase-detecting gates. Any deviation of the incoming multiplex signal from the demultiplexer sweep frequency will be detected and the appropriate ADD or SUBTRACT gate action initiated.

When the receiving terminal is in exact synchronism with the incoming multiplex signal, the leading edge of the on time of element 1 of ring 4 will coincide with the trailing edges of the incoming multiplex signal code pulses. The incoming multiplex signal is inverted and applied to a signal pulse univibrator whose function is to generate a pulse for each negative-going transition in the multiplex signal. This pulse, adjusted to have a width slightly less than the on time of one element of ring 4, is applied to the six phase-detection AND gates. The second input to each phase-detection gate is obtained from ring 4.

No connection to a phase detection gate is made from element 1 of ring 4. Therefore, when the receiving terminal is in synchronism with the incoming multiplex signal, the output of the signal pulse univibrator will fall within the boundaries of the output of element 1, and there will be no output derived from any of the phase-detection gates.

If the incoming multiplex signal should be lagging, the output pulse from the signal pulse univibrator overlaps with elements 2, 3 or 4 of ring 4. This coincidence will be

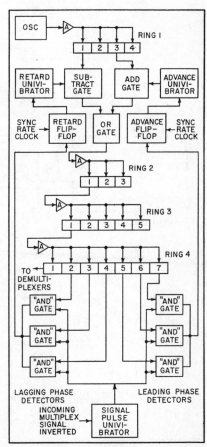

FIG. 7—Receiving terminal frequency dividers and synchronizer

detected by the lagging phase detectors, resulting in the RETARD flip-flop being turned on. The need for retard action is thus stored. In like manner, if the incoming multiplex signal is leading, detection will be made by one of the leading phase detectors, and the need for advance action will be stored in the ADVANCE flip-flop.

Amount of Correction

The ADD and SUBTRACT gates are each controlled by a univibrator. Thus when the RETARD univibrator is triggered, the SUBTRACT gate is closed for a period of time equal to the univibrator relaxation cycle. When the ADVANCE univibrator is triggered, the ADD gate will be opened for a length of time as determined by the univibrator.

The amount of correction provided per correction rate cycle is controlled by adjusting the operating time of the ADVANCE and RETARD univibrators. By varying the univibrator operating period, one

or more drive pulses can be added or subtracted per cycle. Each pulse added or subtracted will advance or retard the succeeding rings approximately 63 microseconds, or 1 percent of the on time of a signal ring element.

Rate of Correction

The correction rate cycle is established through the action of the RETARD and ADVANCE flip-flops. An externally generated sync rate clock pulse is applied at periodic intervals to the RETARD and ADVANCE flip-flops. If one of the flip-flops has been previously turned on, due to the detection of an out-of-phase condition, the arrival of the next sync rate clock pulse will reset the flip-flop and cause the associated univibrator to be triggered. Thus, the rate of correction can be controlled by adjusting the frequency of the sync rate clock input. The rate of correction can be varied from 6.25 pulses per second to one pulse per 20 seconds.

A slow correction rate is generally used to slow down the action of the synchronizer circuits during poor signal periods. Since the phase detectors will respond to every change in the phase relationship between the incoming multiplex signal and the receiving demultiplexer signal ring driving frequency, the synchronizer circuits will attempt to establish an in-phase relationship, not only with normal signals, but with all extraneous bursts of noise and distortion which may be received.

The multiplex equipment described was developed under the direction of T. A. Hansen, project engineer, with circuit design assistance from F. D. Biggam, R. J. Reek, R. A. Slusser and the author.

REFERENCES

(1) F. D. Biggam, A Transistorized Time Division Multiplex Telegraph Set, *AIEE Trans Paper No. 56-987.*

(2) T. A. Hansen and R. D. Slayton, An Electronic Time Division Multiplex Terminal Set, *AIEE Trans 70,* 1951, p 1.

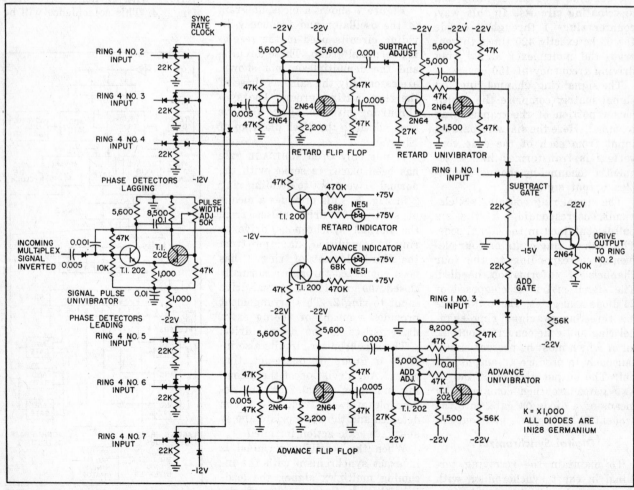

FIG. 8—Synchronizer circuit compensates for effects of oscillator drift and long-term signal distortion resulting from multipath radio transmission

Chapter 9

AUDIO COMMUNICATIONS CIRCUITS

FORK-DRIVEN DUAL-TONE STANDARD

By BYRON H. KRETZMAN
Erco Radio Laboratories, Garden City, N. Y.

IN TUNING the selective amplifiers of an a-f type radioteleprinter converter designed for 850-cycle shift, the usual method is to use a tunable audio oscillator as a signal generator. Some means of accurately setting the oscillator to either of the standard mark and space frequencies, namely 2,125 and 2,975 cycles, is necessary.

Common practice is to compare the frequency of the tunable oscillator with a 425-cycle tuning fork standard. Lissajous figures on an oscilloscope are then obtained in the 5-to-1 and 7-to-1 ratios as required. This procedure may prove unreliable if the a-f oscillator is unstable. Described below is a fork-driven standard that provides 2,125 and 2,975 cycles directly. Using this standard, it is unnecessary to employ an a-f oscillator or even an oscilloscope to align a converter. Operation is simplified by using transistors and a self-contained battery power supply.

Construction is simplified by using an etched or printed-circuit wiring board. The transistor oscillator is a pulse-generator driven by a 425-cycle tuning fork. Transistor amplifiers tuned to the fifth and seventh harmonics amplify the 2,125 and 2,975-cps components.

▶ **Magnetic Circuit** — Originally, the standard was a 435-cycle fork. The softer crotch part was notched with a good round file to move the frequency lower, comparing it with another fork of the correct frequency. Not much filing is necessary to move only 10 cycles.

Exciting coils for the fork were obtained from a high-impedance headphone. Each coil measured about 1,000 ohms d-c resistance. Magnetic bias for the fork is supplied from a ½-in. round by ½-in. long permanent magnet from a small a-c/d-c set speaker. Hex-head ¼-20 steel bolts were soldered to each end of the magnet to extend the magnetic circuit to the coil pole pieces. An air gap of about ¹⁄₆₄-in. is between each of the pole pieces and the tines of the fork. The handle of the fork was threaded with a ¼-28 die and bolted to a triangular mounting bracket.

▶ **Electrical Circuit** — The oscillator uses a 2N107 *pnp* transistor. The circuit is that of a pulse generator whose repetition rate is controlled by the mechanical resonance of the tuning fork. Resistance-capacitance coupling is used to couple to the amplifiers, each of which also uses a 2N107 transistor in a grounded-emitter circuit.

Toroidal inductors paralleled with capacitors tune each collector output circuit to the correct harmonic. Each collector circuit, in turn, is capacitance coupled to another similar tuned circuit to further peak the desired frequency.

With the nominal 88 mh toroids used, it was found that approximately 0.053 μf was required for resonance at 2,125 cycles. At 2,975 cycles, approximately 0.028 μf was required for resonance.

Output is high impedance, in the order of two to three volts rms. Distortion, measured at 2,125 and 2,975 cycles is less than 3 percent, which is a usable value. By careful selection of the capacitors tuning the toroids to resonate the circuits as closely as possible, distortion could be further reduced to less than 1 percent.

Loading the output circuit increases distortion. Wherever possible, coupling to the amplifier being aligned should be to a grid, through a blocking capacitor.

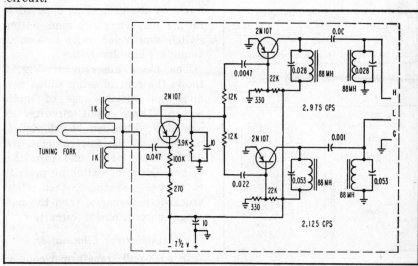

Circuit of the tuning-fork controlled standard with two outputs

PUBLIC-ADDRESS SYSTEM ADJUSTS TO AIRCRAFT NOISE

By J. M. TEWKSBURY

Research Engineer, Bendix Radio Division, Bendix Aviation Corporation, Baltimore, Maryland

Aircraft passenger-address system uses single preamplifier and up to five power amplifiers and speakers for uniform audio distribution throughout seating area. Differences in ambient noise level in the air and on the ground are compensated by switching. Transistors are used as matching devices to supplant transformers at input and output, and as electronic filter for noise

SUCCESSFUL DESIGN of aircraft passenger address systems requires flexibility enough to compensate for differences in noise level from one type of aircraft to another, for the changing levels of ambient noise encountered in various positions in any given aircraft, and for differences in the noise patterns aloft and on the ground. Fig. 1 shows how the noise curve changes from the nose to the tail of a typical aircraft. Improved aircraft design has reduced ambient noise levels to a point where the use of the passenger address system for radio programs, frequent routine communications from crew to passengers, tape recordings and other forms of entertainment has become practical.

Transistor Modules

The amplifier to be described is the first-known completely transistorized aircraft p-a system amplifier available to the aviation industry. The key to its flexibility and efficiency is the use of modular construction.

In a passenger address system using this amplifier the seating area is divided into from one to five audio zones, depending on the type of craft. A plug-in amplifier module producing eight watts of audio power is provided for each zone. Individual potentiometers for each amplifier provide independent control of the sound level. To compensate for the difference in audio levels on the ground as opposed to in the air, an air-ground output switch is provided which acts on all amplifiers simultaneously.

The block diagram of Fig. 2 shows the control setup which permits the selective use of audio power. Three input circuits are available which can be adapted to use with carbon microphones, dynamic microphones and tape machines. Order of switching priority is pilot-to-passengers, then stewardess-to-passengers, then the passenger-entertainment circuit.

Transformer Elimination

In push-pull transformer-coupled output circuits there is an auto-

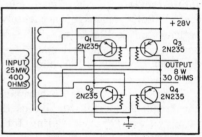

FIG. 3—Power amplifier module schematic

transformer action which places a voltage equal to double the supply potential on the transistors. Hence with a 28-v supply there would be 56 v on the transistors at maximum output. This would necessitate the use of transistors with a high breakdown voltage. But transistors of this type are relatively expensive and their current linearity is generally not satisfactory for this application.

Circuit Details

These difficulties are eliminated by the use of the bridge circuit shown in Fig. 3. Transistors Q_1 and Q_4 are conductive for half a cycle while Q_2 and Q_3 supply the other half-cycle to the load. The maximum peak voltage applied to the load is about 26 v, while the peak current reaches 0.93 amp at 8 w output. Although the 2N235 transistor is capable of 26-w peak output in class B, the 1,200-ohm bias resistors bring about conservative operation in class AB.

One of the inputs required is 200 ohms push-pull at — 50 dbm. The

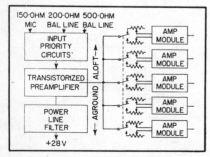

FIG. 1—Noise levels aboard aircraft

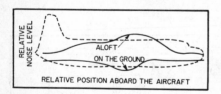

FIG. 2—Functional diagram of 5-station passenger-address system for airliners

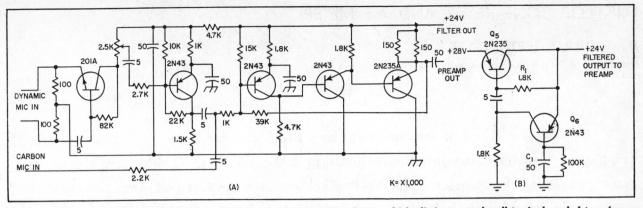

FIG. 4—Schematic of system preamplifier (A). First transistor is coupling device which eliminates push-pull to single-ended transformer. At (B) is electronic filter for d-c line

conventional way to connect such an input is to employ a push-pull to single-ended transformer. An *npn* transistor, however, connected as shown in Fig. 4A is completely equivalent, and this circuit accomplishes the input function with a saving in weight and volume.

Noise Filtering

Another problem to be overcome in the development of p-a amplifiers involves the prevalence of noise components on the 28-v lines of virtually all commercial aircraft. This may be as high as two volts rms on the 28-v bus. This noise must be filtered to prevent it from appearing in the preamplifier stages.

This filtering is accomplished by the special electronic filter shown in the schematic diagram of Fig. 4B.

Referring to the diagram, R_1 biases Q_5 so that, with the current normally present in the circuit, there is a 24-v output and C_1 charges up to approximately 24 v. Any change in the voltage at the output side of the filter is amplified by Q_6 and applied to Q_5 in such a way as to oppose the change. Experimental results indicate that C_1 is amplified 1,000 times in value, thus having an apparent capacity of 50,000 μf.

It is not at all necessary to filter the A+ supply to the bridge amplifiers because the collector impedance is extremely high. Consequently little a-c noise is transferred to the low-impedance bridge output load.

The collector current in the am-

plifier is essentially independent of collector voltage. The base current is produced by the signal input voltage. The current which flows in the collector circuit is many times greater than the base current once saturation has been reached. Thus the output load is almost completely isolated from any a-c components on the d-c input line.

Heat Dissipation

Perhaps the most serious limiting factor in the use of transistors in this application is their inability to withstand high temperatures. A study of the heat-dissipation problem resulted in the establishment of certain basic data.

The data shows that the average power dissipated can be calculated by integration and the equation reduced to $P_{av} = 1/2\pi \, [E \, (2P/Z)^{1/2} - (\pi P/2)]$, where E is the supply voltage, P is the power output of the bridge and Z is the load impedance.

With a supply of 28 v and a load impedance of 30 ohms it then becomes a simple matter to plot the dissipation of each transistor against the power output of the bridge.

This amplifier is the result of growing demands for lighter, smaller, less complex and more reliable airborne units. It has been made possible largely by the use of transistors and the application of new design techniques which transistors permit.

Modular 5-station amplifier. Two sets of potentiometers, one for air and one for ground, permit audio level balance

FLIP-FLOP, TONE KEYER AND A-F METER

By NORRIS HEKIMIAN

Department of Defense, National Security Agency, Washington, D. C.

Temperature-stabilized flip-flop, tone keyer and audio-frequency meter using junction transistors make use of favorable large-signal properties of transistors. The flip-flop achieves temperature stability through diode switching in the emitter circuits. The tone keyer switches tones ranging in frequency from 100 cps to 200 kc at rates up to 10 kc. The frequency meter uses silicon transistors to achieve operation up to 100 C

LARGE-SIGNAL OPERATION of junction transistors results in many useful and interesting circuits.

Improved high-temperature operation of a junction transistor flip-flop is achieved with diode switching in the emitter circuits. Stability factor is improved from 20 or more to about 2 or 3. Operation up to 70 C is obtained using transistors rated for 50 C.

Also stable against temperature variation is a tone keyer that controls tone frequencies from 100 cps to 200 kc at keying speeds up to 10 kc. Keying may be by a d-c signal or by contacts. Supply voltages from 6 to 35 v can be used.

A frequency meter employing silicon transistors withstands temperatures up to 100 C. The meter is designed for audio frequencies. It has a substantially linear frequency scale. There is no zero offset.

Stabilized Flip-Flop

A limiting factor in the use of germanium junction transistors is collector saturation current I_{co}. The saturation current increases exponentially with temperature and doubles in magnitude for every 10 to 12 C increase. Even at low junction temperatures the effects of this current are important.

Since the saturation current flows in the emitter and base as well as in the collector circuit, the effects of I_{co} changes in the output depend on the type of circuit. The grounded-emitter circuit is exceptionally subject to such changes since it is usually designed for maximum current and voltage amplification and for large source resistances driving the base.

Such design is inimical to best high-temperature operation. However, since the grounded-emitter circuit yields both current and voltage gain with junction transistors, it is desirable for bistable flip-flops and some economical means of high-temperature stabilization must be achieved.

Temperature Compensation

Temperature compensation of an amplifier is usually achieved by cancellation of the transistor I_{co} with that of another transistor or by use of some thermally sensitive resistance. The former method is usually preferred since cancellation over a wider temperature range will be most effective for similar types of units. However, for a bistable switching circuit such as a flip-flop it is not possible to achieve compensation without introducing more transistors in each stage.

These compensating transistors must adequately match the active transistors in the flip-flop stages, especially at the highest temperature anticipated. This is particularly true because of the expo-

nential dependence on temperature.

Considerable improvement in temperature stability can be achieved by appropriate resistances in the emitter and base circuits. This improvement is at the sacrifice of efficiency because of losses in the resistances and a reduction in signal swing for switching circuits such as the flip-flop.

A-C Amplifiers

The most useful application of this method is for a-c amplifiers where the swing and gain reductions can be eliminated by adequate bypassing. For bistable circuits and d-c amplifiers such bypassing results in large low-frequency attenuation that is usually unacceptable.

By operating the transistor with large collector currents and low collector voltages, the effects of relatively large I_{co} at elevated temperatures can be kept within the limitations of dissipation and maximum collector current ratings. Further, since in low-voltage high-current circuits, resistance values are low, the larger I_{co} at high temperature results in a smaller effect on output signal voltage.

This method is not economical of supply current but need not require excessive supply power because of low supply voltages. However, the requirement for large current is undesirable because of the requirements imposed on power-supply filtering when using a-c supplies as

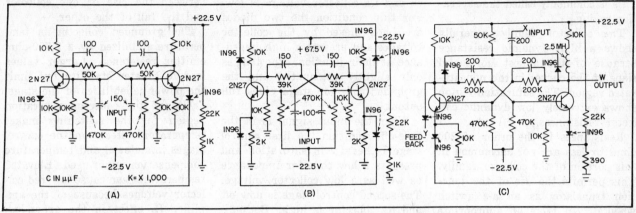

FIG. 1—Transistorized flip-flop uses diode switching in emitter circuit for temperature stabilization (A). Collector clamping improves high-frequency operation (B). Output inductance (C) improves interstage switching in binary counters

well as the increased power losses in conductors and connections. For systems where total power consumption and current drain are small this method is often the best.

Silicon Transistors

The higher work function involved in transporting thermally created carriers across the collector-base junction in silicon transistors results in I_{co} currents several orders of magnitude less than in germanium devices of similar ratings. This difference in work function results in a less perfect switch in the on condition since the collector voltage in saturation cannot drop as low as for germanium. Representative values are 1 volt for silicon and 0.1 volt for germanium.

Furthermore prices for established types of silicon transistors are higher than for germanium units of comparable rating. Eventually, however, the silicon transistor or a silicon-germanium alloy transistor may become the best solution to the temperature problem.

Circuit Discussion

The circuit shown in Fig. 1A employs bias stabilization in the emitter of a grounded-emitter amplifier. In addition it provides diode gating of the emitter stabilizing resistance to reduce the d-c degeneration in the on condition so

that almost full output swing is provided. Since the stabilization is only required in the off condition, the on condition being aided by the saturation current, this arrangement makes full use of the stabilization.

The basic limitation is the use of two diodes, two emitter resistors, and a voltage return for the diodes for each flip-flop stage. The voltage return can be common for several stages, depending upon the permissible drain and impedance of the return voltage source. A separate voltage supply can be used or, if common battery operation is desired, a voltage divider can be used as shown in the circuits of Fig. 1.

Since the flip-flop is symmetrical the regulation of the divider is usually adequate. Calculation shows that for the transistor in the off state the stability factor is in the order of 2 to 3 depending upon transistor parameters. This compares favorably with the ultimate limit of unity for the stability factor. An unstabilized stage would generally have a factor in excess of 20.

Since the stabilization is applied only to the off transistor, the on transistor is not inhibited from going deeply into saturation. Since this effect is more pronounced at elevated temperatures, the transistor goes even further on and

deeper into saturation. The resulting increased carrier storage slows down the transistor in changing from on to off and reduces the upper limit of switching speed.

This effect can be reduced by preventing the transistor from going into saturation. The price paid is another voltage source and another pair of diodes for clamping. Such a system incorporating both the bias stabilization and collector clamping for high-speed operation is shown in Fig. 1B. Results for both circuits are tabulated in Table I. An unstabilized circuit of similar design and with similar transistors consistently failed in operation at about 40 to 45 C.

The input trigger requirements are about 8 to 10 v but trigger rise time may alter the requirements. Improvement in triggering between stages can be achieved by peaking the output collector with an inductance as in Fig. 1C. The value of inductance is noncritical and can be 1 mh or less.

The transistors used were type 2N27. Representative data by the manufacturer show alpha = 0.95 to 0.995; alpha cut-off frequency = 1.0 mc, minimum; I_{co} at 30 v V_c = 30×10^{-6} amp.

Tone Keyer

Typical grounded-emitter characteristic families for junction transistors show a portion of the family merging at low collector voltages and high collector currents. This is commonly called the saturation region.

The collector family generally shows a high dynamic resistance because of the almost horizontal slope of the curves in the nonsaturated region. The saturation region shows extremely low dynamic collector resistances.

Resistances in the order of 200 ohms or less are not uncommon in this portion of the collector family. This permits the use of the junction transistor as an a-c switch when driven from an appropriate impedance source. This property, not unlike the output impedance changes of a vacuum-tube cathode follower with bias, is used in the tone keyer.

The tone keyer shown in Fig. 2 consists of a source impedance, the shunting impedance of the control transistor and a grounded collector buffer amplifier. A second control transistor is included.

The source impedance is kept above 10,000 ohms by the series resistance. A blocking capacitor effectively prevents possible d-c components or return paths in the tone generator from effecting the keyer operation.

Pinching Diodes

Two 1N67A's are pinching diodes. Consider the input as a d-c signal. When the input is in the tone-on condition the first control transistor is off.

The second control transistor is on and its collector voltage is low. For this condition the two diodes are back biased by the collector voltages and present a high impedance at their junction. The tone is only slightly attenuated and the grounded collector stage gives full output.

When the input signal is in the tone-off condition the first control stage is heavily in saturation and presents a low collector impedance as well as a low collector voltage. The second control stage is now off and its collector is high. Both diodes are now conducting and are themselves low impedances.

The diode connected to the collector of the first control transistor ties the signal line to the low collector-to-ground resistance of the first stage and shorts the signal. The second control stage provides a path for d-c through the two diodes.

The first diode effectively ties the signal to the collector of the first control stage by assuring that the diode is forward biased throughout the a-c signal swing.

This prevents d-c transients through rectification of the signal in the pinching operation. The second control stage provides constant B+ drain so that use of the tone keyer in existing circuits does not upset voltage regulation. A cross-coupling capacitor between the collectors of the control transistors suppresses transients which could occur because of the time difference between the rise of one collector and the fall of the other.

The grounded collector is temperature stabilized by a 2,700-ohm emitter resistance. Larger values lead to greater stability but limit the power capabilities of the stage.

Use of the collector saturation region for the switching makes temperature effects on the control stages immaterial and temperature compensation is not used. Elevated temperatures can cause lowered collector voltages because of the amplifier I_{co} effect in the grounded-emitter stages but only extreme temperatures can cause saturation in the absence of an on signal. Overall operation was satisfactory to 58 C.

Leakage Effects

Since the switching is not perfect because the collector and diode impedances do not go to zero, there is tone-off leakage but this leakage is approximately 30 db below the tone-on state. For a 10,000-ohm source impedance this indicates a diode plus collector impedance of about 300 ohms.

The off leakage signal can be bucked out by injecting a portion of the tone-generator signal in opposition to the output signal. This method of cancellation of the residual level will result in only about 3-percent loss of tone-on signal

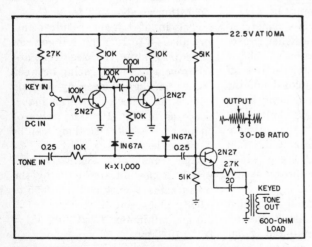

FIG. 2—Tone keyer consists of two transistor control stages and grounded-collector buffer amplifier. Either a d-c control signal or mechanical switching can be employed

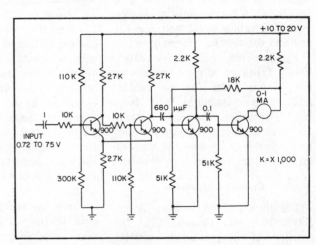

FIG. 3—Audio frequency depends upon triggered single-shot multivibrators to supply standardized pulses into integrating circuit at rates depending upon unknown frequency

when the initial on/off ratio is 30 db.

The buffer output was made grounded collector so that the high input impedance from the tone source plus the 10,000-ohm resistor would not be appreciably loaded during the tone-on state and yet would supply about 0 dbm to a 600-ohm line. This stage is conventional.

Operation is linear for inputs up to 1.0 volt. The base bias resistors were a compromise between high impedance and good temperature stability.

Total power consumption depends on the supply voltage but for the nominal design center value of 22.5 v, the drain is less than 10 ma from the signal battery. The major part of this power is taken by the buffer amplifier to allow it to supply 0 dbm to a 600-ohm load. Reduced output requirements would allow an increase in the 2,000-ohm emitter resistor and a proportionate decrease in current.

Signal input impedance to the tone is dependent on the keying state and varies from 10,000 to 40,000 ohms as determined by the base biasing resistors of the buffer amplifier and the 10,000-ohm resistor. Keying signal input resistance is a minimum of 100,000 ohms in the tone-off state. For applications requiring higher output levels, a stage of power amplification can be added to the grounded collector buffer.

Frequency Meter

Measurement of frequency may be accomplished by measuring the time average of a standardized pulse triggered from an incoming signal. Since the pulses are standardized, the time average is directly proportional to the occurrence rate or frequency of the pulses. For any given frequency the meter reading is proportional to the standardized pulse width and amplitude as well as the input frequency.

In the frequency meter shown in Fig. 3 a single-shot or monostable multivibrator generates the standardized pulses. A milliammeter in series with the normally off transistor collector provides a reading

proportional to the input frequency. By inserting the milliammeter in the normally on collector lead a reading proportional to the period of the input signal could be developed.

Silicon transistors eliminate the effects of changes in ambient temperature normally found in germanium transistors. Further, silicon transistors have low reverse collector currents denoted by I_{co} so that there is virtually no zero current and thus no need for a zero-set adjustment.

Because the transistor makes an excellent switch, the amplitude of the single-shot pulse used as the standard pulse is essentially proportional to the supply voltage and the milliammeter reading for any given input frequency is nearly proportional to supply voltage. Since the circuitry is symmetrical insofar as the emitter and collector circuit loads are concerned, the supply drain is independent of the duty cycle of the single shot and hence is independent of input frequency or pulse width of the single shot.

This is desirable since the power-supply drain is steady, except for transient conditions. Hence power-supply regulation is not affected by input frequency changes or changes in standard pulse-width variations. Also, by having supply regulation independent of input frequency, while the meter calibration is almost directly dependent on frequency, a convenient means of meter full-scale calibration is obtained either by power-supply variation or by use of a series rheostat.

This method of scale calibration is effective for small changes and not for changes of 5 or 10 to 1 because of the possibility of excessive meter currents as well as pos-

sible excessive dissipation in the transistors.

For changes of 5 or 10 to 1 or more, the single-shot coupling capacitor can be switched to different values. The values shown are for full-scale meter readings of 300 to 1,000 cycles. The standard pulse widths should be less than half of the input signal period so that each input signal cycle is effective in triggering the multivibrator. The pulse width should be wide enough so the duty cycle of the lowest frequency will result in measurable average current flow through the meter.

In the event that other than a straight line frequency-scale reading is desired, a meter movement with shaped pole pieces could be used. Such movements can yield, for example, logarithmic scales expanded about a given region.

A two-transistor preamplifier provides adequate sensitivity and allows operation on sinusoidal waves. The preamplifier is essentially a transistorized Schmitt trigger. This circuit requires only 0.75 v rms or an equivalent peak input.

The minimum input impedance is no less than 10,000 ohms as determined by the series base resistance. With this resistance shorted out the sensitivity is increased so that only 0.3 v rms is required but the circuit presents nonlinear input impedances to the signal source.

Elimination of the input blocking capacitor allows the preamplifier to trigger on slow level changes and with the addition of a series adjustment-voltage for setting of the reference level, allows the circuit to measure frequency as referred to transitions about any arbitrary level.

Table 1—Characteristics of Transistor Circuits

Circuit	Frequency	Temp	Power Supply	Output
Stabilized, unclamped.......	20 kc	70 C	22.5 v @ 2 ma	15 v
	30 kc	25 C		15 v
Stabilized, clamped........	50 kc	60 C	22.5 v @ 5 ma & 67.5 v @ 7 ma	16 v
	100 kc	25 C		16 v

VLF OSCILLATOR KEYS VHF GENERATOR

By LEON H. DULBERGER

Test Equipment Engineer, Electronic Division, Stromberg-Carlson Co.,
Div. of General Dynamics Corp., Rochester, N. Y.

Two transistor multivibrators control relay system to key signal generator at either 4 or 6 pps. Either modulation or carrier can be controlled with accuracy of ± 5 percent by simple system requiring only minor modification of signal generator

DEVELOPED to replace an electro-mechanical keying arrangement, the transistor multivibrator described here drives a grounded-emitter direct-coupled relay amplifier to pulse either the modulation or carrier of a vhf signal generator at very low frequencies. Figure 1 shows pulse widths and repetition rates required in this application.

A transistor multivibrator was chosen for long trouble-free equipment life and the other attendant advantages of transistors, small size, low power needs and low heat dissipation. Type 2N104 junction units were chosen. These are small-signal *pnp* audio-frequency types. The basic circuit developed is shown in Fig. 2.

Operation

Considering Q_1 conducting and Q_2 cut off, C_2 is charged to the supply voltage while C_1 is discharging through R_1. When the charge on C_1 is low enough to allow Q_2 to conduct, the decrease in voltage at the collector will be coupled by C_2 to Q_1.

This will cause Q_1 to conduct less heavily bringing its collector to a more negative value. The action is cumulative and results in Q_1 cut off and Q_2 conducting. At this point the charge begins to leak off C_2 and at the proper point the action repeats itself providing a series of pulses.

The time of discharge of either capacitor, and the resultant off time of the transistor it affects is equal to

$$T_1 = R_1 C_1 \log \epsilon \frac{E_u + E_i}{E_u + E_{co}}$$

Where E_u = voltage base resistor R_1 is returned to, E_i = voltage

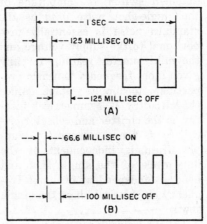

FIG. 1—Pulse times for four (A) and six pps repetition rates (B)

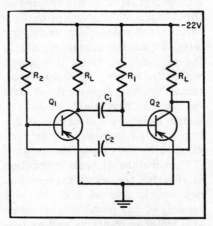

FIG. 2—Basic transistor multivibrator circuit for generator keyer

base must recover from (equal to battery E minus collector E) and E_{co} = voltage level of base when transition occurs. This may be taken as zero.

Twice this time is the total period of a symmetrical multivibrator where $R_1 = R_2$ and $C_1 = C_2$.

For the asymmetrical case, each off time is calculated separately. Emitter resistors were added to provide convenient output points for direct coupling the relay amplifier. If resistors of small value are chosen they raise the natural frequency only slightly and can be neglected in the equation. To achieve the required results and still maintain temperature stability large values of capacitance and small collector and base resistors are required. A compromise was sought that would allow one circuit to produce both of the required outputs with simple resistance adjustments. The arrangement developed is shown in Fig. 3.

Stability Control

Maximum stability is retained by restricting the range of adjustment of the collector load when using it for frequency adjustment. While a ganged potentiometer can be used, fine balance of the transistors is obtained through single unit setting. Symmetry of waveform is controlled by the 100,000-ohm potentiometer in the base return.

These controls allow a minimum of a 2-to-1 frequency range, with

Compact keying unit contains all components except relays, which are mounted in signal generator cabinet

a four-to-one asymmetrical setting, without changing coupling capacitors. Drift of frequency with temperature is ±5 percent. Short-term stability can be improved through the use of temperature controls.

Direct coupling of the output transistor amplifier saves components, and provides sufficient amplification to operate an ordinary 5,000-ohm plate relay. While a small residual current flows at all times through the relay coil, it is below the level that will provide pull-in. A negative voltage applied to the amplifier base initiates a current gain in the output stage closing the relay. The relay contacts pulse the carrier or modulation. All transistors are operated well within their dissipation ratings. Total current drain for each multivibrator-amplifier circuit is about 5.6 ma.

Adjustment

Adjustment to either repetition rate is accomplished on a calibrated oscilloscope. The symmetry potentiometer is set midway and the required frequency is obtained by adjusting first one collector resistor and then the other. Symmetry is obtained by balancing the base return.

There is slight interaction between the symmetry control and the frequency potentiometers, but setting is achieved with little effort.

Final adjustment is made on a time-interval meter. The multivibrator not in use is allowed to continue operation to maintain junction temperatures as maximum frequency drift was noted when the circuit was switched on.

The final circuit was adjusted to the relay contact on-off time shown in Fig. 1. This allows relay operate time to be compensated. Packaging was planned to allow the full advantage of the transistor's small size.

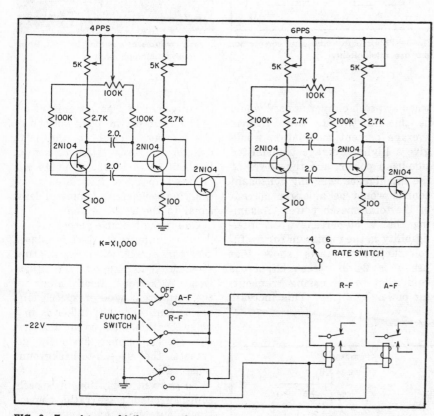

FIG. 3—Transistor multivibrators pulse r-f or a-f modulation at either 4 or 6 pps rate through relays installed in signal generator

INSTANTANEOUS SPEECH COMPRESSOR

By CHARLES R. RUTHERFORD
Temco Aircraft Corp., Dallas, Texas

Maximum range of air-to-ground communications systems is extended up to 100 percent by use of speech compressor combining advantages of clipper and agc types. Use of transistors permits construction of unit that plugs into carbon microphone jack

MILITARY tactical requirements have increased the required distance for communication by 25 to 50 percent over the maximum range obtainable with existing communication equipment.

Communication range with an amplitude-modulated transmitter is a function of the transmitted sideband power and is independent of the transmitted carrier power. Because of the wide variations in amplitude of normal speech, voice peaks producing 100-percent modulation will give an average modulation of only 30 percent. A 30-percent modulation level results in the sideband power being down 10.5 db from the 100-percent modulation level.

Since intelligibility, not quality of speech, is the primary requirement for field communications, a transistorized speech clipper would increase maximum range on many existing communication sets.

The increase in maximum range

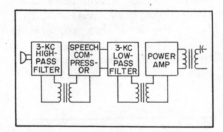

Fig. 1—Filters before and after compressor increase intelligibility

Speech compressor mounted beneath transmitter feeds microphone input

when a speech clipper is used would be due to two factors: Higher average percent modulation would give a higher average transmitted side-band power, and the power of the intelligence carrying consonant sounds would be a larger percent of the total speech power. Assuming that a 50-percent word intelligibility is the minimum for useful communication, tests show that use of a 24-db speech clipper is equivalent to increasing transmitter power by 14 db. This increase

is equivalent to a power gain of 25.

The ratio of the peak voltage of the average vowel to the peak voltage of the average consonant is about 12 to 15 db. Clipping can be used to make the intelligence carrying consonant sounds at least equal to the vowel sounds.

One type of simple speech clipper is a vacuum tube that is biased such that speech will drive the tube from cutoff to plate current saturation. A serious disadvantage of the saturation type of speech clipper is that when the talker is in a noisy location, such as an airplane, the speech clipper seriously decreases the speech-to-background noise ratio.

To prevent degration of speech-to-background noise ratio, conventional automatic gain control circuits have been used to control average percent modulation. This type of speech compressor will hold the average level of speech, and therefore the average percent modulation, constant. However, the circuit will not hold constant the peak values of the voice signal. With the agc type of speech com-

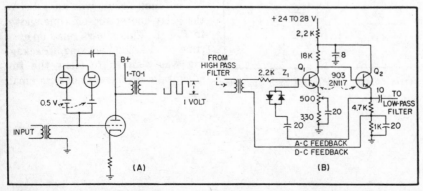

FIG. 2—Vacuum tube a-c feedback type compressor (A) operates in manner similar to the transistorized unit (B) used for air-to-ground communications

pression, normal speech will have peak voltages that are 12 db above the average voltage.

Circuit Operation

The instantaneous speech compressor system, shown in block form in Fig. 1, combines the advantages of both the clipper and the compressor and does not appear to have the disadvantages of either system.

As a-c feedback would decrease the background noise by the same factor as the voice signal, the circuit uses a delayed type of a-c feedback. Figure 2A uses a vacuum tube to illustrate the system principle. For signals with an a-c value at the plate of the tube less than 0.5 v the circuit will act like a conventional amplifier. For an instantaneous voltage that is greater than 0.5 v, one of the diodes will conduct and the signal will be fed back to the grid with a 180-deg phase reversal. If resistances of the diodes and batteries are low and gain of the tube is high, the waveform in the output circuit will be the same as the waveform obtained with a conventional speech clipper. However, as both signal and noise are fed back when the diodes are conducting, the speech-to-background noise ratio remains about the same.

A schematic of the transistorized instantaneous compressor circuit is shown in Fig. 2B. Basically, this circuit is a common-emitter amplifier direct coupled to a common collector stage. The bias for the first transistor is obtained from a d-c voltage divided in the emitter of the second transistor. As a common-emitter circuit gives a phase reversal, while a common collector circuit gives an in-phase signal, this method of obtaining bias for the first transistor results in a circuit with a high degree of d-c stability. The common-collector

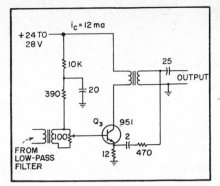

FIG. 3—Output amplifier feeds into aircraft radio microphone jack

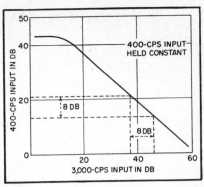

FIG. 4—Tests show ratio of voice to noise signal remains constant

stage gives a low-impedance source for driving a low-pass filter in the output and the a-c feedback circuit to the base of the first transistor. The transistors used are low-gain silicon type 903 having an alpha of 0.9 to 0.95.

The unique feature of this circuit is the use of silicon diodes in the a-c feedback circuit. The diodes will not conduct in the forward direction until a voltage of about 0.5 or larger is impressed across the diode. As a result, the forward resistance of the diode is very high until the instantaneous value of the a-c feedback signal is greater than 0.5 v. Above this level, the impedance of the diode is low. In addition to the compression obtained by feedback, compression is also caused by the sudden decrease in input impedance of the first transistor. When the diodes are not conducting, the impedance Z_1 is about 4,000 ohms as compared to 300 ohms when conducting.

Output Amplifier

The output amplifier, shown in Fig. 3, uses type 951 silicon power transistor. A feedback network from the secondary of the output transformer to the emitter resistor minimizes phase shift in the output transformer. The 25-μf capacitor in the secondary of the output transformer isolates the transformer winding from the d-c voltage applied to the microphone.

To check the effect of the instantaneous speech compressor on the speech-to-background noise

ratio, a 400-cps signal of fixed amplitude and a 3,000-cps signal whose amplitude could be varied were fed to the input of the compressor. Figure 4 is a plot showing how the 400-cps output from the compressor varied as the amplitude of the 3,000-cps input signal was increased. When the 3,000-cps input is increased the 400-cps output must decrease if the speech-to-background noise ratio is to remain constant.

Intelligibility of speech in the presence of low-frequency noise can be greatly increased by using proper filtering before and after a speech clipper. The filter preceding the clipper attenuates the low frequencies at the rate of 6 to 12 db per octave. The filter following the clipper attenuates the high frequencies at the rate of 18 to 24 db per octave.

Although controlled intelligibility tests have not been made, short-range tests on two pieces of military equipment have shown that sideband power is increased at least by a factor of 4 when the speech compressor is used.

A disadvantage of the instantaneous speech compressor is that when the talker is in a noisy location, the amplitude of the noise between speech will be practically the same as the amplitude of the speech. However, tests at the Harvard Psycho-Acoustic Laboratory have shown that noise that is present only between words does not affect the intelligibility of the message.

AUDIO INDUCTION PAGING SYSTEM

By RAY ZUCK

Government and Industrial Div., Philco Corp., Philadelphia, Pa.

Transistor device smaller than a cigarette lighter and weighing one ounce provides sound reception without wires connected to the user or radio-frequency transmission requiring a license. Audio amplifier serves as the paging source, eliminating customary lights, gongs or loudspeakers

ENERGY flowing through a loop of wire around the listening area is picked up by the miniature loop of a receiver, then amplified and converted to sound for the wearer of the instrument in this inductive paging system.

Because the sending loop is low impedance, it can be fed directly from the output of an audio amplifier, a tv set or a radio. Input to the audio amplifier can be a microphone for paging purposes or an f-m or a-m tuner for radio reception on planes, buses or trains. In the home, the tv set, record player or tape recorder could be used for the signal source.

Signal Propagation

The equation in the box indicates that radiated field strength is proportional to I, inversely proportional to λ and inversely proportional to distance d from the antenna. At the antenna, electric and magnetic fields are greater in magnitude and different in phase from the radiated field calculated from Maxwell's equations. The induction field diminishes at a faster rate with distance from the antenna than the inverse relationship of the radiation field and becomes negligible at a distance of a few wavelengths from the antenna or loop.

At a small fraction of a wavelength, the induction field is much stronger than the radiation field. For a loop antenna, the induction magnetic field is much stronger than the induction electrostatic field, while the reverse is true of the straight-wire induction field. The stronger induction magnetic field of the loop antenna is used in this system.

Interference from 60-cycle equipment such as transformers is limited to the induction source and this area is directly proportional to current. For 60-cycle equipment not over 1,000 watts in size, this interference distance is limited to a few feet.

Circuit Description

Low-impedance pickup coil (Fig. 1) is coupled to an audio taper volume control at the input of T_1. Output of T_1 is loaded by the input impedance of T_2 and T_1 output voltage conveniently provides bias for T_2 which, in turn, provides bias for T_3. Output stage, T_3 is biased to provide maximum undistorted current swing across the earphone.

The earphone acts as a d-c load for the stabilizing energy, which is fed back to the input resulting in low-frequency degradation and

---

Radiation from a Wire

The strength E of the radiated field from an elementary antenna is given by

$$E = \frac{60\,\pi L}{d\,\lambda} I \cos \omega \left(t - \frac{d}{c} \right) \cos \theta$$

E = field strength in millivolts per meter
L = length of wire in meters
d = distance from antenna in meters
λ = wavelength in meters
I = current magnitude
t = time
C = velocity of light = 3×10^8 meters per sec
θ = angle of elevation from plane of antenna

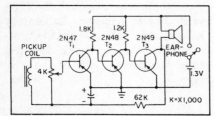

FIG. 1—Circuit of three-transistor portable receiver. Direct-coupled amplifier has feedback from output to input

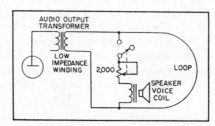

FIG. 2—Parallel operation of loudspeaker and paging system with separate volume control for speaker

Table I
Wire Size vs Impedance

AWG	Ohms per 1,000-ft length
16	3.8
14	2.7
12	1.9
10	1.2
8	0.75

25-db gain per stage at higher audio frequencies. With V_{ce} of T_1 equal to V_{be} of T_2, the point of quiescent operation is near the knee of the E_c-I_c curves. However, since the output of T_1 and T_2 is fed into the low impedance of T_2 and T_3 respectively the shunting effect of R_c $(1-\alpha)$ is negligible and allows operation of the amplifier down into and below the knee region of the E_c-I_c curves.

As the collector-to-emitter voltage falls toward zero from this knee region, alpha also approaches zero. Keeping alpha near normal and consequently maintaining the circuit gain limits the operating point to this knee region or above.

Power gain G_P of a common emitter transistor amplifier is expressed as

$$G_P = \frac{R_L}{R_{IN}} \times \left(\frac{\alpha}{1-\alpha}\right)^2$$

when R_L is the load resistor and R_{IN} is input impedance.

When $R_L = R_{IN}$, that is, T_2 provides the load for T_1 and T_2 and T_3, the gain becomes

$$G_P = \left(\frac{\alpha}{1-\alpha}\right)^2 = 1{,}520 = 31.8 \text{ db}$$

where $\alpha = 0.975$

The above represents the maximum available gain of a direct-cascaded common-emitter configuration. Approximately 25 db per stage is realized in this direct-coupled circuit. Approximately the same gain results from a common-base amplifier configuration but represents only one component per stage (a resistor) for amplification.

Installation

The loop of wire that sets up the magnetic field is low impedance and is fed from the output of an audio amplifier. Any size of wire from No. 20 to No. 8 awg can be used for the loop. For best reception, the loop impedance should be matched to the output impedance of the driving transformer. Typical impedances for a loop length of 1,000 feet are shown in the table.

A small radio, battery portable or tape recorder provides sufficient

drive for a home or office area 20 by 30 feet. For larger areas the power required is calculated on the basis of 50 to 500 milliwatts per 1,000 sq ft depending upon the ambient background noise of the paging area. For areas larger than 5,000 sq ft, multiloops fed in series and phased so that the magnetic flux is additive, will improve performance.

Multiloop Use

In multifloor buildings, loops on different floors must also be in phase to prevent cancellation of flux. For widely separated areas, one amplifier can be used with distribution lines to step-down matching transformers for each loop. The area of adequate listening signal level outside and in the plane of the loop is equal to approximately three times the area of the loop.

The distance perpendicular to the plane of the loop for adequate pickup is approximately equal to plus and minus the smallest dimension of the loop.

AUDIO INDUCTION RECEIVER

INDUCTION pickup from an audio loop placed in the region for listening provides headphone audibility with a self-contained receiver. For home use, a loop is connected in series with the speaker voice coil and the receiver volume is turned up.

The amplifier driving the loop must have an output impedance of 4 to 8 ohms. The amplifier should have an output rating of about 2 watts for normal listening. Low-frequency boost of about 15 db near 150 cycles improves fidelity.

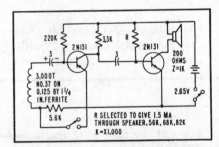

Circuit diagram of induction receiver

If a loop comprising 6 turns of No. 18 wire is made to encircle an area 10 by 40 feet adequate field

will be created using the 2-watt amplifier.

The circuit diagram of the induction-pickup device and the photograph show the nature of the equipment employed. The antenna coil is encapsulated in the clamp that fits behind the ear. The receiver proper contains the miniature loudspeaker behind a punched grill and, on the other side, an off-on switch.

Information about the equipment was furnished by Norris Electronics Corp., Scottsdale, Arizona.

SENSITIVE DYNAMIC MICROPHONE

DESIGNED to increase the intelligibility of mobile communications, a dynamic microphone employing a transistor audio amplifier between microphone unit and cable has recently been described at Wescon by A. A. Macdonald of Motorola.

The response characteristic of the dynamic element was shaped to eliminate undesired vehicle and wind noises. It is essentially flat from 300 to 4,000 cycles. The dynamic unit is relatively insen-

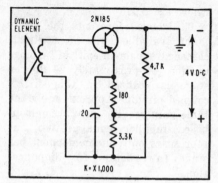

Circuit of the Motorola dynamic microphone using a single transistor

sitive to hum fields.

Using the carbon-button current source for the transistor preamplifier, the unit becomes completely interchangeable with the conventional carbon microphone for which most communications transmitters are designed.

TELEPHONE RINGING CONVERTER

SIGNALING over typical local battery telephone lines is generally accomplished with a low-voltage 20-cps hand generator. Bells and other indicators respond to such impulses and voice communication is then initiated. Such methods are particularly true of simple military telephone systems.

Recent more complex military systems provide 12 speech channels over so-called spiral-four cable. Electronic repeaters are required about every six miles, but many of these devices require no operating personnel.

Maintenance tests often require establishing contact from a repeater point. Since the newer military systems employ ringdown calling methods that operate at

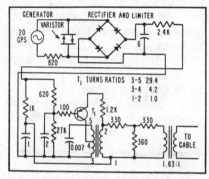

Circuit shows transistor converter to supply 1,600-cps ringing tone

1,600 cps, the hand-cranked generator at 20 cps cannot be used here.

▶ Oscillator—Test set telephones are now being equipped with a point-contact transistor circuit

that oscillates at 1,600 cps when d-c potential resulting from rectification of the 20-cps signal is applied. At the receiving location, visual and audible signals call the attendant.

It is necessary to rotate the hand crank only a couple of turns to produce the signal. Power from the generator is sufficient to operate the transistor oscillator circuit, so no batteries are required. There is no warmup time and the circuit shown is so compact that it adds little weight or bulk to the test set.

Information on this transistor application has been provided through the courtesy of Bell Telephone Laboratories.

CARRIER PHASE DETECTOR

THE transistor demodulator described here provides considerable amplification in addition to operation as phase-sensitive rectifier. The half-wave circuit is shown in the diagram. Carrier reference voltage turns the emitter currents on and off in alternate half periods of the carrier waveform. If the input signal from Q_1 is zero, the emitter-to-base voltages applied to Q_1 and Q_2 would be equal and there would be no output.

If a signal is fed from T_1 in such a sense that a negative voltage ap-

pears at the base of Q_1 and a positive voltage on the base of Q_2, the emitter-to-base voltage of Q_1 will be greater than that of Q_2 and a positive output will be produced.

A full-wave demodulator would employ two circuits of the type shown with additional secondary windings on the transformers. This information has been abstracted from the article "A Transistor Demodulator" by H. Sutcliffe appearing in *Electronic Engineering*, Mar. 1957, p 140.

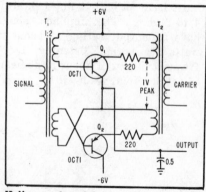

Half-wave demodulator for 2-kc carrier using two transistors. Full-wave version would use two-circuits of this type and additional transformer windings

MEASURING PARAMETERS OF JUNCTION TRANSISTORS

By ROY W. HENDRICK, JR.
Cornell Aeronautical Laboratory, Inc., Buffalo, New York

ALTHOUGH transistor quality may be roughly evaluated by its base-to-collector leakage current and base-to-collector gain, temporarily baffling results may be faced in dealing with the transistor circuits if these are the only data available. For instance, a two-stage amplifier with an expected gain of 5,000 produced a meager 40 until the static transistor currents were increased. The large increase in input resistance due to lower collector currents had shunted the signals away from the transistor bases, dropping the overall gain.

This experience prompted the development of a test unit which measures all the hybrid parameters of a junction transistor. The instrument can measure the follow-ing characteristics at any static collector current from 0.15 to 15 ma: collector leakage with the emitter open, I_{co}; collector current with base open, $I_c(I_B=0)$; base-to-collector current gain, β; input impedance r_i; output impedance with the base at a high impedance to the signal r_o; output impedance with the base terminated in 1,000 ohms, r_o'. The resistances designated are for the grounded-emitter connection.

Except for leakage measurements, the tester determines dynamic characteristics using a 60-cps test signal. Little difference has been encountered between parameters measured with this frequency or the usually specified 1,000-cps test frequency.

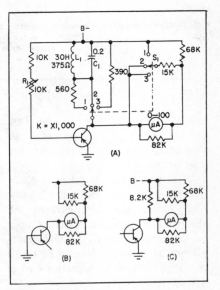

FIG. 2—Collector-current monitoring circuit (A), collector-leakage monitoring system for open emitter (B) and open base (C)

The input resistance r_i is h_{11}, β is h_{21} or h_{cb} and the reciprocal of the open-base output resistance $1/r_o$ is h_{22} or h_c. The inverse feedback ratio h_{12} is difficult to measure directly but is easily calculated from

$$h_{12}=[(R+r_i)/\beta]\,[(1/r_o)-(1/r_o')]$$

where R is the 1,000-ohm resistance terminating the base when measuring r_o'.

Circuit

Figure 1 is the complete schematic diagram of the tester. Its operation is easily understood by considering how it is used to measure each individual transistor parameter.

A series of five lever switches each having three positions determines the particular transistor

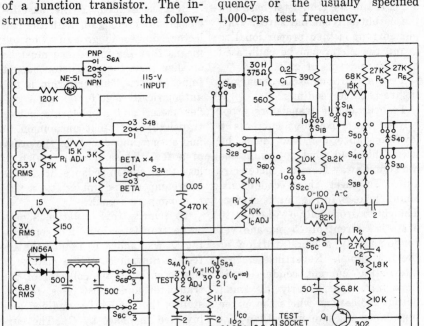

FIG. 1—Tester uses 60-cps signal to determine four common-emitter hybrid parameters

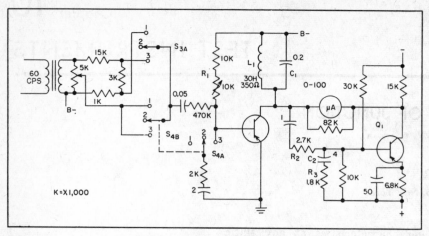

FIG. 3—Circuit used for measuring transistor beta and input impedance

Transistor is attached to tester with spring clips (at right)

characteristic being measured. One switch turns the meter on and chooses power supply polarity for *npn* or *pnp* types.

With the other switches in their neutral positions the tester monitors static collector current as shown in Figure 2A. Current may be adjusted with R_1 to test transistors at currents from 0.15 to 15 ma with full-scale deflections of 2 and 20 ma.

The full-wave rectifier a-c meter responds equally well to d-c currents of either polarity, hence requires no lead reversal with change in transistor type. Collector-current range switch S_1 alters the meter shunt and adds a voltage divider. This arrangement provides a constant driving impedance for the meter so the scale non-linearity will be identical on low and high-current ranges. In position 3, it also places a purely resistive load on the transistor so auxiliary tests of cut-off frequency may be carried out using the tester as a device for furnishing power and controlling the operating point of the transistor.

Collector Leakage

The emitter-open transistor leakage, I_{co}, is measured with S_2 in Fig. 1 in position 1. The meter is placed in series with the transistor with only a high-impedance shunt, as shown in Fig. 2B. The shunt in Fig. 2C was chosen to adjust the a-c meter to read 100 μa d-c full scale.

Base-open leakage I_o ($I_B = 0$) is measured similarly, but S_2 in position 3 adds a shunt across the meter to reduce full scale sensitivity to 2 ma.

Current Gain

To measure current gain a known a-c is fed in the transistor base and the output current measured. Output current is then proportional to β and the meter can be calibrated directly in terms of gain.

Two precautions are necessary to insure reliable readings.

To operate the transistors properly at large collector currents a low-impedance d-c load must be used; this load must not shunt the a-c signal away from the meter circuit. To meet these conditions, shunt choke L_1 in Fig. 3 was chosen. The choke drops only 5 v d-c due to a static current of 15 ma and yet conducts only a small portion of the a-c signal. Nevertheless, this alternating current conduction is important (particularly at small output indications where the meter impedance rises to more than 10,000 ohms) in accurate measurements because the inductance varies with static current and thus the fraction of the signal shunted may

be a function of the current in the choke.

Capacitor C_1 minimizes the variation in signal shunting by tuning the inductor below the 60-cps signal frequency. Thus, as the inductor current increases, the inductance drops but the resonant frequency increases closer to the signal frequency. In this way, signal impedance can be maintained nearly constant.

Beta is defined as the base-to-collector current gain with the collector shorted to ground. Shorted means that the impedance must be smaller than the transistor output impedance r_o. This criterion is not automatically met for an a-c rectifier meter and a moderate power transistor; it is not uncommon to find output impedances of the order of a few thousand ohms which is less than the impedance of a 100-microampere a-c meter to a 25-microampere signal.

To remedy this situation a one-stage transistor amplifier comprising Q_1 was added to reduce the effective impedance of the meter. If the potential at the collector of the transistor under test rises, for instance, the change is amplified and inverted in phase by Q_1. The current which flows back through the meter to the collector of the test transistor due to this inverted signal is nearly the opposite of the

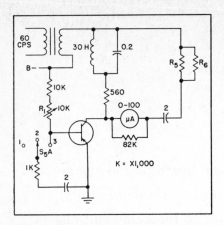

FIG. 4—Output resistance measurement ohmmeter circuit

collector current of the tested unit. Hence, little net unbalanced current results at the collector terminal.

The detector-unit impedance is approximately the total resistance in the base circuit Q_1 divided by its current gain. For the circuit in this tester the effective impedance is about 100 ohms, much less than most transistor output impedances. Resistors R_2, R_3 and capacitor C_2 in the base circuit of Q_1 form a phase-shift correcting network to prevent self oscillation of this feedback amplifier.

The base-collector current gain is measured by injecting a known 60-cps current from a high-impedance source into the base and reading the output current. Two choices of input current are available by switch S_{3A}, roughly $\frac{1}{2}$ or 2 μa. These correspond to full-scale current-gain indications of 200 or 50.

Input Resistance

Transistor input resistance r_i is measured with a circuit arrangement similar to that used in determining β. An unknown base signal current is adjusted to make the output meter read full scale and then a portion of this base current is shunted away from the base through resistor R_4 in Fig. 3 by S_{4A}. The fraction of current still entering the base is

$$i/i_o = R_4/(R_4+r_i) = (1+r_i/R_4)^{-1}$$

from which the input resistance scale may be calibrated. For example, since R_4 is 2,000 ohms, a half-scale reading after adjustment and shunting means the input resistance is 2,000 ohms, a two-thirds full-scale reading means 1,000 ohms and a one-third, 4,000 ohms.

An a-c operated ohmmeter is used for measuring the output impedance with the base at a high signal level impedance (r_o) or at 1,000 ohms (r_o'). Figure 4 illustrates how the meter is placed in series with an alternating potential, a fixed resistance and the transistor being tested. The meter current, neglecting the choke shunt, is

$$i_m = e_s/(R_5+R_m+r_o)$$
$$= \frac{e_s}{(R_5+R_m)}\left[1+\frac{r_o}{R_5+R_m}\right]^{-1}$$

where e_s is the applied potential, R_5 is the series resistance, R_m is the meter resistance and r_o is the measured transistor resistance. Metered output current i_m changes with output resistance r_o in the same manner as shown above for the input resistance. Consequently, the same scale divisions can be used. Proper choice of e_s and R_5 multiply the resistance scale by 10 to ensure that the scale center corresponds to a typical output resistance value. Because the meter impedance changes with applied current the given expression is not precise, but if the series resistor and the applied potential are experimentally fixed with R_5 so zero resistance is full scale and 20,000 ohms is half scale, adequate accuracy is available for all but the most precise applications.

Construction Features

The relative position of the power transformer and load inductor can result in an induced potential in the inductor and a spurious bias in the meter reading. The positions used were chosen emperically by connecting the meter across the inductor and moving the transformer to various places until a point was found which produced negligible pickup.

The power transformer used to operate the transistors and furnish the a-c test signal is a modified Stancor 6134 filament transformer. The high-current secondary was removed and in its place a 6.8-v power winding and 5.3 and 3.0-v signal windings were wound. It is not necessary that the two signal windings be separate, a 3-v tap on the 5.3-v secondary would be satisfactory.

Regulation of the transistor power supply will be a little better if a full-wave rectifier is used with a center-tapped 13.6-v winding. This would also aid the filtering possibly circumventing the use of a choke filter element—and increasing the ripple frequency well above the signal frequency.

The inductive kick of the transistor-load choke when the current is abruptly changed with 15-ma flowing in the choke produces over a 100-v pulse, enough to ruin the majority of transistor types. Because of this pulse, a transistor should never be pulled from the socket with the power on. Switch S_{6A}, a section of the power switch, shorts the inductor as the power is turned off, safeguarding the transistor.

When the meter is switched to read leakage current, the d-c collector current is rapidly interrupted, making control of the switch opening sequence important if the resultant inductive pulse is not to ruin all transistors tested. Switch S_{2B} shorts the inductor and going from position 2 to 1 it closes before S_{2A} grounds the transistor base interrupting collector current.

It is also important that S_{2B} break contact going from position 2 to 3 and cut off the base current to the transistor before S_{2C} removes the low-impedance meter shunt or else the meter will be an a sensitive current scale while still carrying the full static collector current. To insure these contact sequences, S_2 must be modified.

Battery Operation

In instances where self-contained operation would be an advantage, battery life would be excellent since instrument warm-up is not required and current is not drawn except when actually testing. A Burgess B5 battery should furnish 500 to 1,000 hours of testing under normal circumstances.

This modification necessitates an internal oscillator that could be a neon-bulb type operating from $67\frac{1}{2}$-v battery or a transistor audio oscillator. A reduction in size of the transistor load choke and coupling capacitors can be achieved by running the oscillator at a higher operating frequency.

TEST EQUIPMENT FOR TRANSISTOR PRODUCTION

By ANDREW B. JACOBSEN
Senior Staff Engineer
and CARL G. TINSLEY
Electrical Engineer, Motorola, Inc., Phoenix, Arizona

Separate parametric and application test equipments are used for production-line testing of several-thousand power transistors a day. Parameter measurement results are indicated by go-no go indicator and also by conventional meter. Application test measures power gain and distortion. Use of semiconductor components in circuitry assures dependable operation

MASS PRODUCTION of power transistors requires rapid and efficient testing procedures and instruments. The equipment to be described uses transistors to test several thousand Motorola 2N176 power transistors per day.

Test Specifications

There are two types of transistor test specifications. The parametric test determines the magnitude of a particular transistor parameter, for example the small-signal current gain β.

The other, an application test, determines the performance of the transistor in a circuit having certain fixed operating conditions, for example the power gain for a given circuit.

Parametric Tests

The parametric test circuits provide the necessary d-c and a-c electrical conditions for measuring collector and emitter diode saturation currents L_{co} and I_{co}, small signal current gain β, input impedance h_{11e} and collector saturation voltage V_{os}.

The manual switching arrangement shown in Fig. 1 connects the transistor to the test circuits. All test circuits ground the collector of the power transistor to reduce electric shock hazard, provide adequate heat transfer and allow for automatic transistor feed, if desired.

The test circuits and d-c regulated power supply are completely independent plug in units.

Each test circuit is simultaneously connected to the test transistor, the indicating meter and the go-no go lamp.

Prime power source is 117-v, 60-cps through a line-voltage regulator. The 35-v d-c power supply in Fig. 2 incorporates a bridge rectifier and open-ended regulator with a Zener diode for reference. The output voltage drops about 0.5 v when the current varies from no load to full load of 3 amp. Ripple

Operator uses parametrical test equipment to measure I_{co}, I_{co} and current gain of power transistors

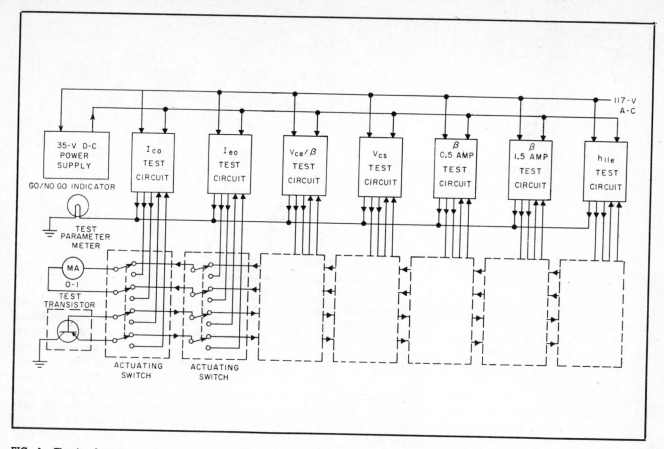

FIG. 1—Circuit-selecting switching and indicator system for parametric test equipment. Individual test circuits are plug-in type

voltage is less than 0.1 percent at full-load current.

Collector Back-Current Test

Collector back current I_{co}, is measured with a base-to-collector voltage of 35 v and the emitter open. Figure 3 shows the details of this circuit.

The meter which indicates I_{co} is in series with the test transistor and is calibrated to read 10 ma at full scale. Transistor Q_2 is on and Q_1 is off for all transistors with I_{co} less than 3 ma. For this condition the relay K_1 is energized and the indicator light is off. If I_{co} exceeds 3 ma, Q_1 is turned on and Q_2 is turned off dropping out the relay

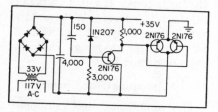

FIG. 2—Thirty-five-volt power supply for parametric test equipment

and turning on the indicator light.

The resistor in the collector of Q_1 limits its collector dissipation to a safe value for shorted transistors or those with high I_{co} values. This protective circuit causes a voltage error for I_{co} values above 3 ma (maximum allowable for the 2N176).

Emitter Back-Current Test

Emitter back-current I_{eo} is measured with a base-to-emitter voltage of 12 v and the collector open using the circuit of Fig. 4. A separate floating 12-v power supply keeps the collector at chassis ground.

The meter circuit is the same as for the I_{co} test and is calibrated for a full-scale value of 10 ma.

The Zener diode gives a constant voltage source for test transistors with I_{eo} less than 3.5 ma. With a short circuit in the test socket, for example, the Zener diode drops out and the circuit is current limited to about 6 ma.

Transistor Q_1 is biased to ener-

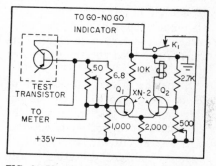

FIG. 3—Plug-in circuit for collector back current test

gize K_1, lighting the indicator lamp, if I_{eo} exceeds 2 ma. This circuit provides accurate measurements up to the rated value at 2 ma.

Current-Gain Test

The complete schematic circuit for measurement of current gain is shown in Fig. 5. For this test, the transistor is operated with 12-v d-c collector-to-emitter voltage and 500-ma d-c collector current. The d-c conditions are established with transistor Q_1 and resistor R_1. A Zener diode provides a voltage

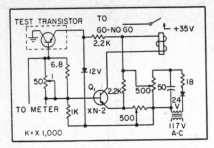

FIG. 4—Test circuit for emitter back current

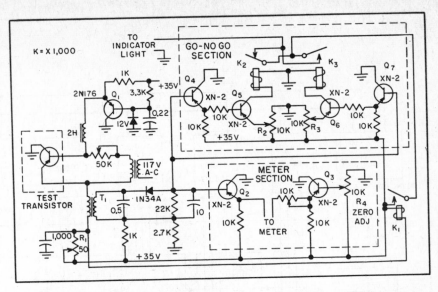

FIG. 5—Test circuit for measuring current gain at 0.5-amp collector current

reference and Q_1 provides the base current.

Resistor R_1 from the 35-v supply provides a constant-current supply for the emitter. By holding the emitter current constant, the collector current cannot vary more than ± 1 percent with variations in base current, depending on the current gain of the transistor under test.

For the test signal a 1-ma 60-cps constant current is applied between the base and emitter of the test transistor. Since the 2-henry choke is a high impedance compared to the input impedance of the transistor, a constant base signal current is provided for the transistor under test.

Since the primary impedance of T_1 is low compared to the output impedance of the transistor, the transistor under test is operated in a bootstrap circuit, equivalent to common-emitter operation, with shorted output. The voltage developed across the secondary of T_1 is rectified and the resulting d-c voltage is proportional to the current gain β of the test transistor. The output of the rectifier is fed to the meter section and the go-no go section.

Emitter-follower inputs reduce loading of the 1N34A rectifier.

The go-no go section provides a light indication for a β value less than 30 and for a β value more than 70. The voltage applied to the bases of Q_4 and Q_7 is proportional to β. This voltage is zero for $\beta = 0$ and increases in a negative direction.

For negative potential inputs to Q_4 representing a β less than 30, Q_5 is nonconducting, relay K_2 is not energized and the indicator lamp is on. For β more than 30, K_2 is energized and the lamp contacts are open.

For a potential at Q_4 and Q_7, representing $\beta = 70$ or more, Q_6 conducts energizing K_3 and the indicator lamp is on. Potentiometers R_2 and R_3 set the emitters of Q_5 and Q_6 for operation at $\beta = 30$ and $\beta = 70$.

In the meter section, the emitter voltage of Q_3 is nearly equal to the setting of R_4 and the emitter voltage of Q_2 follows the output of the rectifier. The 1-ma meter is connected between these two emitters with a series calibrating resistor adjusted for a full-scale reading at $\beta = 100$.

Relay K_1 removes the voltage from the light indicator when no transistor is under test.

Input-Impedance Test

The circuit for measuring input impedance h_{11e} of the transistor under tests is similar to the β cir-

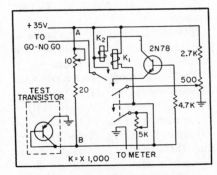

FIG. 6—Saturation voltage test circuit

cuit in Fig. 5. The transistor is operated at 12-v collector-to-emitter voltage and 500-ma collector current. A constant 1-ma alternating current is applied between base and emitter. For this circuit T_1 has a higher input impedance and samples the voltage between base and emitter measured by the metering circuit. The go-no go circuit is also similar to that used in Fig. 5.

Saturation-Voltage Test

The saturation voltage test is made with 1.5-amp emitter current, the base connected to the collector; the maximum acceptable collector-to-emitter voltage is 1.2 v using the test circuit shown in Fig. 6. The resistance from point A to point B is adjusted for 1.5-amp collector current.

To protect the meter circuit and the npn transistor, K_1 removes the supply voltage from these circuits when no transistor is in the test socket. With a transistor in the test socket, the meter functions as a voltmeter reading the emitter-to-collector voltage.

The series resistor is adjusted for a full scale meter reading of 5 v. The emitter of the npn transistor is connected to a voltage-divider network and this voltage is adjusted to energize K_2 if saturation voltage exceeds 1.2 v. With K_2 energized, the go-no go indicator light is turned on.

Application-Test Circuit

The application test automatically provides the following operating conditions for the power tran-

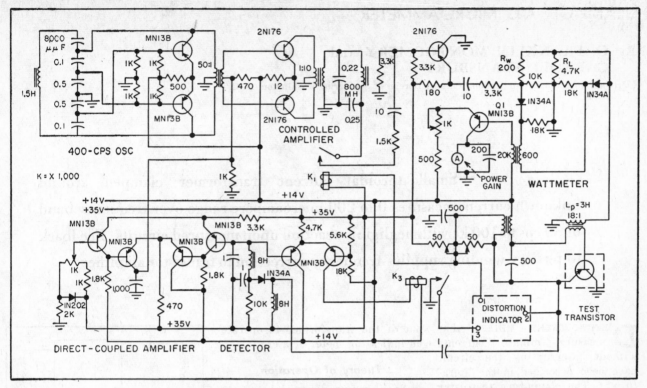

FIG. 7—Circuit for measuring power gain and distortion of transistors with input impedance of 10 to 15 ohms and power gains of 30 db

sistor: 14-v collector-to-emitter voltage, 500-ma collector current and an audio output of 2 w. This circuit gives a direct meter reading of power gain in db and a light indication for amplifier distortion in excess of 9 percent.

The transistor power gain for these conditions is obtained from a calibrated input power meter with the output automatically held constant at 2 w.

Two regulated power supplies, 35 v and 14 v, similar in design to that in Fig. 2, accommodate the test transistor, an oscillator, various amplifiers, a metering circuit and a distortion indicator.

Operation of the automatic power gain test and distortion indicating circuit may be followed by

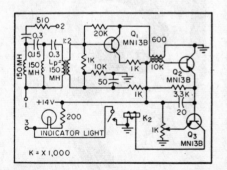

FIG. 8—Distortion indicator circuit plugs into circuit shown in Fig. 7

refering to the Fig. 7.

The 400-cps oscillator runs continuously and the gain of the control amplifier is set by an amplified error signal to provide the necessary input signal for 2-w output signal power.

The test transistor output is rectified by the detector circuit and is compared with the Zener reference voltage to provide an error signal to the differential direct-coupled amplifier. This amplifier provides enough gain so the output for test transistors having a power gain of 26 to 36 db is 2 w ±0.3 db.

Relay K_1 removes the input signal to the wattmeter when no transistor is in the test socket.

Input Wattmeter

Signal power input to the test transistor is a measure of the transistor power gain in this circuit because the signal power output is held constant. Input power to the test transistor is measured by the wattmeter circuit in Fig. 7, which, when calibrated is accurate within ±0.1 db for transistors with input impedances between 8 and 22 ohms. Outside this range the error increases to ±1 db for 4 to 40-ohm input impedances. Nearly all the transistors of the type to be tested

are within the 8 to 22-ohm range of input impedances giving the power gain tester an overall accuracy of ±0.3 db for power gains of 26 to 36 db.

The input to the transistor under test has less than 1-percent harmonic distortion. This low value is obtained by using a 400-cps filter which attenuates the second and higher harmonics.

Distortion Indicator

The distortion indication circuit is shown in Fig. 8. The output of the test transistor is passed through a high-pass filter that attenuates the 400-cps fundamental 36 db compared with the second and third harmonics.

The output of the filter is amplified by transistor Q_1, rectified by Q_2 and the resulting d-c is applied to the base of Q_3.

The emitter of Q_3 is biased to a potential equivalent to 9-percent distortion. When the distortion exceeds 9-percent, Q_3 energizes relay K_2, lighting the distortion indicator lamp.

A negative feedback circuit stabilizes the gain of Q_1. The relay turns on distortion indicator lamp when power output cannot be maintained at 2 w.

CLAMP-ON A-C MICROAMMETER

By G. FRANKLIN MONTGOMERY and CARROLL STANSBURY

Electronic Instrumentation Section, National Bureau of Standards, Washington, D. C.

Small toroidal current transformer clamped around unknown current measures 0 to 200-microampere range over frequency band of 50 cps to 100 kc with negligible reaction upon measured circuits. Feedback to tertiary winding supplies frequency correction to transistor amplifier

DIRECT-CURRENT METERS that measure minute currents without interrupting the circuit have been described in the literature.[1,2] The clamp-type transistor microammeter to be described was developed to demonstrate the practicality of similar a-c instruments.[3]

This meter uses a small current transformer which can be clamped about the unknown current. An amplifier increases the electrical output of the transformer to operate an indicating meter or a cro.

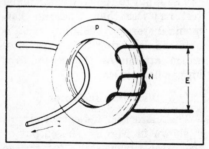

FIG. 1—Secondary winding and unknown current interlace closed magnetic core

Theory of Operation

An unknown sinusoidal current of I amperes surrounded by a closed magnetic core with a secondary winding of N turns is shown in Fig. 1. The amplitude of the induced secondary voltage is

$$E = \omega N P I \qquad (1)$$

where P is the core permeance in webers per ampere-turn.

If the secondary is connected to the input terminals of a zero input-impedance current amplifier as shown in Fig. 2, the input current will be

$$I_s = (I/N)/[1 + (f_c/f)^2]^{1/2} \qquad (2)$$

where $f_c = R_s/(2\pi L)$, R_s is the effective a-c resistance of the secondary and L is its self-inductance. The output current of the amplifier

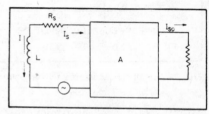

FIG. 2—Equivalent circuit of transformer probe and preamplifier

is substantially independent of f in the range where $f >> f_c$. When the transformer's physical dimensions are small, f_c may be several kc.

The method of effectively reducing f_c for this instrument is to supply negative feedback from the amplifier output to a tertiary winding on the core. When this is done, the output current of the amplifier becomes $I_{so} = (I/N)/[1 + (f_c/$

$$I_{so} = (I/N)/[1 + (f_c/Af)^2]^{1/2} \qquad (4)$$

and f_c is reduced by amplifier current gain A.

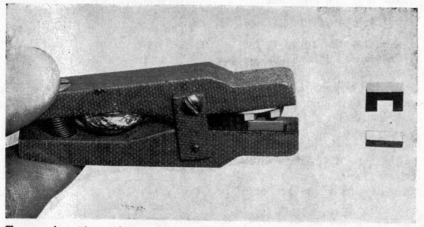

Clamp probe with toroid core shown at right

Complete microammeter with clamp

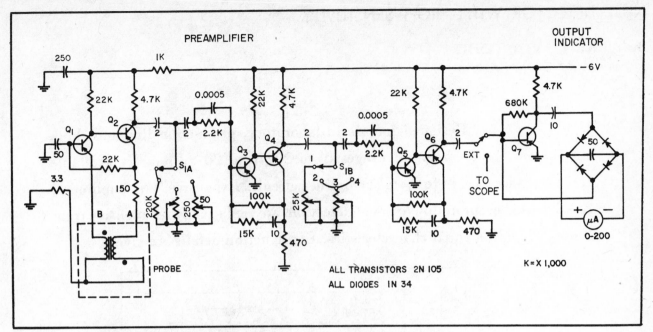

PREAMPLIFIER

OUTPUT INDICATOR

ALL TRANSISTORS 2N 105
ALL DIODES IN 34

K = X 1,000

FIG. 3—Clamp-type transistor microammeter has four full-scale ranges from 200 μa to 200 ma

Preamplifier $Q_1 Q_2$ in Fig. 3 supplies frequency correction for the pickup transformer whose signal winding terminates at B. The feedback winding terminates at A.

Output Indicator

The 200-microampere output indicator is driven by Q_7 through a full-wave bridge rectifier; 100-percent negative feedback is used around Q_7 to improve the rectifier linearity. The ratio of d-c output to a-c input of Q_7 is therefore slightly less than unity and an input of about 220 microamperes a-c deflects the meter full scale.

If the transformer secondary winding consists of 250 turns, the output current of preamplifier $Q_1 Q_2$ will be about 0.8 microamperes a-c for an unknown current of $I = 200$ microamperes. Consequently, a current amplification of $220/0.8 = 280$

is needed between the preamplifier output and the input to Q_7. Feedback pairs $Q_3 Q_4$ and $Q_5 Q_6$ furnish gain of 20 per pair. Adjustable

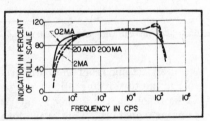

FIG. 4—Frequency response curves for four ranges of operation of transistor microammeter

shunt attenuators permit calibration of each current range.

The transformer probe consists of a silicon iron core of 0.014-in. laminations with $\frac{5}{16}$-in. outside dimensions and a thickness of $\frac{1}{8}$ in. The stack of C-shaped laminations fixed in the lower jaw of the clamp

is wound with two 250-turn windings of number 44 wire. The I-sloped stack in the upper jaw is mounted on a pivot so it will seat properly on the rising legs of the C.

The response of the completed microammeter is shown in Fig. 4. The most sensitive current range, 200 microamperes full scale, is useful in the frequency range from 50 cps to 100 kc. The internally generated noise is about 20 μs.

This work was done at NBS as part of a research program sponsored jointly by the ONR, the AFOSR, and the AEC.

REFERENCES

(1) E. H. Frei and D. Treves, Electronic Current Meter with Clip-on Probe, 12th National Electronics Conference, Chicago, Ill. Oct. 1, 1956.
(2) E. H. Frei and D. Treves, Clip-On Milliammeter Uses Magnetic Amplifier, ELECTRONICS, p 204, Jan. 1957.
(3) G. F. Montgomery and C. Stansbury, The Clamp-Type Alternating-Current Microammeter, AIEE Winter General Meeting, New York, Jan. 21, 1957.

NULL DETECTOR WITH HIGH SENSITIVITY

By CARL DAVID TODD

Semiconductor Products Department, General Electric Co, Syracuse, New York

Designed for use in laboratory, production line and shop, instrument has seven input ranges from 20 μv to 20 v. Readily available transistors are used in four-stage amplifier circuit. Expensive junction diodes, required for accurate low-voltage detection, are replaced by inexpensive transistors providing equal characteristics. Construction details are given

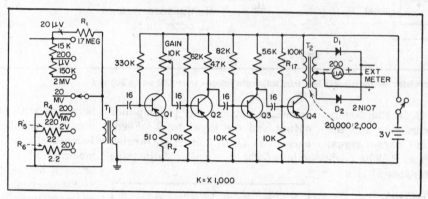

FIG. 1—Four-stage amplifier uses low current-gain transistors

IMPORTANT CHARACTERISTICS of a null detector are high sensitivity, low internal noise, moderately fast response to changes in input levels, and an ability to withstand a large overload.

Circuit Description

The complete schematic diagram of a null detector with these characteristics is given in Fig. 1. It is basically a high-frequency audio amplifier whose output is rectified and fed into a sensitive d-c microammeter.

The input impedance varies

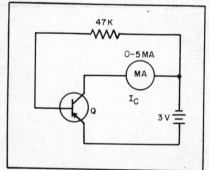

FIG. 2—Test circuit measures relative current gain h_{fe} of transistors

when switching from the 20-microvolt range to the 20-millivolt range. After reaching the 20-mv range, it remains constant. This was done to avoid high resistance values normally required for the higher voltage ranges and also partially to eliminate noise in the input.

The input transformer is a modified push-pull driver transformer. The center-tapped secondary was modified to give two identical windings. By breaking the center lead and bringing out both wires separately a unity ratio transformer is obtained.

The input impedance to the null-detector amplifier was approximately 20,000 ohms. With the transformer connected, the input impedance dropped to 2,000 ohms.

The four-stage transistor amplifier uses low current-gain transistors. The amplification is sufficient. High-gain transistors may be used with slightly better results accompanied by a higher price tag.

The amplifier output is fed into a transformer originally designed as the driver interchange trans-

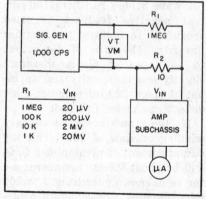

FIG. 3—Test set-up to determine values of detector input attenuator resistors

former for a push-pull amplifier. The centertap allows full-wave rectification without need of a bridge.

To achieve good results at low voltages, junction diodes are required for the rectifiers. At the present time, junction diodes are expensive. Type 2N107 transistors sell for much less and make excellent junction diodes. Either the emitter-base diode or the collector-base diode may be used. Both diode sections can not be used since they

144

Null detector fits in 5 by 7-in. case

Internal view shows parts layout

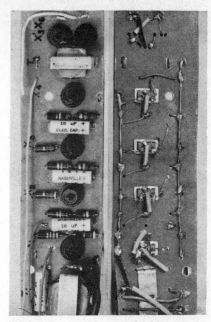
Subchassis parts layout and wiring

are not independent.

Terminals are provided so the 200-microampere meter may be used for external functions. No switching is necessary to disconnect the output circuit from the meter since the diodes isolate the meter from the null detector whenever current is supplied to the external terminals.

Construction

The null detector was built in a 5 by 7 by 2-in. aluminum chassis. The amplifier and rectifier circuitry are mounted on a Textolite board subchassis. This is in turn mounted to the side of the main chassis. Printed circuit techniques are not used, but rather, small wires are employed to connect the various components together.

The current gain h_{fe} of the transistors used as diodes is not important. It may be wise, therefore, to use the two lowest h_{fe} units for this function saving the others for the amplifier.

The test circuit shown in Fig. 2 may be used to measure relative gains of the transistors. The transistor that has the highest collector current will, in general, have the highest h_{fe}.

The transistors had an h_{fe} ranging from 18 to 40. If transistors with current gains greatly different from this are used, it will be necessary to change the resistors supplying the d-c base-bias current. As a general rule, the base-to-supply resistor should give a collector current of about 0.5 ma.

Tests

A 1,000-cps signal applied to the amplifier input as shown in Fig. 3 allows measurement of the input voltage in the microvolt range. The exact values of the resistors used in the attenuator depend on the 1,000-cps voltage amplitude and the lowest voltage that may be measured by a vtvm or an oscilloscope.

The values shown are for use with a signal generator capable of delivering 2 volts rms. When two volts are applied to the input attenuator, 20 microvolts will appear across the 10-ohm resistor. This

voltage should produce a deflection on the meter. If there is no deflection, the gain of the amplifier may be too low. To determine this, increase the input voltage by using a smaller series resistor R_1. Several values are given in Fig. 3. Once a deflection is obtained, the transistors should be swapped back and forth to maximize the gain.

Final gain adjustment is done by varying the value of resistor R_7.

The input-attenuator resistors are best determined by using a potentiometer for the resistor, applying the desired full-scale voltage to the input as shown in Fig. 3, then adjusting the potentiometer to give a full-scale meter reading.

The resistors should be determined in numerical order. With resistor R_1 connected in place, resistor R_2 is selected to give full-scale deflection with 200 microvolts applied. Resistor R_3 is selected in a similar manner.

Resistors R_4, R_5 and R_6 when placed across the amplifier input should produce full-scale readings at the desired voltages.

PORTABLE FREQUENCY STANDARD

By DONALD S. BEYER

General Electric Co., Syracuse, N. Y.

IN EVERY PHASE of electronics, the need often arises for a convenient signal source of known frequency, voltage and waveform. When using an oscilloscope, for instance, it would be most useful to have a reference-frequency signal source to connect to the vertical input terminals with the object of establishing an accurate time base on the X axis.

The device to be described is small enough to be always handy and easy to use for performing all the above services. In addition, it has many uses, such as its use as a calibration signal source for alignment of radio receivers.

▶ **Circuit**—A low-priced *pnp* transistor in common base configuration is used in a modified Colpitts oscillator. It is frequency stabilized by a 200-kc crystal in a parallel resonant circuit. For battery economy, only $4\frac{1}{2}$ v of collector bias is used. For the same reason, emitter bias is obtained without a resistor voltage-dividing network by deriving this bias directly from one cell of the battery.

Common emitter configuration is used for the other *pnp* transistor. This stage is overdriven by output from the oscillator so the output will be rich in harmonics owing to the clipping that results. This stage also serves as a buffer, to prevent loading by external circuits.

Output signal obtained is 3.5 v peak-to-peak of clipped sine wave at 220 kc, having a rise time of approx. 0.1 μsec. The output impedance is 1,000 ohms. Harmonics are present to over 10 mc. Stability is such that when the output is shorted, the oscillator does not shift more than 5 cps.

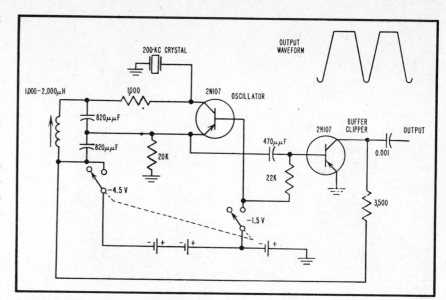

Schematic circuit diagram of the portable frequency standard

The oscillator is adjusted as follows. Move the slug so it is nearly out of the coil (low-inductance end) and loosely couple the output to a broadcast receiver tuned to 600 kc. Correct adjustment is obtained when the S meter or tuning eye indicates the strongest harmonic signal of 200 kc.

For more precise adjustment, couple the output to a short-wave receiver tuned to standard-frequency station WWV on 5 mc. Adjust the slug until zero beat is heard.

▶ **Use**—In aligning radio receivers, every 200-kc point on the dial can be marked. When connected to the horizontal or vertical inputs of an oscilloscope, a convenient frequency reference is available for measuring signals of unknown frequency.

By counting the number of 200-kc square waves that appear with a given horizontal sweep-frequency setting and then by counting the number of cycles of unknown input signal when using the same sweep frequency, the frequency of the unknown input signal can be determined. It can be used as a complex waveform reference signal to indicate oscilloscope amplifier response, and rise time errors.

OSCILLOSCOPE PREAMPLIFIER

PREAMPLIFICATION of signals into an oscilloscope is often necessary but frequently awkward when conventional amplifier units must be employed. A commericial amplifier employing transistors uses the circuit shown.

Voltage gain is 1,000. Noise is 1 μv rms referred to input and the

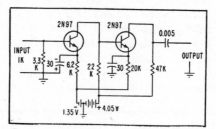

Circuit of Du Mont preamplifier

response is 3-db down in the range from 20 cps to 30 kc. The unit is powered with mercury cells that have a life of 1,000 hours in this type of service. Input resistance is approximately 1,000 ohms.

Chapter 11

INDUSTRIAL CONTROL CIRCUITS

VELOCITY-TYPE SERVO SYSTEM

By **HERBERT L. ARONSON** Project Engineer
WILLIAM R. LAMB Electronics Engineer
Baird Associates—Atomic Instrument Co., Cambridge, Mass.

Velocity-type servo system uses single-rate feedback loop and d-c stabilization. Network design is based on constant-current driving source and low-impedance load conditions imposed by transistor operation. Double common-collector power stage with inverse feedback lowers output impedance by factor of 200 and cuts motor corner frequency in half. Power resistors mounted on copper-bar heat sink minimize stabilization needs

DEMAND for a miniature servo system of lower power consumption has resulted in the development of a unit having the advantages of transistorized design, greater life expectancy, smaller size, simpler thermal design and lower operating voltages.

The servo amplifier and associated motor-tachometer form a highly stable and accurate rate servo system, the operation of which is insensitive to carrier frequency variations. The desired 5-cps bandwidth is achieved by d-c stabilization specifically designed for operation at transistor impedances.

Power Levels

Maximum power output of six watts is developed from a 15-mv input signal. Overall power gain is thus 98 db. A maximum of 400 ma is required from the 28-v supply to maintain this power level only for brief moments of rapid motor acceleration. Full-speed motor operation at 4,000 rpm causes a 200-ma current drain. Under standby conditions the drain falls to 28 ma.

The block diagram of Fig. 1 shows the fundamental system. An a-c preamplifier supplies driving power for a phase-sensitive demodulator. The resulting low-frequency error signal passes through the stabilization network to a phase modulator. Efficiency of the stabilization and conversion circuitry

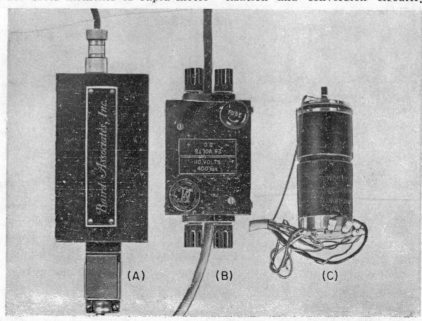

FIG. 1—Servo system employs velocity feedback overall as well as negative feedback locally in power amplifier

Three units of system comprise servo amplifier (A), output transformer (B) and motor-generator (C). Amplifier occupies 17 cu. in.

makes d-c amplification unnecessary. Power stages following the modulator consist of a driver and push-pull output.

Preamplifier Stages

Preamplification is required to raise the signal level at the output of the summing circuit to a value suitable for demodulation. Approximately 90 db of power gain is achieved with three R-C coupled common-emitter stages shown in Fig. 2. The high-impedance output of the summing circuit is matched to the low-impedance input of the first transistor by a subminiature transformer. Magnetic shielding is used to prevent coupling between this and other transformers.

The first two stages employ silicon transistors which, due to their low leakage currents, require no emitter stabilization over the desired operating temperature range. A germanium transistor is used in the output stage to supply up to 60 mw drive to the demodulator. Loop gain of the system is controlled by changing the value of R_1 and thereby varying the amount of current feedback in this stage. Some feedback must be retained to ensure proper operation of the d-c stabilization circuitry since it is designed on the basis of a constant-current driving circuit of high output impedance.

Phase-sensitive demodulation of

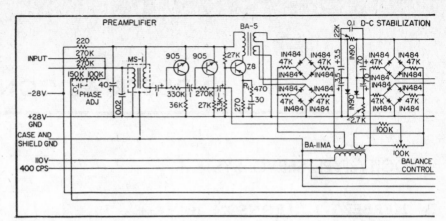

FIG. 2—Schematic of servo amplifier system. Value of C_1 is chosen to correct tacho

the preamplifier output supplies the input signal for the stabilization network. Network output is converted to suppressed-carrier a-c by means of a phase modulator. Both demodulator and modulator consist of full-wave switching circuits utilizing subminiature silicon diodes. The circuit has a 50-db dynamic range characterized by extremely low drift and residual output.

Impedance Functions

Because of the switch-type nature of the circuit, both the modulator and demodulator appear as impedances which are functions of their driving and load impedances respectively. Thus the demodulator acts as a current source for the stabilization network while

the modulator is made to appear as a low-impedance load for the stabilization network.

Conventional designs are developed on a voltage-transfer basis which assumes a constant voltage source and an infinite-impedance load. Neither of these assumptions is true in the system under discussion. A power loss of 50 db could be anticipated if the output of such a network were loaded by the input impedance of a transistor. But if current is taken as the dependent variable the design will be based on a constant-current driving source and a low-impedance load. Reasonable impedance matching is then possible and the only power loss is that due to any series resistance in the network itself.

An R-C network designed on a

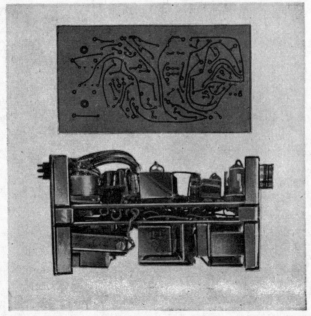

Epoxy glass circuit board (top) measures 1¾ x 3½ in. Two-board assembly shown in side view (below)

Tranformers mounted underside (left). Separate printed circuit board holds other parts (right)

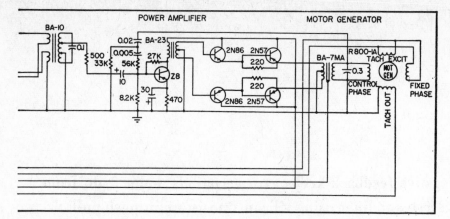

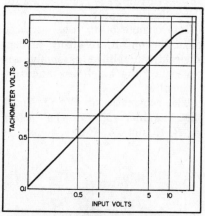

meter phase shift. Value of R₁ controls current feedback and loop gain

current basis allows for a realistic value of output termination. An L-C circuit would require 5,000 henrys to obtain the desired characteristics, and while such a choke is obtainable in miniature size the high series resistance would negate the advantages. The R-C circuit is therefore used because of its simplicity, reasonably small components and moderate efficiency.

The network is designed on the basis of a 10,000-ohm termination; the value presented by the input impedance of the modulator when loaded by the first power-amplifier stage. Network calculation results in a value of 20,000 ohms for the series resistance with a power loss of only 6 db. Capacitors of 3.5 and 70 μf provide the required low-frequency time constants. Because the tantalytic capacitors in the network are polarized units, two of them are connected back to back to permit the flowing of bilateral currents. Germanium diodes are connected across each capacitor to shunt the reverse currents.

Power Amplification

The power amplifier section is designed to develop the required six watts with low distortion and the low output impedance necessary to reduce the servo motor corner frequency. When the motor is driven from a high-impedance source, the corner frequency is 2.5 cps. This frequency increases to 5 cps when the motor is driven from a 200-ohm source, which is less than 1/10 the stalled motor impedance.

The common-collector stage is a linear voltage amplifier for all voltages over a few tenths of a volt. Thus no biasing is necessary and a standby current drain for the four transistors of only a few milliamperes is caused by the transistor leakage current. Inasmuch as this current is thermally generated within the transistors its

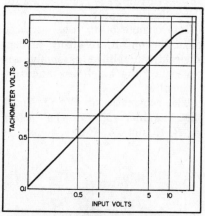

FIG. 3—Transfer characteristic curve is linear over range of better than 100 to 1

magnitude will vary with temperature. When the transistors are at room temperature it will be 1 or 2 ma. But when the transistors are heated by internal power dissipation or increased ambient temperature it may temporarily increase to 50 ma.

The output impedance of a common-collector stage is approximately equal to the driving impedance of the preceding stage divided by the current gain of the transistor. Since two common-collector stages are cascaded together in this circuit the output impedance is equal to the driving impedance seen by the first transistor divided by the product of the current gains of both transistors. This is an im-

pedance reduction of about 200 for the transistors used. Since the voltage gain is nearly one this results in a power gain of about 22 db.

Two resistors of 220 ohms each are connected from base to emitter of each power transistor to improve the thermal stability of the circuit. If they were not present the thermal current generated within the transistor would cause its base to become slightly positive with respect to the emitter. This in turn would cause the medium-power driver transistor to be cut off, eliminating the stabilizing effect of its normally low output impedance. The result at best would be reduced overall efficiency and it could result in thermal runaway and destruction of the transistor.

A medium-power class-A stage supplies power for the output. It has a maximum output power of 50 mw and a standby dissipation of 150 mw. Negative feedback reduces the output impedance and improves stability. Low output impedance is desirable to prevent changes in motor impedance from reflecting back through the common-collector stages and influencing the gain.

The output impedance with feedback is 2,400 ohms at the driver transformer primary as compared to 9,000 ohms without feedback. The power gain of this stage is 28 db resulting in a total power amplifier section gain of 52 db.

Negative Feedback

A negative feedback loop around the power amplifier section further reduces the system output impedance and provides part of the 90-deg phase shift required for operation of the motor.

Satisfactory stability is obtained by designing the feedback loop to supply a 60-deg phase shift and detuning the output of the modulator transformer to obtain the remaining 30-deg shift. The 6-db feedback reduces the output impedance from 700 to 200 ohms as seen at the 110-v secondary of the output transformer. This is a driving impedance of less than one-tenth the stalled motor impedance and represents a sufficiently stiff source for proper motor operation. The complete system transfer characteristic is shown in Fig. 3.

PUSH-PULL SERVO AMPLIFIER

By R. T. HENSZEY

Lear, Inc., Learcal Division, Los Angeles, Calif.

Negative feedback keeps gain variations under 1 db from —55 C to 100 C in servo amplifier using silicon transistors in push-pull circuit that delivers 0.75 watt into transformer or 7.5 watts to saturable reactor. Signal source impedance is low compared with input impedance. Unfiltered, rectified a-c is adequate for collector supply

SILICON TRANSISTORS are used in a preamplifier which may be coupled to a push-pull transformer for 0.75 watt output or to a saturable transformer for 7.5 watts output.

Negative feedback is the major factor contributing to the performance of the amplifier about to be described. Other features aiding the overall performance are low signal-source impedance compared with input impedance, direct coupling of transistors and the use of unfiltered full-wave rectified a-c for the collector supply.

Interchangeability was stressed from the outset, resulting in considerably simplified assembly and testing of the amplifiers and subassemblies.

The schematic of Fig. 1 shows the amplifier up to the points A, B, C and D. Here, it may be connected either to a push-pull transformer or to a discriminator saturable transformer if higher gain and output are required. The connections are shown in Figs. 2A and 2B respectively.

The amplifier is designed to operate from a low-impedance E-I type pickoff. The pickoff tuning is integral with the pickoff. The loads for both amplifiers are two-phase torque motors operated with locked

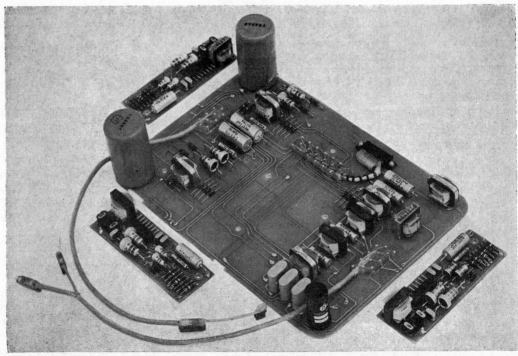

Main assembly board contains two complete 7.5-watt servo amplifier channels and a 0.75-watt channel plus related switching circuits. Subassembly boards contain complete amplifier except for output transformers

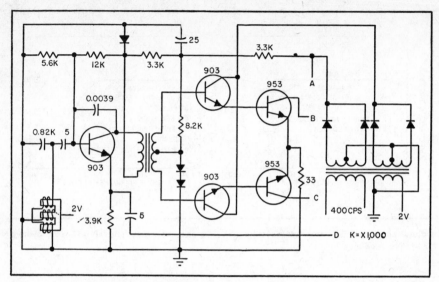

FIG. 1—Amplifier, supplied by unfiltered, rectifier a-c, drives either push-pull transformer or saturable reactor shown in Fig. 2

Table I—Characteristics of Amplifiers.

	Class B Transformer Output	Saturable Transformer Output
Load Z	790 $\angle$ 48 untuned 1,200 $\angle$ 0 tuned	1,150 $\angle$ 49 untuned 1,770 $\angle$ 0 tuned
Signal Source Z	500 ohms	500 ohms
Input Z	500 ohms without feedback 3000 ohms with feedback	500 ohms without feedback 3000 ohms with feedback
Power Out	0.75 watt max (30 v out)	7.5 watts max (115 v out)
Power Gain	60 db without feedback 45 db with feedback	70 db without feedback 55 db with feedback
Negative Feedback	15 db at max output	15 db at max output
Power In for max Out	0.75 microwatt without feedback 25.0 microwatts with feedback	0.75 microwatt without feedback 25.0 microwatts with feedback
Gain-Power Out	Constant	Constant

rotors. Since there is little motion of the motor rotors the load impedance is constant. All of the data were obtained with the actual operating sources and loads.

The input stage is a single-ended class A stage with emitter-bias stabilization. This stage is transformer-coupled to a balance driver stage. The driver stage is direct-coupled to the balanced output stage which is biased for class B operation. Proper bias is maintained at the output over the temperature range by the two diodes in the driver base circuit.

Feedback

The major advantage of the negative feedback incorporated in the circuits is temperature stability. Figure 3A shows the gain variation from −55 C to 100 C. The curves were prepared from averages of six tests for each amplifier type. Transistors with h_{fe} ranging from 7 to 35 at 400 cps for the common-emitter circuit were used. The maximum variation for any individual test was 1 db.

Another advantage of the negative feedback is that parts interchangeability without adjustment is possible. Transistors with h_{fe} ranging from 7 to 50 have been tested in the circuits. In most cases, the operation is completely satisfactory. However, when transistors which have a high h_{fe} are used simultaneously, there is a tendency toward oscillation at a low signal level.

The transistors in the balanced stages were matched to within 20 percent for h_{fe} at 25 C. An additional advantage of the feedback is the increased frequency response. The effect of feedback on the response to a sinusoidal input is shown in Fig. 3B and 3C.

The overall negative feedback in both circuits is approximately 15 db at maximum output. At small signal levels, the feedback increases because of the increase in open-loop gain. Necessary stabilization of the amplifiers is obtained by proper attenuation of high and low frequencies.

The high-frequency response is limited by the 0.0039-μf capacitor from collector to base of the first stage. This capacitor also tends to lower the input and output impedance of the first stage.

The low-frequency response is limited by the 5-μf capacitor in the feedback circuit to the emitter of the first stage. This capacitor is small enough to produce degeneration in the first stage below 400 cps. Also, the low-frequency feedback is attenuated by this capacitor.

The variation of capacitance

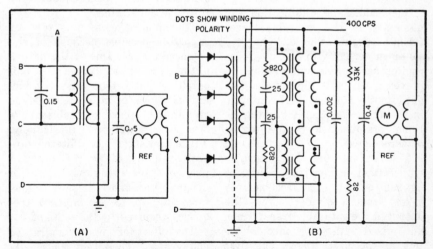

FIG. 2—Diagram shows method of connecting servo amplifier to transformer (A) and to saturable reactor (B) shown in Fig. 1

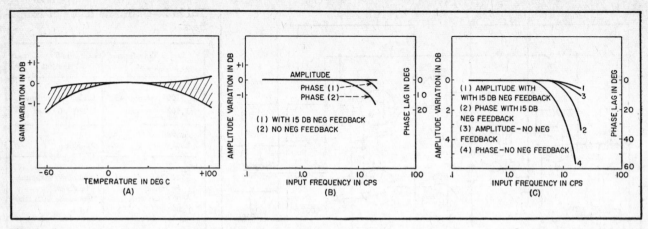

FIG. 3—Curves show spread of gain-temperature characteristic over twelve tests (A), response characteristic of 0.75-watt system (B) and 7.5-watt system (C) with and without negative feedback

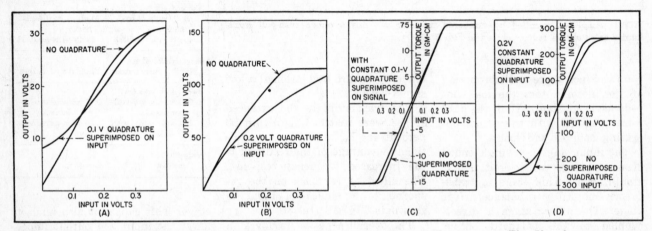

FIG. 4—Curves show output-input voltage characteristic of 0.75-watt system (A) and 7.5-watt system (B). Also shown are output torque-input voltage characteristics for the 0.75-watt (C) and 7.5-watt (D) systems

with temperature is such as to aid the overall stabilization. The open-loop frequency response (point *D* grounded) is approximately 300 to 1,200 cps. To make this measurement it was necessary to use a filtered collector supply

Saturable Transformer

Some additional stabilization was necessary in the saturable transformer output. The 820-ohm resistor in series with the 25-μf capacitor across the control winding prevents self oscillation of the output circuit. The 0.0033-μf capacitor in the feedback circuit gives additional stabilization in the region of zero signal.

The output of the saturable transformer is nonsinsoidal except at full output. The feedback is taken from a divider across the output at the saturable transformer and mixed with the sinusoidal input in the first stage. The first stage has sufficient dynamic range

to handle the resulting complex waveform.

Power Supply

The collector supply is somewhat unconventional. The unfiltered full-wave rectified a-c supply reduces transistor dissipation considerably. Also, some quadrature rejection is attained.

The application of this type of supply is possible because of the balanced stages. Due to the nature of the collector supply, the push-pull output is not a true class B stage but is more of a switching-type circuit. No phasing problem has been encountered with this supply.

The first stage has a filtered supply consisting of an R-C network plus a Zener diode which has a dynamic impedance below 50 ohms from d-c up to the highest frequency required by the system.

The effect of input quadrature voltages on the output voltage is shown in Fig. 4A and 4B. The

amount of quadrature superimposed was the maximum which would not saturate the first stage when combined with full input signal. The voltages presented on the curve were measured with an ordinary rectifier voltmeter and represent the total of in-phase, quadrature and harmonics.

Performance

The system performance is much better than indicated by the output voltage curves alone. Since the two-phase motor load acts as an additional discriminator against quadrature, the effect of quadrature superimposed on the input is further reduced. The composite system performance is presented in Fig. 4C and 4D. Good linearity is maintained from input to output.

The average characteristics of the two amplifier types are presented in Table I. These values represent test data from several amplifiers at room temperature.

RELAYS WITH LOW IDLING CURRENT

By D. W. R. McKINLEY

Radio and Electrical Engineering Div., National Research Council, Ottawa, Ontario, Canada

Electronic relays of remote-control devices operate electro-mechanical relays requiring 2 or 3 watts; consume few microamperes when idling. Circuits for c-w audio and pulsed-video control signals are shown

REMOTE radio control equipments require receivers which will run continously with the lowest possible standby battery drain and which will deliver, on demand, 2 or 3 watts of power to a rugged and reliable heavy-current relay to turn on gasoline-electric generators or other equipment.

This article describes two types of transistor relays which can be used at the output end of the receiver to convert the signal waveform into useful d-c power. The control signal may be a carrier modulated at 1,000 cps or it may consist of short pulses from a microwave transmitter.

Basic Circuit

The basic circuit shown in Fig. 1 follows a cascade complementary arrangement. Input transistor, Q_1 should be of silicon to ensure that its idling current (I_{co}) at zero bias will be about 1 or 2 μa. A germanium transistor can be used with -1.5-v base bias, but the I_{co} will usually be higher than that of the silicon types with no bias.

Input transistor, Q_1 acts as a single-ended class-B amplifier for the signal voltage. The 2-μf capacitor from collector to ground filters the half-sine wave output of Q_1 and provides a smooth d-c bias for Q_2.

The effective input impedance is about 8,000 ohms and an input signal of 2 to 3 vrms will develop 8 v d-c across a 15-ohm output load in the emitter circuit of Q_2. Power gain of the relay is 30 to 40 db, depending on the input level. Effi-

ciency of the output transistor varies with the signal level and ranges up to 70 percent.

A 6-v d-c 0.5-amp dpdt relay operates satisfactorily when substituted for R_L. Idling power is $\frac{1}{2}$ mw, compared to 4 w delivered power when the relay trips.

Pulse-Sensitive Circuit

When pulses of 1 or 2-μsec duration are used for signaling, the circuit of Fig. 1 is not satisfactory and a pulse-stretching technique should be used. In Fig. 2, feed back transistor Q_3 and network R_1, R_2, C are added for this purpose.

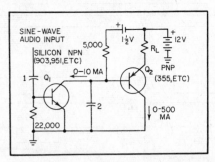

FIG. 1—Relay operates on c-w audio input

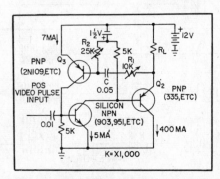

FIG. 2—Pulse-stretching technique adapts circuit for pulsed-video input

The rising edge of an input positive pulse turns on Q_1, Q_2 and Q_3, in that order, and the regeneration around the loop speeds up the action. The transistors remain turned on full until the charge on C is dissipated causing a sharp fall in the transistor currents.

The width of the square wave produced across R_L may be controlled by varying either R_1 or R_2. The duty cycle is adjusted to be as high as possible (up to 70 percent ON time can be obtained) at the lowest pulse repetition frequency used; for higher repetition frequencies the relay will simply count down.

Switching Efficiency

Since the transistors are acting as switches in Fig. 2, efficiencies are higher than in the circuit of Fig. 1. For example, with 12 v on the collector of Q_2 and 50 percent duty cycle, a 10.5-v square-wave is developed across a 15-ohm load, with an average emitter current of 340 ma. Average d-c output power developed across R_L is then 3.6 w.

The efficiency of Q_2 is nearly 90 percent and is independent of signal level since the transistor is either full on or shut off.

A pulse amplitude of 1 v is sufficient to trigger the relay. A 6-v d-c relay may again be used in place of R_L. When zero-bias silicon transistors were used, total idling current was less than 5 μa, but useful d-c power was limited to 0.5 w with the transistors available.

NAVIGATION TRAINER CONTROLS SHIP MODELS

By ALAN L. RICH

Senior Project Engineer, Teletronic's Laboratory, Inc., Westbury, L. I., N. Y.

> Two self-powered radio-controlled ship models, a control center and a water tank, scaled 75 to 1, train naval personnel in the art of ship handling. Ship characteristics such as response to helm, engine telegraph, acceleration and deceleration have time lags similar to full-scale ships. Device also simulates wind and water currents. Control is proportional, utilizing three audio channels modulating a single carrier frequency in the 30 to 42-mc band

INTENDED for use at naval training schools and reserve training centers for demonstration and exercise in the art of ship handling, the ships'-characteristics demonstrator consists of a control unit, two self-powered radio-controlled ship models and a tank.

Only one model is used at a time. The operator steers the ship and actuates the engines by three knobs on the control-unit panel which correspond to the helm and the engine telegraphs. Rudder control is continuous. Engine control is in the following steps: flank speed ahead, full speed ahead, standard speed ahead, 2/3 speed ahead, 1/3 speed ahead, stop, 1/3 speed back, 2/3 speed back and full speed back.

The control system is the proportional type and utilizes pulse-duration modulation of three audio channels modulating a single r-f carrier.

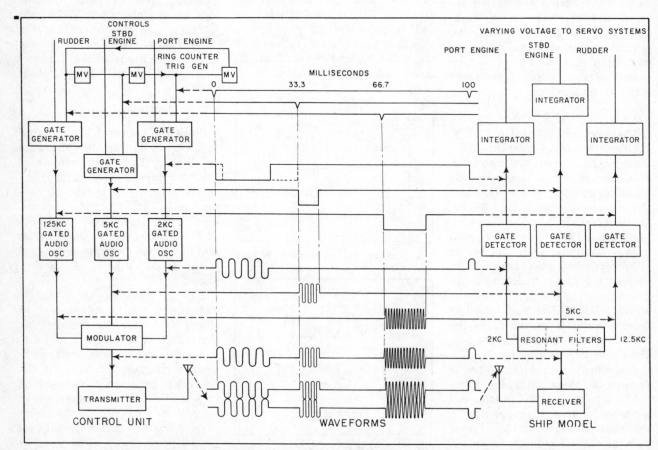

FIG. 1—Information transmission link of naval trainer showing waveforms. Control information is transmitted at 10 cps

The control unit contains the modulator circuitry, transmitter, power supply and the ships' battery charger. The transmitter is low powered, crystal controlled and conventional. It is presently operating on an assigned frequency of 34.54 mc, but may be tuned to any frequency in the 30 to 42-mc band with suitable crystals.

Single-screw ships are represented by an AK cargo vessel model. The other model, a destroyer escort DE 51 class, has twin screws independently controlled and twin rudders. Both models have molded fiberglass hulls four feet long and are scaled approximately 75 to 1 in length with corresponding scaling of displacement, speed, turning radii and advance and transfer.

Time lags similar to those of a full-scale ship are incorporated into all responses; response of rudder to helm, response of ship to rudder, response of engines to engine telegraphs and response of ship to engines.

The ships' circuitry is transistorized, with servomechanisms to position the rudder and actuate the drive motors. The batteries are easily removable, provide 6 hours of continuous operation and can be recharged in less than 2 hours.

The tank has a 20-ft-square water surface corresponding to scaled dimensions of 1,500 ft by 1,500 ft. When filled to a depth of 6 in. it has a scale depth of 37 ft.

Associated with the tank are a wind generator and a water current generator. Either or both may be introduced into the demonstration.

Information Control

The information transmission link includes the generation, transmission and reception of the signals used to control the operation of the ship models. Block diagrams and waveforms of the system are shown in Fig. 1. Control information is transmitted to the models at a basic rate of 10 cps. Timing triggers are generated in a ring-of-three counter, and drive three variable-delay gate generators in sequence. The gate widths are determined on the front panel of the control unit by the port engine, starboard engine and rudder Controls. Each gate may be varied from virtually zero to one-third of the repetition period. The three channels time-share the repetition period.

Each gate actuates an audio oscillator. The oscillator generates an audio pulse of the same duration as, and coincident with, its initiating gate. Audio frequencies generated are 2 kc, 5 kc and 12.5 kc for the port engine, starboard engine, and rudder channels respectively. Outputs of the three oscillators are combined and amplitude modulate the radio transmitter.

In each receiver, Fig. 2, a diode detector drives a high-gain three-stage transistor audio amplifier. With the transmitter located adjacent to the tank there is a 40-db variation in received signal-strength. In lieu of agc, the audio amplifier limits at a low input level so that the output amplitude remains constant over a large range of signal strength. Distortion introduced by clipping results in negligible crosstalk since the audio frequencies of the 2 kc, 5 kc and 12.5 kc—channels bear no low-order harmonic relationship.

Channel Subassemblies

The resonant filter, gate detector, limiting amplifier, integrator and the electronic portion of the servo system of each of the channels comprise a small subassembly. The schematic is shown in Fig. 3.

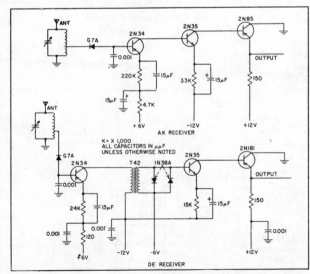

FIG. 2—Receivers in the ship models. In each receiver, a diode detector drives a high-gain three-stage transistor audio amplifier

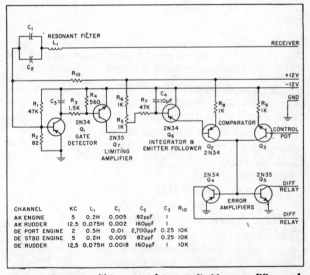

FIG. 3—Resonant filter, gate detector, limiting amplifier and the electronics of the servo-system in the ship model

155

ROLL STABILIZER FOR MISSILE SHIPS

By RICHARD SCHEIB, JR.

Marine Division, Sperry Gyroscope Company, Division of Sperry Rand Corporation,
Garden City, New York

Measuring, computing and servo techniques are combined to control underwater fins. Up to 90 percent of ship's roll is eliminated thus providing stable platform for missile launching. Servos combine transistor and magnetic amplifiers obtaining 15-watts output for fin orders. Automatic overload protection of fins in heavy seas also is provided

AMERICAN SHIPS are again being equipped with stabilizers to attenuate their rolling motion. They have been installed on the SS Mariposa and SS Monterey in Pacific cruise service and on the Navy's USS Compass Island, a missile ship now in service in connection with the US Navy's Polaris missile program.

To the commercial ship owner, stabilization offers improved passenger comfort and safety as well as economies afforded by maintain-

Stabilization Problem

Frequency and amplitude of a ship's rolling motion depend on the frequency spectrum of the waves as well as the ship's natural roll frequency and damping characteristics. The largest amplitude and greatest regularity of roll motion occurs at frequencies near resonance, typically 4 cpm for a 20,000 ton ocean liner.

Reduced but more random amplitudes occur at frequencies below and above the resonant fre-

quency. An ocean liner stabilizer must, therefore, cope with this wide range of roll frequencies, typically from 1 to 10 cpm. A stabilizer theoretically capable of maintaining a 5-degree list reduces rolling to negligible amplitudes in all but most violent seas.

Of the many methods available, ships are stabilized most effectively by actively controlled underwater fins. Typically, two fins only 7 by 14 feet in size will stabilize a 550-foot, 20,000 ton ship. Fins such as these are capable of applying stabilizing moments indefinitely to cope with low-frequency wave disturbances. By making the fin operate rapidly enough, the stabilizer is able to counteract the highest frequencies of wave disturbance to which the ship will respond. When operated within the capacity of its fins, the Gyrofin stabilizer is capable of attenuating the roll amplitude by as much as 90 percent, reducing a roll of ±20 deg to less than ±2 deg.

Two 100-sq ft fins, located about

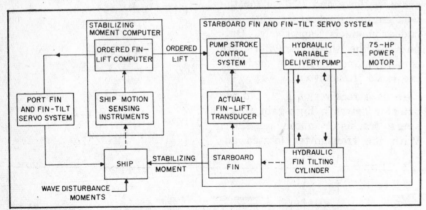

FIG. 1—Complete ship stabilizer system consists of port and starboard fin-tilt servos and stabilizing-moment compute. Hydraulic system provides huge torque

ing speed and schedule in rough weather. To the Navy, ship stabilization also offers improved missile launching conditions and navigation accuracy under adverse sea conditions.

Computer studies facilitated the development of the new stabilizer, called the Gyrofin Ship Stabilizer. The dominant role played by the application of electronic and servo techniques to the stabilizer control system will be discussed in this article.

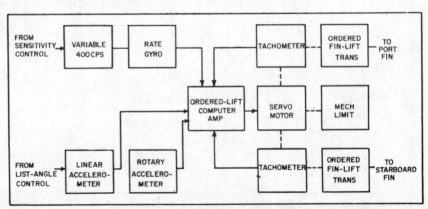

FIG. 2—Stabilizing-moment computer is located in control console on ship's bridge

the middle of the ship, 20 feet below the water line, provide the stabilization. When not in use they are folded back into recesses in the hull. The fins cooperate, one by tilting downward, the other upward, to apply a stabilizing moment to the ship. They are capable of lift forces up to 70 tons each, resulting in a maximum stabilizing moment of 6,000 ton-ft. The fins operate rapidly at speeds up to 30 degrees per second and are capable of accelerating to that speed in 0.3 second. Thus a fin may travel between the 25 degree extremes of its travel in less than 2 seconds.

Hydraulics

To impart these accelerations and speeds to a 20-ton fin moving in water at 20 knots, hydraulic servo controls are used to apply torques up to 100,000 lb-ft to the fin shaft. The hydraulic piston and cylinder that operates the fin shaft through a simple crank arrangement, is controlled by a variable-delivery, reversible-flow pump that can transfer oil from one side of the cylinder to the other at rates up to 100 gpm. This pump, with a peak output of 100 hp, incorporates a hydraulic pump stroke amplifier which requires a mechanical input control power level of only 2 watts.

Since this heavy fin actuating machinery can be controlled by only a few watts of power, small electro-magnetic transducers are used to obtain signals indicating the ship's motion as well as fin action, low-level transistor and magnetic amplifiers are used in computation and special electronic circuits were

developed to assure the efficiency and safety of stabilizer operation.

A 400-cps power frequency is used exclusively in the control system to take advantage of the high signal gradient characteristic of 400-cps transducers and the availability of standard 400-cps magnetic amplifier and servo components.

Operating Principles

Basic principles of stabilizer operation are illustrated in Fig. 1. The stabilization system has two major divisions, the stabilizing moment computer which detects the roll motion and orders the proper stabilizing moment and two identical fin tilt servo systems which cause the fins to apply the ordered stabilizing moment to the ship.

The stabilizing moment computer deduces the disturbing moment being applied to the ship by wave action from measurements of the ship's response to these disturbances. Sensing instruments detect and measure several functions of the ship's roll motion from which the stabilizing moment is computed. The fin-lift computer orders both fin servos to apply equal fin-lift forces to the ship which cooperate to produce a stabilizing moment equal and opposite to the wave-disturbing moment.

The desirable equality of fin-lift forces is assured first by the equality of fin-lift orders to each fin servo and second by the unique lift-control feature of the fin servo system. As shown in Fig. 1 the hydraulic servo system responds to the difference between the ordered and actual fin-lift signals; the

latter derived from a strain gage which measures the actual lift force exerted by the fin.

The stabilizing-moment computer system, located in the control console on the ship's bridge, is illustrated in Fig. 2. The ship's roll motion is measured by a linear acceletometer that senses the ship's roll angle, a conventional rate gyro that senses the roll velocity and a rotary accelerometer that senses the roll acceleration. Each instrument incorporates a small electromagnetic pickoff that supplies a 400-cps signal proportional to the function being measured.

The linear accelerometer can be tilted physically to compensate for any steady list of the ship so that stabilizer capacity will not be wasted in combating small list angles. Excitation of the roll-velocity transducer can be varied to adjust the stabilizer for optimum performance as sea conditions vary.

Computer Servo

The sensing instrument signals are combined by a servo system that continuously computes fin lift orders proportional to the combination of the three roll motion functions. As shown in Fig. 2, the computer servo system consists of a resolver measuring the servo position and two transmitters that send ordered lift signals to the fin servo units. Mechanical limits on the computer servo prevent the ordering of a fin lift exceeding 70 tons.

The servo amplifier combines the 400-cps signals linearly to control the servo motor. In response to signals from the sensing instruments

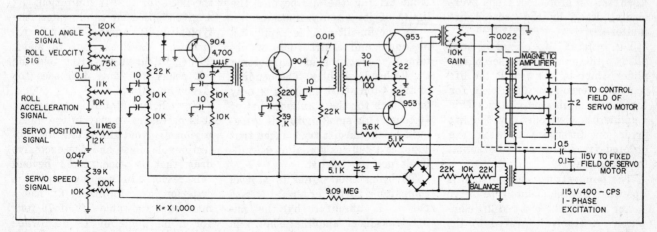

FIG. 3—Low input impedance of first transistor stage assures accurate signal mixing at input to ordered-lift computer servo amplifier

the servo nulls when the feedback signal cancels the sensing instrument signals. The tachometer signal attenuates the servo response to high frequencies from the ship's vibrations.

Computer Amplifier

The computer servo amplifier shown in Fig. 3 combines transistor and magnetic amplifiers. The five input signals are attenuated, phase shifted and mixed by a network of resistors and capacitors. The low input impedance of the grounded-base input transistor is used to assure accurate signal mixing. The three-stage transistor amplifier section features local degeneration in each stage, transformer coupling between stages, tuning of interstage transformers to the 400-cps signal frequency and adjustable gain control in the negative feedback loop around two stages.

The rectifier in the input circuit limits input signals to avoid exceeding the peak rating of the first transistor. The balance control cancels any in-phase residual null signals from the sensing instruments.

The transistor amplifier section supplies about 0.3 watt to drive a half-wave magnetic amplifier that can supply 15 watts to the control field of the two-phase servo motor. The servo motor fixed-field excitation is obtained from the supply line through a phase-shift network.

Two identical fin-tilt servo systems receive equal ordered fin-lift signals from the computer system. A schematic of one complete fin-tilt servo system is shown in Fig. 4. Each fin is protected from overloading itself by the ordered lift limiter.

The ordered lift signal is supplied to the pump-stroke servo amplifier which is identical to the lift computer servo amplifier except for the input mixing circuitry. This amplifier is controlled by the lift error or difference between the ordered-lift signal and the actual fin lift signal.

The pump-stroke servo system is operated by the same type servo motor used in the ordered-lift computer. A resolver, geared to the servo motor, feeds a signal propor-

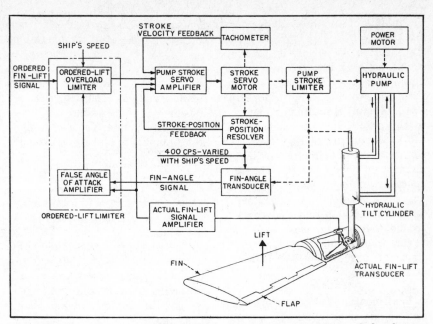

FIG. 4—Strain gage in shaft supporting fin provides actual fin-lift signal that is compared with ordered fin lift to determine lift error to drive pump-stroke servo

tional to pump stroke (fin speed) back to the stroke servo amplifier. Thus, for small values of lift error, pump stroke is proportional to lift error up to 15 tons.

Hydraulic Stability

To maintain hydraulic system stability the excitation of the stroke-position resolver must be varied with ship's speed to compensate for the variation of the actual fin lift. The tachometer geared to the servo motor stabilizes the stroke servo and the overall hydraulic servo system.

An auxiliary mechanical coupling to the fin-tilt cylinder automatically returns the pump stroke to zero as the mechanical angular limits of fin motion are approached. This prevents the pump from driving the fin beyond the limits of its angular motion.

As the fin-tilt cylinder applies a fin angle and a lift force is obtained, the stub shaft supporting the fin bends slightly. The bending of this 4-ton shaft amounts to only 0.012 inch for the maximum lift of 70 tons. This small motion is measured by the fin-lift transducer that supplies a 400-cps signal of 0.4 millivolt per ton of lift.

To bring this signal to a level comparable to the ordered-lift signal it is amplified by the fin-lift transducer amplifier shown in Fig. 5. Grounded-emitter transistor

stages are used to obtain high gain. Gain stability is assured by local degenerative feedback and feedback around the entire amplifier.

The push-pull power stage supplies an amplified lift signal to the stroke servo amplifier and the ordered-lift limiter circuit. Amplifier gain is adjusted by varying the amount of feedback around the three stages.

To achieve the most efficient operation from a stabilizer of given size and to assure safety of the ship and its stabilization equipment under the most adverse sea conditions the ordered-lift limiter system is employed.

Fin Lift Limitation

The characteristics of the fin must be considered to understand the need for fin-lift limitation. At a given speed in calm water, the typical fin-lift against fin-angle relationship is linear out to about 20 deg of fin angle, after that the lift peaks and then decreases for larger fin angles.

For efficient stabilizer operation, it is necessary to stay in the linear region to avoid the cavitation, objectionable noise and the excessive drag that is encountered beyond this region. In keeping with this criterion, the stabilizing fin has a maximum allowable lift of 70 tons for a fin angle of 18 deg when moving at 20 knots in calm water.

Figure 6 shows the maximum allowable lift as a function of speed. One function of the lift limiter is to keep lift orders within the appropriate boundaries of Fig. 6.

The fins rarely operate in undisturbed water. The angle of water flow with respect to the fin differs from the angle of the fin with respect to the ship by the so-called false angle of attack. False angles arise from the penetration of wave action to the fin depth and from various ship motions including pitching, heaving (vertical translatory motion) and residual ship rolling. In the extreme, false angles of 15 to 20 deg are possible. The fins are permitted an angular travel of ±25 deg so that maximum permissible lift forces may be obtained in spite of false angles.

ship's speed so the two signals cancel each other when no false angle is present. Output of the false-angle amplifier is proportional to the false angle while its phase indicates the direction of the false angle.

The false angle of attack amplifier shown in Fig. 7 is a single-stage transistor amplifier. The actual fin-lift and fin-angle signals are mixed in the primary of the input transformer. The push-pull transistor stage with transformer output supplies the signal proportional to the false angle of attack.

The ordered-lift signal input to the limiter section of this circuit is a 400-cps sine wave. But for the attenuation of the 10,000-ohm series resistor and the shunting effect of the biased back-to-back diodes, the output or limited-lift

biasing voltage is full-wave rectified to avoid lift signal waveform distortion.

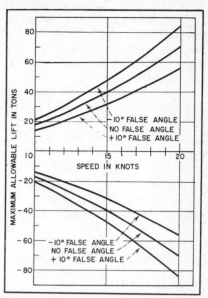

FIG. 6—Curves show maximum allowable lift as a function of speed. Lift orders must remain within boundaries shown

The common connection of the 160-ohm resistors in the limiter circuit is connected to ground through the output transformer of

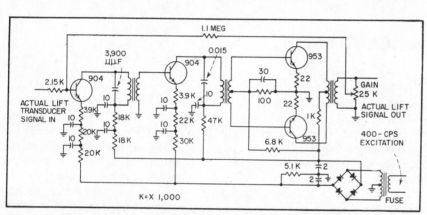

FIG. 5—Fin-lift transducer amplifier supplies actual-lift signal to stroke servo amplifier and also to ordered-lift limiter circuit which assures safety of the ship

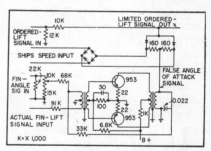

FIG. 7—Ordered-lift limiter circuit determines false angle of attack of fins and uses this to modify ordered-lift signals

At 20 knots and with a false angle of +10 deg, the linear region limits are +56 tons and −84 tons.

The second function of the lift-limiter system is to recognize the magnitude and direction of false angles and to limit the lift orders asymmetrically in accordance with the curves of Fig. 6.

False-Angle Determination

The false angle of attack of a fin at any given instant may be found with reasonable accuracy from the algebraic difference between the actual fin lift and the lift expected from the fin based on its angle with respect to the ship. The ordered-lift limiter system shown in Fig. 4 consists of a false-angle amplifier and the overload limiter proper.

The false-angle amplifier receives the fin-lift signal and a fin-angle signal. The latter is corrected for

signal would follow the input signal. The variation of the lift limit with ship's speed is accomplished by the back-to-back diodes which shunt the ordered-lift signal. Their

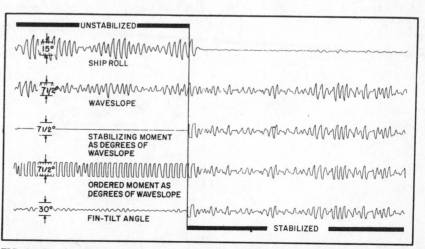

FIG. 8—Curves obtained from analog computer investigation under accurately simulated sea conditions show system response to ship's roll when unstabilized and stabilized

the false-angle amplifier. The false-angle signal now biases the entire limiter circuit by an amount proportional to the magnitude of the false-angle signal. This bias is asymmetrical as required by fin characteristics and is in a direction which depends on the relative phases of the false-angle signal and ordered-lift signal. Thus this circuit achieves the ordered-lift limitation as a function of ship speed, modified asymmetrically by the magnitude and direction of the false angle of attack experienced by the fin.

Fin lift is prevented from exceeding limits of efficient operation and the fin is protected against damage by sea action for all operating speeds of the ship. A separate limiter system is required for each fin since each fin generally experiences different false angles of attack.

Performance

Typical stabilizer performance is illustrated in Fig. 8. These records were obtained from an analog computer investigation conducted under accurately simulated sea conditions. In the unstabilized portion of the record, wave disturbances are causing near resonant rolling with amplitudes up to 13 deg. Ordered stabilizing moments, expressed as equivalent steady list of the ship, are limited to 4 deg. The stabilizing moment is constantly zero although there is appreciable fin motion as required by the lift control feature to maintain zero actual lift in the presence of false angles of attack caused by the ship's roll.

The average characteristics of the disturbing wave slopes are continued through the stabilized portion of the record. The ship's roll motion averages less than 1 deg. The ordered and actual disturbing wave slopes are within stabilizer capacity. The record shows the effect of an occasional disturbing wave slope which exceeds the stabilizing capacity.

CONTROLS FOR AIRBORNE SEARCHLIGHT

OPTICAL methods are ultimately needed in nighttime operations of Navy patrol aircraft that initially make contact with targets by radar. A new carbon-arc searchlight has recently been developed by American Bosch Arma Corp. that employs transistors.

Typical of the newer approach is the positive carbon drive mechanism for which a simplified schematic diagram is shown. This is essentially a focus control system in which a photodiode feeds a single transistor stage to operate a sensitive relay. The relay contacts normally shunt a field-control resistor in the drive motor.

Mechanically, the system comprises a small rhodium-plated mirror set at 45 degrees to the carbon axis so the image of the crater is reflected through the center hole of the main reflector to the photodiode.

▶ Speed Change—Gearing between motor and positive carbon normally moves the carbon forward at a fixed rate. When the crater is at the focal point, the small reflector is set such that the reflected beam strikes the photodiode at a point just off its sensitive area. As the carbon burns back from the focal point, the reflected beam moves onto the sensitive area of the photodiode, changing its impedance and unbalancing the bridge circuit.

A small signal applied to the collector of the transistor is amplified to a level sufficient to operate the relay. Contacts of the relay remove the shunt across the field resistor, decreasing field current and increasing motor speed. The carbon then moves forward at greater than the burning rate and the projected beam is thus removed from the photodiode.

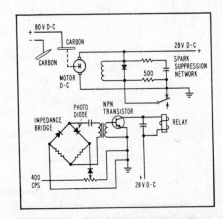

Schematic of the positive carbon drive

SENSITIVE CONTROL RELAY

By HARRY A. GILL

Development Laboratories, Fisher Scientific Co., Pittsburgh, Pa.

Control device with almost indefinite life operated from socket power is mounted on printed-circuit board. Semiconductor diodes and transistors permit miniaturization with size determined only by a-c output fixtures, transformer and relay. Controlled circuits can be normally on or off

ADVANTAGE of inherent long-life characteristics of transistors is exploited in the design of a new transistorized relay. Two matched 2N109 transistors are used in balanced input and one 2N44 in the relay coil circuit.

Balanced input compensates for ambient temperature changes; it also permits greater versatility in the control circuits. The relay can be energized by a normally open or normally closed contact, by illumination on a cadmium-sulfide cell or photocell and by a control current of 12 microamperes more or less. With the current input feature, a photovoltaic cell, for example, can be used to control the opening and closing of the relay.

The schematic circuit diagram is shown in Fig. 1. With S_1 in normally open position and no input signal, the bias currents of Q_1 and Q_2 are equal. The voltage drops across R_9 and R_{10} are also equal and opposite; therefore, the resultant voltage at the base of Q_3 is zero.

If J_1 and J_2 are shorted with a resistance up to 400,000 ohms, the increased bias on Q_1 upsets the balance of voltage on R_9 and R_{10}, producing a negative voltage at the base of Q_3. Sufficient collector current then flows through Q_3 to energize relay K_1.

With S_1 in the normally closed position and J_1, J_2 open, the collector currents through Q_1 and Q_2 are unbalanced to the point where Q_3 conducts sufficiently to energize K_1. Shorting J_1 and J_2 balances the collector currents again and K_1 opens.

If a photocell of the cadmium-sulfide type is connected across J_1 and J_2, changes in illumination will cause the cell resistance to vary, thus producing the same effect as opening and closing the circuit across J_1 and J_2.

Reverse Operation

It is sometimes more convenient to actuate the relay by opening a contact rather than closing it. Using the normally-closed control option permits this to be accomplished and yet if the relay fails, it will fail safe, shutting off the load.

If it is desired to operate the relay from a source of low current, such as a photovoltaic cell, the terminals of the cell are connected to J_2 and J_3 with the positive terminal on J_3; S_1 must be on normally open. The action is now push-pull; the Q_1 collector current is increased while Q_2 collector is decreased. The resulting voltage across R_9 and R_{10} causes the relay to energize.

A slight delay in opening and closing the relay is obtained by closing S_2. The delay time varies depending on the resistance of the input circuit but it is intended primarily only to prevent relay chatter when a slowly moving, light contact is being made or broken.

Sensitivity control R_3 provides a convenient means of limiting the input current if, for instance, it is desired to reduce the output of a photovoltaic cell because of high ambient light.

Equipment side of printed chassis

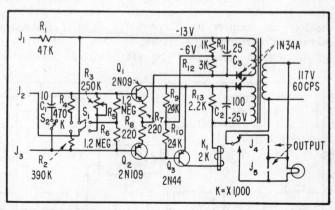

FIG. 1—Circuit diagram of the three-transistor control device

GUIDANCE SYSTEM FOR FEED CARTS

By SHELDON KNIGHT

K & R Development Co., Hughson, California

Semiautomatic equipment for large chicken farm uses carts following buried wire for distributing feed and collecting eggs along rows of coops. Transistor amplifiers feed position data to steering motors from magnetic pickup coils at front of cart. A third amplifier automatically stops cart if it goes off path

Pickup coils at front wheels guide feeding carts along rows of coops following buried wire layout

STEERING of a small, slowly moving cart over a desired path is achieved by the method to be described.

The path is marked out by a buried wire carrying approximately one ampere of 60-cps alternating current. The guidance unit senses the position of the cart relative to the wire by the magnetic field surrounding the wire. If the cart is off course, an error signal is sent to the power steering unit to bring about the necessary correction.

Use of transistors allows operation directly from the 12-v storage battery which supplies motive power for the vehicle, and also eliminates the maintenance problems associated with vibrator type high-voltage supplies. Transistors also allow the unit to give reliable operation in spite of the vibration experienced on the moving cart.

The advantages of this type of guidance over a rail system are obvious. The path is simple and inexpensive to lay out or change. Also, switching can be done electrically from a fixed point by energizing the proper branch at a junction.

Basic Operation

Two small pickup coils are suspended about six inches over the ground in front of the vehicle, and are placed about one foot apart. As shown in the photograph of the

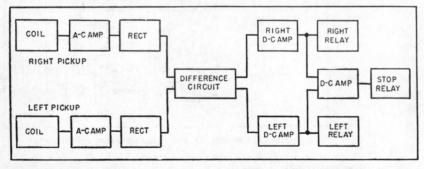

FIG. 1—Block arrangement of circuits used in cart control system

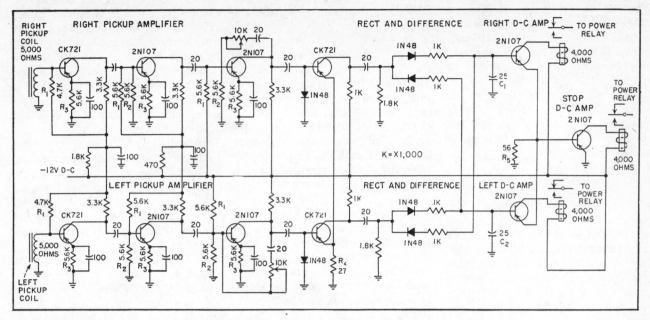

FIG. 2—Four-stage transistor amplifiers boost signals picked up by coils to control steering motors for cart

cart, the two coils are mechanically connected to the front wheels in such a way that they move from side to side as the wheels turn with steering. When the vehicle is on course, the buried wire is midway between the two coils, and so the voltages induced in the coils by the magnetic field surrounding the wire will be equal. If the vehicle drifts off course, one coil will move slightly closer to the wire, producing a difference in the induced voltages.

Each coil feeds a four-stage transistor a-c amplifier as shown in block form in Fig. 1. The output of each amplifier is rectified and the two d-c signals combined to produce a difference output to the d-c amplifiers. The magnitudes of the outputs are proportional to the difference between the rectified outputs of · amplifiers. But the polarities are opposite.

As the vehicle moves off course to the left, the rectified output from right-pickup amplifier increases and the output from the left-pickup amplifier decreases, causing a negative d-c voltage to be developed in the right side final output and a positive d-c voltage to be developed in the left channel.

When the vehicle is on course the output of both amplifiers will be equal and the final outputs will be zero.

Each final output drives a single

stage d-c amplifier which controls a sensitive relay. The transistors in these d-c amplifiers are connected in the grounded-emitter configuration, and collector current is low in the absence of an input signal to the base. When the base is driven negative the collector current increases and closes the relay.

The relays actuate power-control relays which operate the electric steering motor to make the proper correction.

Since the pickup coils are moved laterally as the front wheels turn, they will be in a position of balance over the buried wire after the wheels have turned a few degrees towards the course. Thus, the steering motor will be stopped leaving a small correction to intercept the course.

As the vehicle approaches the course again, the coils which are turned with the wheels, give an error indication in the opposite direction. This starts the steering motor to straighten the wheels, moving the coils back to their normal position, restoring balance and completing the correction cycle.

This is far from the ideal error-actuated servo system. However, it has the advantage of simplicity since no error-time derivative signal is required and since the follow up is purely mechanical. At the one to two miles per hour speed at which this cart operates, the system has proved entirely satisfactory.

A third d-c amplifier and relay is connected with the steering control d-c amplifiers in such a way

All control circuit components including relays are mounted on polystyrene board

163

that if the collector current of either steering amplifier exceeds a preset amount it will close and, through a power control relay, cut off the main motive power to the vehicle. This is a safety measure which will automatically stop the cart should it get more than a maximum distance off course due to some malfunction of the guidance system.

Electronic Details

The complete diagram of the electronic unit is shown in Fig. 2. The four-stage a-c amplifiers use conventional grounded-emitter configurations. A high degree of temperature stability in each stage is provided by stabilizing networks R_1 and R_2 together with emitter resistors R_3.

A slight increase in stability is provided by the method of obtaining gain control. A potentiometer does this by providing variable negative-current feedback around the third stage.

There is a considerable excess of gain provided by the four stages, so during normal operation considerable negative feedback is used. Extending the feedback loop to include more stages is not practical because of low-frequency stability problems which arise from the coupling and decoupling capacitors and the low impedances associated with the transistors.

Because the desired final output is the difference between the two amplifier outputs, and also because there is some loss in the difference circuit, the fourth stage must operate at a fairly high power level. To avoid excessive dissipation in this stage, the bias current is developed by rectifying the signal feeding the stage with the diode in the base input circuit. This maintains the bias and hence the collector current at the lowest pos-

sible level consistent with the signal level.

At first glance it would appear that the low forward resistance of the base-emitter section of the fourth-stage transistor would serve this purpose; however this proved to be too high to allow adequate charging of the input capacitor and so the diode was added. This variable bias circuit also eliminates the need for the temperature stabilizing resistors used in the first three stages.

A small amount of cross coupling is provided between channels by the common-emitter resistor R_4 in the fourth stage of the two amplifiers. The input coils are connected to the amplifiers so that the amplifiers are excited in phase. Thus the signal developed across R_4 due to either amplifier is of such polarity as to reduce the output of the other amplifier. If the output of either amplifier increases, the output of the other amplifier will decrease even if the input to the other amplifier has not changed. This cross coupling feature further increases the differential in the amplifier outputs. The rectifier and difference circuit is a conventional diode system.

The emitters of both d-c amplifiers are grounded through the common resistor R_5. These emitters are also connected to the base of the stop d-c amplifier. Thus a part of the emitter current of either d-c amplifier must flow through the base input circuit of the stop amplifier. If this base current exceeds a certain maximum, the collector current of the stop amplifier will increase sufficiently to close the relay.

Possible Modifications

This unit was designed and built for one specific application. In this capacity it worked well, so no further development effort has been

expended on it. However, usage has pointed out some possible improvements that could be incorporated in subsequent designs.

Bias stabilizing networks, R_1 and R_2, produce more stability than is really necessary, and so introduce an unnecessary loss of gain. These resistors could possibly be increased to as high as 27,000 ohms, depending upon the range of temperature variation expected, and the resulting increase in gain would allow the unit to operate satisfactorily with only three instead of four stages of a-c amplification.

There is a slight lag in the response of the unit due to the time required for charging capacitors C_1 and C_2. This lag places a low limit, in the neighborhood of three miles per hour, on the forward speed of the vehicle. With 60-cps current in the buried wire, the capacitance cannot be reduced below 25 μf and still maintain adequate filtering. For higher forward speeds the wire should be fed from a higher frequency source so that the capacitor values and the resulting lag could be reduced proportionately.

The unit was built on a 4 by 8 inch sheet of $\frac{1}{4}$-inch polystyrene. Holes were drilled in the sheet to allow wire leads to the various components to pass through from the component side of the board to the wiring side.

Components were temporarily mounted with plastic cement. The wiring itself serves to permanently mount the components. This type of construction provides a compact unit in which all parts and wiring are readily available for rapid servicing.

The unit was mounted in a small aluminum box for protection. Gain controls are mounted on the aluminum box, and so do not show in the photograph.

DEMODULATOR-LIMITER FOR REMOTE CONTROL

By N. L. JOHANSON
Boeing Airplane Company, Seattle, Washington

Transistor circuit for carrier-based control systems limits while modulating or demodulating. Circuit operates at high signal-conversion efficiency, produces hard limit and has excellent linearity

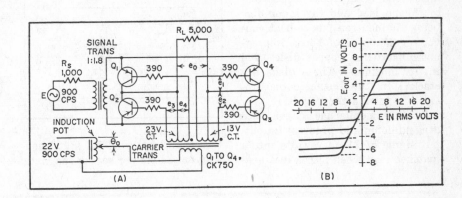

FIG. 1—Autopilot demodulator-limiter has nonsymmetrical and adjustable limit levels (A). Commanded limit levels based on shaft angle θ_0 of induction potentiometer are shown as dashed lines (B)

MANY CONTROL SYSTEMS require hard limiting of control signals to predetermined levels. It is sometimes also desirable to have independent control of the plus and minus, or phased, signal limit levels, such that either or both can be fixed or varied according to some desired function.

In systems which employ medium-level phase-sensitive modulators or demodulators, this flexibility of limit control can be achieved without additional components by using the modulator or demodulator circuit shown in Fig. 1. Where the signal source is isolated from the load, this circuit provides full-wave operation at high-signal conversion efficiency without requiring a center-tapped signal transformer.

Circuit

Assuming a demodulator application for the circuit of Fig. 1A, operation in the linear range is as follows: during one-half cycle of the carrier frequency the bases of Q_1 and Q_3 are negative with respect to their collectors, offering a low impedance to signal current flow through their emitter-collector paths. During this same half-cycle, the bases of Q_2 and Q_4 are positive with respect to their collectors, offering a high impedance to signal current flow through their emitter-collector paths.

During the next half cycle, Q_1 and Q_3 act as open switches and Q_2 and Q_4 as closed switches, resulting in full-wave rectification of the signal input.

As the signal level is increased the mode of operation changes at a predetermined level, causing the output to limit.

During one-half cycle Q_2 presents a high impedance to current flow through its collector-emitter path only as long as its base is positive with respect to both its emitter and collector.

When output voltage e_0 exceeds switching voltage e_1 current flows through the base-emitter path of Q_4 creating a low impedance path for signal current through its collector and emitter. This shunt across the load automatically regulates the voltage at the input terminals of the circuit by virtue of its decreasing impedance as a function of increasing signal. During the next half cycle switching voltage e_2 determines the output limit level.

If the signal phase is reversed, switching voltages e_3 and e_4 replace e_1 and e_2 in determining the limit level on alternate half-cycles of signal voltage. Thus, nonsymmetrical, symmetrical, fixed or variable limits can be attained by the proper choice or variation of the two center-tapped switching voltages. To obtain a hard limit the signal voltage and the switching voltage must have the same wave-form.

Application

Figure 1B shows the limiting characteristics for the circuit of Fig. 1A which was designed for an autopilot system. This application required nonsymmetrical and variable limit levels for the acceleration commands to an elevator servo. Since the two limit levels were to be varied by the same function, the circuit requires only one carrier transformer. In this case, the transformer primary voltage is varied by a servo-driven potentiometer.

Acknowledgement is due Kenneth D. Johanson for assistance in preparing this article and Earling Johnson for constructing and testing the unit.

NUCLEAR REACTOR CONTROL

A SYSTEM using a neutron-sensitive thermopile, a magnetic modulator, an excitation source for the modulator and a transistor amplifier is used to measure neutron flux density in a pile control system. The thermopile is made of individual thermocouple disks consisting of a sequence of three $\frac{5}{8}$-in. diameter disks, the outer disks being bismuth and lead and the inner disk being the neutron flux-to-heat converting material of boron impregnated powdered copper disk 0.006 in. in thickness. These disks are stacked in 100 thermocouple sections to form a thermopile.

The thermopile has a resistance of $\frac{3}{4}$ milliohm and produces an output signal of 21 microwatts into a matched load when placed in a neu-

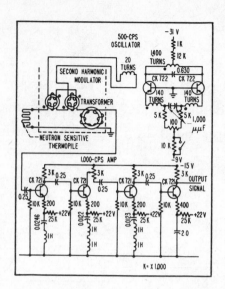

tron flux density of 3.8 x 10¹³ neutron per sq cm per sec. A magnetic modulator designed to match the thermopile uses Mo Permalloy cores.

The transistor amplifier uses common emitter degenerative stages. Frequency selective L-C tuned circuits between emitter and ground are tuned to a frequency near the second harmonic. The fourth stage of the amplifier is untuned. The output of the amplifier is adequate to excite electronic drivers for control-rod actuation. The system using transistors increases reliability of reactor control equipment.

This material was abstracted from "Solid State Neutron Flux Measuring System", *Elec Eng,* p 678,, Aug, 1957.

Chapter 12

INDUSTRIAL MEASURING INSTRUMENTS

INDICATOR MEASURES JET-AIRCRAFT EXHAUST

By GEORGE H. COLE

Assistant Project Engineer, Aeronautical Division, Minneapolis-Honeywell
Regulator Co., Minneapolis, Minnesota

Thermocouple bridge at jet engine exhaust feeds temperature indicating system, including chopper, reference source and a-c amplifier, mounted directly behind 2-inch indicator dial on pilot's instrument panel. Weight and space requirements are about one-fifth that of vacuum-tube or magnetic-amplifier equipment

ACCURATE MEASUREMENT of jet engine exhaust gas temperature permits protection against excessive engine temperatures without sacrificing performance because of conservative operation.

Thermocouples are capable of withstanding high temperatures and corrosive atmospheres and permit temperature averaging around the exhaust area. However, their use in precision systems produces problems in amplifying low level d-c voltage.

The most feasible approach has been to compare the thermocouple signal to a d-c reference voltage and to convert the difference signal to a-c for amplification.

When this technique has been used in conjunction with a null-rebalance servo loop, accuracies in the order of ±0.5 percent for temperature ranges of 1,000 C. have been achieved.

A typical electron-tube or magnetic-amplifier system, miniaturized and ruggedized for airborn use, consists of a 2-in. diameter, 4-in. long panel indicator and a remotely located amplifier about 4 by 4 by 4 inches in size. The volume of this two-unit system is about 75 cubic inches and total weight is on the order of 5 pounds.

Transistorized temperature indicator occupies space of 17 cu in. behind instrument panel as compared to 75 cu in. for vacuum tube unit. Power consumption is about 7 watts and weight is less than a pound

The equipment described here uses semiconductor devices for reference voltage supply, thermocouple bridge, modulator and amplifier. The entire circuit is mounted in a 2-in. diameter case, 5½-in. long.

The repeatability and stability of the reference voltage directly affects system accuracy. For an accuracy of ±0.5 percent, the reference voltage requires a ±0.1-percent stability.

Use of subminiature voltage-reference tubes was considered, but study and tests of the Zener voltage characteristic of silicon diodes showed that the required stability could be obtained using these components.

The relatively low ranges of Zener voltages available with diodes also assured that little power would be wasted in dividing the reference voltage to thermocouple levels.

Diode Circuit

A silicon-diode circuit and compensation network were designed requiring only a single diode. A simplified schematic is shown in Fig. 1.

Only moderate filtering of the rectified supply is necessary because the regulation properties of the diode effectively remove some of the ripple. In Fig. 1, resistors R_1, R_2 and R_3 are used for line-voltage compensation, temperature sensitive resistor R_c for temperature compensation and R_p for calibration purposes.

The circuit actually represents an unbalanced bridge, part of the output of which is used to energize the temperature bridge. The diode is operated with the anode nega-

tive and no current is passed until the voltage exceeds the Zener voltage of about 8 volts. At this point, breakdown occurs and the current increases very rapidly with voltage. The operating point of the diode is set at about 1 ma by current limiting resistor R_1.

In the other leg of the bridge, R_3 is chosen so that the voltage-current slope matches that of the diode. The difference between the two voltages then remains practically constant even though the supply voltage varies as much as ±15 percent. The diode, however, is temperature sensitive which would cause thermocouple-bridge voltage to increase directly with temperature. As the increase is practically linear it is relatively easy to compensate with a copper resistor, R_c, in series with the output to the thermocouple bridge. Adjustable resistor R_p adjusts bridge current to compensate for variations in Zener voltage of the diodes and also resistor tolerances.

The entire circuit can be contained within approximately one cubic inch and offers indefinite life. A unit which has been operated continuously for a period in excess of 10,000 hours has undergone less than 0.03-percent change in output. The reduction in power requirements compared to a reference-tube circuit is about 15 to 1.

Thermocouple Bridge

The thermocouple bridge, shown in Fig. 2, supplies voltage to the rebalance potentiometer for nulling incoming thermocouple signals, and

also furnishes thermocouple cold-junction compensation.

This bridge utilizes large resistors with low temperature coefficients to load the reference voltage and set up constant currents in two bridge legs. One current path is through a temperature-sensitive resistor so that the product of the constant current and the temperature-sensitive resistor provides a millivoltage equal and opposite to that of the chromel-alumel cold junction.

The two voltages cancel each other throughout the ambient temperature range of −85 to +78C within ±2C. error. The compensating elements are mounted together to minimize temperature differentials.

The second bridge leg contains resistors which establish the rebalance voltages at the ends of the rebalance potentiometer. In this case the measured temperature range is 0 to 1,000 C and the potentiometer voltage correspondingly extends from 0 to about 45 millivolts. The resistance of the rebalance potentiometer is such that it does not appreciably load the bridge and at the same time will not insert enough series resistance in the null circuit to cause loss in sensitivity. Potentiometer R_p is used to establish the desired level of bridge current. Total volume is less than one cubic inch excluding the rebalance potentiometer which is contained in the motor section.

Transistor Modulator

The transistor-type modulator, in Fig. 2, converts d-c error signals received from the thermocouple bridge to essentially square-wave a-c before amplification. It will handle signal magnitudes of 40 microvolts without mechanical contacts.

Two transistors act as single-pole switches which close and open on alternate half-cycles of line frequency. As each closes, it provides a low-impedance path for the signal current to flow through one-half of the primary of the output transformer. Opening and closing of the transistor switches is accomplished by the a-c voltages e_1 and e_2 which alternately bias the transistor collector and emitter positively and

FIG. 1—Reference-voltage circuit using silicon diode has less than 0.03-percent change in output in more than 10,000 operating hours

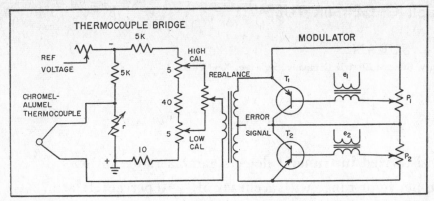

FIG. 2—Thermocouple bridge and modulator unit provide error voltage to transistor servo-amplifier

negatively with respect to the base, thus creating corresponding low and high impedance paths. Potentiometers P_1 and P_2 are high enough in resistance to minimize their shunting effect during the off cycle of each switch.

The circuit actually forms two bridges with e_1 and e_2 as excitation voltages, transistor junctions as two legs, potentiometer halves as the other two legs and each bridge output impressed across one-half of the output transformer primary. So that none of the excitation voltage appears in the output, it is essential that each potentiometer leg ratio be the same as the transistor junction resistance ratio.

Because the ratio of emitter and collector junction resistances in the forward and reverse directions changes as temperature varies, it is necessary to stabilize the bridge balance over a wide temperature range. This was done by a combination of design fixes, some of them based on empirical data. As a result, the circuit has been developed to where less than 120 microvolts are introduced to the modulator output by the excitation voltage through a temperature range of −85 to +72 C. Once adjusted, the circuit has shown no need for readjustment.

Transistor Amplifier

The transistor amplifier was designed specifically for low-level amplification and temperature stabilization. The circuit uses transistors with easily obtainable parameters.

Three low-power transistor stages in cascade are used for signal amplification and an H-4 power-type transistor is used for the servo-motor drive. The three low-power stages are used common emitter and operate class A for small signals. Saturation does not occur until almost 50 millivolts input signal (about a 1,000 C. temperature signal) is reached.

Transformer coupling is used for the inputs to the first and second stages. The transformer input to the first stage provides impedance matching and also isolates the thermocouple circuit from ground. Because of the latter, accidental grounds in the engine thermocouples do not offset system calibration. The second stage is direct coupled to the third stage which is transformer coupled to the output stage.

Power for the first three stages are obtained from a common rectified supply. The output stage has its own power supply to avoid coupling to earlier stages, and also because its larger current requirement would cause undue power loss in the filter section required by the first three stages. Filtering for the output stage is not strong as the 800-cycle ripple is noneffective.

The output stage acts as a discriminator to drive the servo motor in the correct direction to balance out error signals. The H-4 power transistor is operated with grounded collector. This connection simplifies the mounting and improves heat dissipation from the transistor since the collector is grounded to the transistor case.

Indicator Section

The H-4 transistor drives a miniaturized, two-phase servo motor designed for transistor amplifier operation by a low-impedance phase. The line-phase to amplifier-phase power ratio is approximately 8-to-1 to improve torque over balanced power operation.

A 400-to-1 gear train drives the rebalance potentiometer, rate feedback potentiometer and main and vernier pointers.

The entire indicator section including motor, gear train, potentiometers and space for the addition of a switch for actuating external circuits is contained within a 1½-in. long section of the 2-in. diameter case.

Construction

Components are assembled on five circular, stacked cards. Components are preassembled on the five cards which are then mechanically fastened and wired together in the final amplifier assembly. In laying out the circuit and mounting components, it was essential to locate certain components not only in exact locations but also with a specific orientation to reduce pickup.

Because of the use of miniaturized components throughout, the entire assembly is rugged. The case is hermetically sealed after calibration adjustments are made as the low aging properties of the components involved do not require readjustment. The assembly is shown in photograph prior to canning and sealing.

The accuracy of the system has proven to be as good as that of larger vacuum-tube units. Under room temperature conditions, the error is less than ±5 C. across the 1,000 C. range. Under environmental extremes, less than ±12 C. error occurs.

The travel time of the pointer across the 1,000 C. range is less than 3 sec and response to a step input occurs with negligible overshoot. Power consumption of the system is 7 watts.

One result of transistorization is the almost complete lack of sensitivity of the high-gain circuit to pickup from external sources.

Acknowledgement is due William Freeborn and Nathan M. Lawless, senior development engineers, for their contributions in design of the instrument.

GAMMA-RAY MONITOR USING GEIGER TUBE

By J. M. JACKSON and J. J. SURAN

H. M. E. E. Department, Electronics Laboratory, General Electric Company, Syracuse, New York

Transistorized instrument detects gamma radiation over 1 to 1,000 milliroentgen per hour range with accuracy of ± 40 percent. Alarm, which is adjustable from 10 to 900 mr per hr, indicates when radiation exceeds preset value. Reliability is achieved partially by simplicity of circuit design; all active circuits, except for power supply, are mutivibrators

PROBABILITY that a complex circuit will fail is usually a function of the number of components in the circuit. Hence, reasonable objectives for a reliable design are to use a minimum number of components and to make those that are used as individually reliable as possible.

On the basis of these considerations and in the light of component availability, it was decided to transistorize an entire gamma detector with the exception of the detecting element, a Geiger-Muller tube of a conventional and proven type.

System Description

A block diagram of the gamma-radiation monitor is given in Fig. 1 and the circuit diagram is shown in Fig. 2.

The detecting element followed by a pulse amplifier increases the energy level of the signal pulses and also performs a discrimination function to minimize noise effects. The amplified pulses are fed to a binary counter which converts the pulse train to a square wave of constant amplitude and of frequency equal to one-half the pulse repetition rate of the signal. One of the outputs from the binary stage is to a visual-count indicator circuit.

A second output from the binary stage is to an integrating circuit which converts the square wave to a d-c signal whose amplitude is proportional to the frequency of the square wave. A current-sensitive relay in series with the output meter actuates an alarm device when the radiation level exceeds some predetermined safe value.

The detecting tube is sensitive to gamma radiation but has almost no response to alpha, beta or neutron radiations. Its output consists of a train of pulses at a repetition rate proportional to the intensity of the gamma radiation field. An experimental curve of the response of the tube type used in this circuit is illustrated in Fig. 3. The output pulses tend to be uniformly large at low count rates, but as the rate increases a larger percentage of the pulses become smaller in amplitude.

Transistor Selection

Detailed consideration of the circuit specifications for the entire system led to the conclusion that, with a single exception in the power supply, one transistor type could be used throughout.

Transistors with an a_b cutoff frequency of 1 mc are more than adequate in this application.

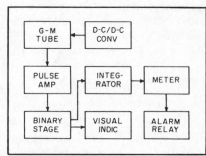

FIG. 1—Block representation of gamma-ray detector

Geiger-Muller tube is housed in external unit partially seen at lower right

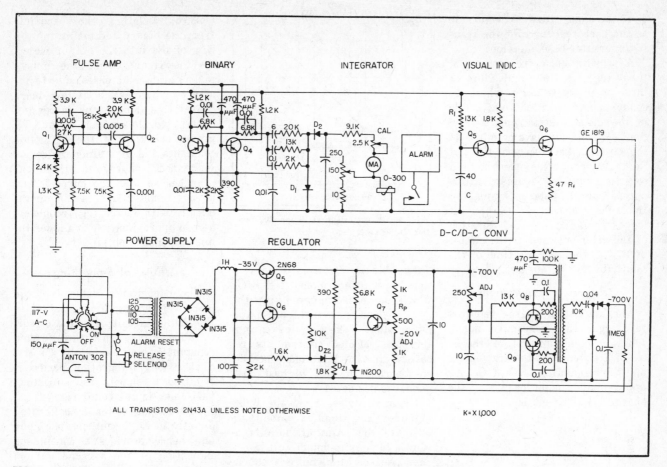

PULSE AMP   BINARY   INTEGRATOR   VISUAL INDIC

POWER SUPPLY   REGULATOR

D-C/D-C CONV

ALL TRANSISTORS 2N43A UNLESS NOTED OTHERWISE   K = X 1,000

FIG. 2—Eleven-transistor unit uses type 2N43A in all circuits except for two 2N68's in regulator

Though current-amplification factors of 30 or greater are desirable, the designs were based on a lower figure to insure reliable operation despite aging effects. Reverse leakage current of less than 2 μa at room temperature were specified to insure adequate performance over the desired ambient temperature range.

An inverse collector-to-emitter voltage rating of at least 20 v was selected to facilitate the design of the power supply. Finally, a power dissipation rating of 100 to 150 mw at room temperature was specified to insure a proper margin for derating.

On the basis of these desired properties, the type 2N43A *pnp*

alloy transistor was selected for all the circuits except the voltage regulator for which a 2N68 *pnp* alloy power transistor was selected.

Pulse Handling

Pulses from the G-M tube feed a monostable multivibrator configuration of the type illustrated in Fig. 4.[1] Inclusion of R_e in the emitter lead of Q_1 unbalances the normally bistable circuit and makes it monostable. In the monostable mode, Q_1 is normally cut off and Q_2 is conducting; Q_2 is unsaturated in its conducting state for maximum circuit sensitivity.

Diode D provides a high-impedance triggering point for the G-M tube, thus eliminating the necessity for an additional buffer stage between the detector and amplifier. Since the cutoff potential developed across R_f divides between the back resistances of D and the emitter-base junction of Q_1, it is desirable to have the back resistance of

Inside view of transistorized gamma-ray monitor shows component location

171

D of the same order of magnitude as the back resistance of the transistor emitter-base junction.

A positive pulse from the G-M tube triggers the multivibrator into an unstable state where Q_1 conducts and Q_2 is cut off. While in its unstable state, pulses from the G-M tube will have no effect on the circuit. The recovery time of the multivibrator may thus be used to discriminate against after-pulses from the detector by making the time constant of the circuit longer than the expected duration of the after-pulses. For the circuit of Fig. 4, the recovery time of the multivibrator is 50 μsec.

Triggering Requirement

The amount of charge required to trigger the monostable multivibrator is a function of the α-cutoff frequency of the transistors and the conducting level of Q_2.[2] An approximate equation for the charge requirement is

$$Q_T \cong \frac{I_{c2}}{\omega \alpha_b} \cong$$

$$\frac{1}{\omega \alpha_b} \left(\frac{R_b}{R_b + R_k + R_1} \right) \left(\frac{E_{bb}}{Rf} \right) \quad (1)$$

where Q_T is the trigger requirement in coulombs, I_{c2} is the collector current of the conducting transistor in amperes prior to triggering and ω_{ab} is the cutoff frequency of the transistors in radians per second. For the circuit of Fig. 4, Q_T is approximately 5 x 10^{-10} coulombs.

Discrimination against low-charge noise pulses is accomplished by adjusting the steady-state conduction current of Q_2 by varying R_f.

The charge available from the G-M tube during a primary discharge is approximately 10^{-9} coulombs, providing a trigger margin of at least 100 percent. The output of the multivibrator consists of a negative pulse 50-μsec wide and approximately 10 v in peak amplitude.

The binary stage can load the circuit appreciably without adversely affecting the operation of the monostable multivibrator during the regenerative cycle. An estimate of the peak pulse power gain of the amplifier circuit is 25 to 30 db.

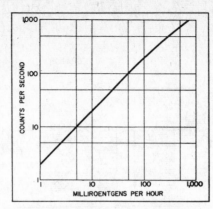

FIG. 3—Characteristic of Anton 302 Geiger-Muller tube

The binary stage is a conventional emitter-coupled flip-flop which is base-triggered by the negative pulses from the pulse amplifier stage. The coupling capacitors are selected so that its recovery time is less than one-half the period of the highest frequency anticipated.

Negative pulses from the pulse amplifier are applied simultaneously to the bases of both transistors in the flip-flop circuit. Since comparatively low pulse frequencies below 10 kc are anticipated, a diode routing gate in the trigger circuit is not required. For optimum trigger action the flip-flop is designed as a nonsaturating circuit.[1]

Logarithmic Integration

The collector swing of the binary transistors is 10 v and 10 ma. The collector of Q_4 is loaded by a logarithmic integrating circuit which delivers a d-c current to the output meter. Since the output meter must cover a range of three decades, 1 to 1,000 mr/hr corresponding to approximately 1 to 1,000 counts per second, the amplitude of the d-c output from the integrator is required to be proportional to the logarithm of frequency.

Logarithmic conversion is obtained by a simple R-C network attached to the collector of Q_4. When the collector of Q_4 swings negative, a quantity of charge is pumped into the 250-μf capacitor and the charge magnitude is determined by the R-C network time constants.

During the positive half of the square wave D_1 conducts and D_2 is

blocked, isolating the capacitor from the binary and maintaining a flow of d-c into the output meter. As the frequency of the binary multivibrator increases, due to increased gamma activity, more charge is pumped thus increasing current to the output meter. By proper selection of the R-C time constants and by scaling the series resistance values of the R-C branches so that the current flow always tends to increase logarithmically with increasing frequency, a good approximation to a logarithmic response is obtained.

Statistical Smoothing

The capacitance value is large to provide a long time constant in the output circuit. For the circuit in Fig. 2, the integrating time constant is approximately 2.5 sec. Such a long integration period is required to smooth the statistical variations in the count rate.

The output meter is calibrated directly in milliroentgens per hour and hence provides a continuous indication of the average gamma radiation intensity. The current-sensitive relay in series with the meter provides a high-level alarm.

The setting of the alarm point, which may be varied from 10 to 900 mr/hr by a variable resistor shunting the relay coil, is complicated by the random nature of detected radiation. If the alarm is set too close to an ambient level, inadvertent alarm triggering may occur due to statistical variations in the count range. Setting the alarm point experimentally by use of a calibrated cobalt-60 source has proved to be the most satisfactory means for determining the proper alarm level.

Visual Count Indicator

The visual-count indicator causes a lamp to flash every time a gamma interaction occurs in the G-M tube. A circuit of this type is useful only in the low-count region since the resolution of the eye is limited to approximately 20 flashes per second. Nevertheless, it is of inestimable value as a psychological channel to indicate that the circuit is operating satisfactorily.

A direct-coupled monostable trigger circuit meets the requirements of the monitor. Transistor Q_5 is normally conducting and Q_6 is held cut off by the low collector potential of Q_5 and by the voltage drop across R_f. Capacitor C is therefore charged to a low potential which is approximately equal to the voltage drop across R_f. Since indicating lamp L is in the collector circuit of Q_6, the lamp is normally off.

Lamp Lighting

If a negative pulse is applied to the base of Q_6, the latter is momentarily pulsed into a conducting state. The flow of current in the emitter of Q_6 will increase the negative potential across R_f, but since the base of Q_5 is momentarily maintained at a constant potential by C, Q_5 is forced into a cutoff state. This action causes the negative potential at the collector of Q_5 to increase which in turn drives Q_6 into saturation, thus lighting L.

At the termination of the trigger action the capacitor begins to charge through resistor R_1. When the negative potential across C becomes equal to the potential drop across R_f, Q_5 begins to conduct causing its collector potential to drop, hence, Q_6 is driven back to a cutoff state. The lamp is then extinguished and remains so until another trigger pulse from the binary stage initiates the circuit action. The resolution of the indicator circuit is

$$ f_r \cong \left[R_1 C \ln \left(\frac{E_{bb} - E_1}{E_{bb} - E_2} \right) \right]^{-1} \quad (2) $$

where E_{bb} is the supply voltage of Q_5, E_1 is the potential drop across R_f under quiescent conditions and E_2 is the potential drop across R_f when the lamp is lighted. For the circuit in Fig. 2, Eq. 2 indicates a maximum resolution of approximately 20 pps. Any higher frequency will result in a constant glow of the indicator lamp.

Reliable Triggering

Separate voltage supplies for Q_5 and Q_6 are used in the indicator circuit since a regulated voltage supply is required only for Q_5 to insure reliable trigger action. Consequently, the watt of power required to flash the lamp may be drawn from the unregulated supply and thus does not impose unnecessary power drain upon the regulator circuit.

The trigger source for the indicator is the common emitter resistance of the binary-stage flip-flop. A negative pulse occurs at this point each time the binary multivibrator is triggered from the pulse amplifier. In addition, the common-emitter terminal of the binary circuit is a low-impedance point and hence is not seriously loaded down by the visual-count circuit.

Power Supply

The power supply requires an accurate voltage regulator and high-voltage generator. Good voltage regulation is required for the G-M tube since a variation of ± 2 percent about 700 v may result in improper tube response. Close regulation is also required for the multivibrator circuits to insure constant output levels.

The regulator circuit maintains

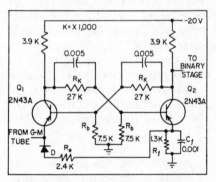

FIG. 4—Monostable multivibrator serves as pulse amplier

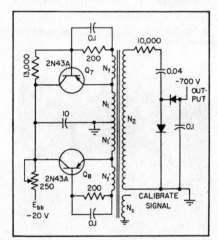

FIG. 5—Voltage-doubler d-c/dc converter supplies 700-v excitation for G-M tube

the voltage level against variations of load impedance as well as against input voltage fluctuations. The reference voltage, from which the difference feedback voltage to a series-regulating element is established, is obtained from zener diode D_{Z1}.[3] The difference voltage is amplified by Q_8 and Q_9 and fed back to the base of power transistor Q_7 which is in series with the line and which constitutes the regulating element. A second zener diode reference, D_{Z2}, stabilizes the base and collector supply voltages of feedback transistors Q_8 and Q_9 respectively, thus making the amplifier gain virtually independent of input voltage changes. Adjustment of the regulated output voltage may be made manually by variation of R_p.

The regulator circuit has an output resistance of approximately 0.2 ohm and regulates the 20-v supply within 1 percent over a temperature range from 0 to 50 C.

High-Voltage Generation

High voltage for the G-M tube is obtained from the d-c/d-c converter of the type illustrated in Figs. 4, 5. A square wave of alternating voltage is generated when the transistors alternately switch supply voltage E_{bb} across transformer windings N_1 and N_1'. Regenerative feedback is supplied by windings N_f and N_f'.

Frequency of operation is governed by the magnitude of the supply voltage and the saturation properties of the square-loop transformer material. If ϕ_s is the saturation flux in webers and if $N_1 = N_1'$, the frequency of oscillation is

$$ f = E_{bb}/(4 \phi_s N_1) \quad (3) $$

Since E_{bb} is a regulated voltage and ϕ_s is fairly temperature insensitive over a wide range of temperature, the frequency of the converter is constant. This property may be used in calibration of the monitor equipment by employing the converter as an internal generator for testing the indicator circuits by taking the calibration signal off winding N_c.

The voltage across secondary N_2, is rectified and filtered in a voltage doubler configuration. The re-

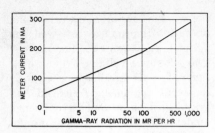

FIG. 6—Output characteristic of complete instrument

sistor in series with the secondary winding limits the maximum current flow in the secondary circuit. A fine adjustment of the d-c output voltage may be made by variation of the 250-ohm potentiometer, which adjusts the collector potential of the transistors within a small margin.

The converter operates at a frequency of 500 cps. The transformer turns ratio (N_2/N_1) is 17.5 and the core is a toroid fabricated from Orthonol square-loop material. Supply voltage E_{bb} is from the 20-volt regulated line.

System Performance

The output response characteristic of the circuit, showing d-c output current as a function of radiation intensity, is illustrated in Fig. 6.

The circuit is calibrated by switching the input of the pulse amplifier from the G-M tube to the differentiated output of the calibration-signal winding on the high-voltage transformer. The meter current under these conditions is known, since the frequency of the converter is constant; therefore, the meter-circuit resistance may be adjusted until a correct reference reading is obtained. Calibration by this means eliminates the effect of component variations which could cause minor differences in the output readings from unit to unit.

Count-rate calibration is not rigorously analogous to radiation calibration. However, the technique does meet the accuracy requirements of the system and provides a simple method for individual adjustments. For calibration in the lower two decades, the circuit depends upon the assumption that the G-M tubes exhibit approximately the same characteristics. Experience shows that this is a reasonably valid assumption and leads to only negligible errors.

Operational Accuracy

Once the system is calibrated, continued accuracy of the output indication depends upon the voltage stability of the regulating circuits. For example, if the output meter is rated as accurate to within 1 percent and has a full-scale deflection of 300 μa, the error anywhere on the scale may be as high as 3 μa.

Since the meter is marked in three logarithmic decades, an error of this magnitude corresponds to one thirty-third of a decade or approximately 7 percent in the indication. Consequently, a 2-percent change in the supply voltage, which leads to the same percentage change in the voltage output of the binary stage, could result in a maximum error of almost 15 percent in the meter indication.

This sensitivity can be reduced by the addition of an emitter-follower amplifier stage between the binary multivibrator and the logarithmic integrator circuit, providing a larger current for the output meter and permitting the use of a meter with a larger full-scale current rating. However, if the voltage regulator circuit is adequate for the required accuracy, as in the present design, the additional transistor stage is not necessary.

Drift

Operation of the monitor over an ambient temperature range of 0 to 50 C meets the given accuracy requirements since the drift in output indication for a single unit is no greater than ±40 percent.

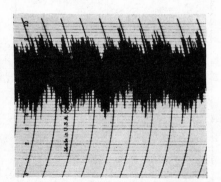

FIG. 7—Five-hour recording of output response to radium watch-dial source. Vertical divisions are 5 μa

The multivibrators in the indicating circuits are stabilized against temperature changes by the d-c bias network so that under the worst possible conditions the amplitude of the output pulse from the amplifier circuit does not vary more than ±10 percent and the output amplitude from the binary stage does not vary more than ±2 percent over the 50 C temperature range.

Frequency stability of the converter in the same temperature range is better than ±2 percent. It is estimated that most of the observed drift is due to component variations in the logarithmic integrating circuit.

Radium Response

A 10-hour pen recording of the gamma monitor output current response to a radium watch-dial source is illustrated in Fig. 7. Each major division along the horizontal axis is a one-half-hour time unit and each major division along the vertical axis is a 5-μa unit. The cyclic variations of 1-hour periods are due to the motion of the minute hand. Added to the hourly variations is a 12-hour cycle due to the motion of the hour hand. Statistical variations from the average radiation level are as much as ±50 percent.

The authors are indebted to W. A. Andrews of the Heavy Military Electronic Equipment Department for his valuable assistance. In addition, the helpful suggestions of H. W. Abbott, E. P. Cleary and D. A. Paynter of the Electronics Laboratory are gratefully acknowledged.

This work was supported by the Bureau of Ships, U. S. Navy, under contract number NObsr-72530.

REFERENCES

(1) J. J. Suran and F. A. Reibert, Two-Terminal Analysis and Synthesis of Junction Transistor Multivibrators, *IRE Trans of PGCT*, Mar. 1956.

(2) J. G. Linvill, Non-Saturating Pulse Circuits Using Two Junction Transistors, *Proc IRE*, July 1955.

(3) H. R. Lowry, Transistorized Regulated Power Supplies, *Elec Des*, Feb. and Mar. 1956.

(4) D. A. Paynter, An Unsymmetrical Square Wave Power Oscillator, *IRE Trans of PGCT*, Mar. 1956.

(5) G. H. Royer, A Switching Transistor DC-to-AC Converter Having an Output Frequency Proportional to DC Input Voltage, *AIEE Winter General Meeting*, Paper No. 55-73, Jan. 1955.

STROBOSCOPE MEASURES SHAFT TORQUE

By JOHN PATRAIKO

Electrical Department, Scientific Laboratories, Ford Motor Co., Dearborn, Mich.

Magnetic pickup for reference pulses feeds transistorized amplifier and shaper that triggers pulser circuit for strobe lamp. Miniature unit has peak light intensity of 2,000 c-p which is equal to average intensity of 10 c-p at 60,000 flashes a minute. One-microsecond pulses at rates between 3,000 and 60,000 fpm are produced by unit designed to determine torque by measuring dynamic shaft twist

STUDY of modern high-speed turbines requires a stroboscope having higher flashing rates, smaller flash durations and reduced jitter than are usually available. Using the unit to be described, an engraved scale can be read indicating shaft twist on a shaft rotating at 60,000 rpm. The shaft twist is used to measure the transmitter torque.

System

Figure 1A shows the physical setup. A block diagram of the strobe system is shown in Fig. 1B.

The reference-pulse pickup, shown in Fig. 1C, is a 1,000-turn magnetic pickup with a permanent-magnet field. A signal of about one volt is generated when the radially-protruding steel pin on the rotating shaft passes through the pickup field. The slope of the signal just as the pin passes through the center of the magnetic field is about 0.2 volt per μsec.

The circuit shown in Fig. 2 amplifies and shapes the pickup signal into a 250 volt, one-μsec positive pulse with an initial slope of 600 volts per μsec.

The output pulse is applied to the pulser and triggers a VC-1258 thyratron within 0.2 μsec from the start of the pulse. The thyratron is employed as a switch to apply a pulse of energy to the strobe lamp. The strobe lamp begins to conduct within 1 μsec of the thyratron firing time.

Pulser Unit

The circuit employed in the pulser is also shown in Fig. 2. A single pulse controlled by the thyratron is used both to trigger and activate the strobe lamp. One ad-

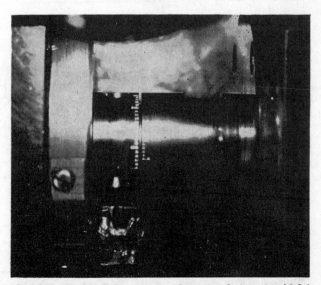

Miniature strobe lamp, magnetic pickup and amplifier, shaper and pulser chassis (left) measure dynamic shaft twist (right)

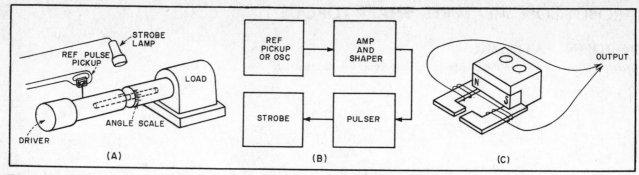

FIG. 1—Test setup to measure shaft twist (A) uses magnetic pickup (B) to provide pulses for strobe circuits (C)

vantage of this circuit is the common ground for the thyratron and strobe lamp. This permits convenient remote operation of the strobe lamp through a single conductor coaxial cable without danger to the operating personnel.

Thyraton Characteristics

Figure 3A shows the idealized waveform of thyratron plate voltage with emphasis on the four phases of pulser operation. Voltages E_T and E_i and times t_1, t_2 and t_3 have been exaggerated. Normally E_T and E_i are a few percent of $2B$ volts and t_3 is a few μsec compared to a millisec for T_R.

With the thyratron open circuited, capacitor C and line N will become charged through resistors R_N, the diode and inductor L. Application of an input pulse at time t_0 drops the thyratron anode to near ground potential E_T and the charged capacitor C is essentially

connected across the strobe lamp. Because of inherent lamp delay, conduction will not occur until time t_1. When the lamp conducts, capacitor C discharges according to the time constant of C and the resistance of the conducting strobe lamp. For circuit values given, the discharge time is about one μsec.

Delay line N in conjunction with R_N produces an inverse voltage E_i at time t_2 by line reflection. The length of line determines the occurance of time, t_2 after the discharge of capacitor C. For this delay line, t_2 is about three μsec.

The inverse voltage present from t_2 to t_3 permits partial thyratron deionization and aids in the prevention of subsequent thyratron reignition. The longer the time t_2 to t_3 the more rapid the subsequent thyratron voltage rise may be without reignition. Thus higher flashing rates can be had with larger values of $t_3 - t_2$.

However for this charging circuit the line time constant ($\tau = t_2 - t_0 = t_3 - t_2$) must be kept much less than the charging circuit time T_R to achieve the required charging circuit isolation during the time the thyratron conducts and the start of cutoff. Also, the line time constant should be slightly larger than the time to discharge C and charging time T_R must necessarily be less than the period of the maximum flashing rate.

Because of charging-circuit resonance the discharge-capacitor voltage attains a value approximately twice the supply voltage at time T_R and retains this value up to the triggering time t_{02} by the unidirectional action of diode D.

Parameter Selection

The flash lamp was selected primarily on the basis of size but certain minimum requirements of firing delay, light output, and dis-

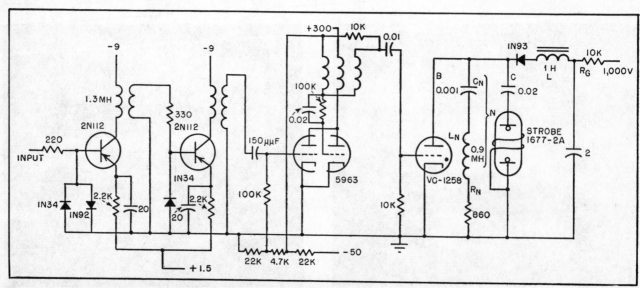

FIG. 2—Amplifier and shaper provide 250-v one-μsec pulse to pulser that acts like radar modulator by discharging stored energy in C through strobe lamp. Common ground for thyratron and strobe lamp permits convenient remote operation

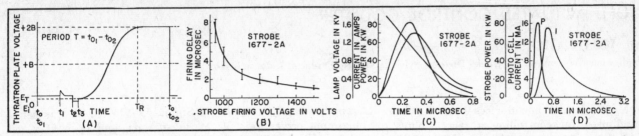

FIG. 3—Idealized thyratron plate waveform (A) shows four phases of pulser operation. Other curves show firing delay of strobe as a function of applied voltage (B), instantaneous strobe voltage, current and power (C) and light output (D)

charge time were necessary. Krypton lamp 1677-2A satisfied these requirements. The curves of Fig. 3 show the results of tests on this lamp. These tests were conducted to aid in selecting some of the operating parameters.

In Fig. 3B the delay is less than one microsecond for voltages greater than 1,500 volts The jitter in firing as indicated by the vertical lines in the figure is also less than a microsec above 1,500 volts.

Strobe Voltage

The curves of Fig. 3C show the instantaneous strobe lamp voltage, current and power using a 0.02 μf capacitor charged to 2,000 volts in series with a hydrogen thyratron acting as a switch. The energy per flash is 26 milliwatt-sec. This value together with the 18-watts rated power input of the lamp permits determination of the maximum flashing rate as 41,000 fpm.

A higher flashing rate may be permitted by lowering the charge voltage or capacitance. The voltage decrease is proportional to the square root of the flashing-rate ratio while capacitance decrease is directly proportional. Lamp dissipation is kept within the rating up to 60,000 fpm by lowering the voltage and maintaining the capacitance.

This is done by inserting a 10,000-ohm resistor R_G in series with the power supply. As power supply current increases because of increased flashing rate the charging-capacitor voltage is reduced by the drop to 1,660 volts at 60,000 fpm.

Light Intensity

Figure 3D shows the shape and width of the light intensity with respect to the instantaneous power curve of Fig. 3C for the same voltage and capacitance. The one-third intensity points are separated by

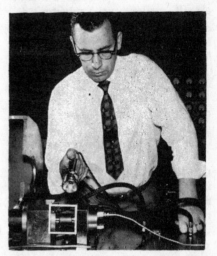

Use of pulser circuit enables operator to use strobe without danger of shock from accidental contact with work surface

one microsecond. The delay in light with respect to instantaneous power is of little consequence since the delay is free of jitter.

Results

Photographs of waveforms at the test points shown in Fig. 4 were taken for a typical flashing rate of 500 fps. These waveforms show that the delay in the firing of the thyratron is negligible and the current for forceful thyratron conduction exists for three microseconds as required. The remaining cycle of current exists as a result of circuit losses, such as an equivalent shunt resistance across inductor L and deionization effects.

The last waveform shows that the light has reached its peak one microsecond after triggering.

Although this article pertains to a particular strobe light system is can be applied to other types of thyratron control devices with increased repetition rates.

REFERENCE

(1) J. Patraiko, A Miniature Strobe Light For a 60,000 RPM Bearing Tester, *IRE Trans* on Industrial Electronics, PGIE-3, March 1956.

FIG. 4—Waveforms taken at 500 fps show input test pulse (A), thyratron plate voltage on scale of one $\mu sec/div$ (B) and 100 μsec div (C) delay network current (D) and light pulse output (E)

LIGHT-ACTUATED CONTROL COUNTER

By JOHN GRANT

Applications Department, Electronics Division, Sylvania Electric Products, Woburn, Mass.

Many circuits and devices have been used for controlling a relay or other actuating mechanism by the light falling on a photocell. This article will describe a simple circuit using semiconductor devices. It incorporates a 1N77A photodiode and two 2N35 germanium junction transistors.

The light-actuated section of the circuit is a bridge in which the 1N77A photodiode is one leg. The diode's reverse resistance is approximately balanced by the 100,000-ohm resistor. The opposite side of the bridge is a 50,000-ohm potentiometer. The potentiometer is used to vary the no-signal current through the relay and to set the overall circuit gain by biasing the transistors into a higher gain region.

The two-stage transistor amplifier is connected in the common-collector configuration to take advantage of the maximum available current gain. The 10,000-ohm resistor increases somewhat the second stage transistor stability and keeps the amplifier input impedance at approximately 100,000 ohms for maximum power gain.

When a light beam strikes the light-sensitive junction of the photodiode, the reverse resistance of the diode decreases, thereby increasing the voltage at the base of the first transistor. The increased voltage allows emitter current to flow from the first transistor into the base of the second transistor. The increased base current of the second transistor increases the emitter current through the relay coil and actuates the relay armature.

If the 50,000-ohm potentiometer is not adjusted correctly, the voltage at the base of the first transistor will be too negative with respect to its emitter for the change in bridge voltage to turn the transistor on. On the other hand, if the potentiometer is set too far in the opposite direction both transistors will stay on. A small amount of experimentation will give the potentiometer setting for the best operation of the amplifier.

The relay used in this setup was a Sigma 5F-1000-G. It was set to close at 4 ma and open at 2 ma. Approximately 5 to 6 ma can be drawn through the relay for positive switching action.

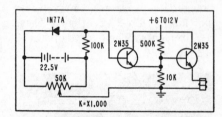

Light operated relay will trigger counter tube at rates better than 60 counts per sec

With a miniature 22.5-v battery and a 6-v battery, the entire device can be built in a box smaller than 6 in. x 2 in. x 2 in. The output of the relay has been used to control a mechanical counter up to 200 counts per minute. If used in conjunction with a glow transfer counter tube, rates better than 60 counts per second could be easily handled. The photodiode frequency response is about 15 kc and the transistors can handle 600 kc.

TRANSISTOR DRIVES CLOCK

By C. HUNTER McSHAN
Great Neck, N. Y.

BALANCE and hairspring assemblies of conventional clocks can be driven by self-switching transistor circuit as shown in the diagram.

The balance wheel is modified by mounting a small cylindrical magnet on its rim and poising with a counterweight. A stator of soft magnetic material with a tapped winding is positioned to receive the magnet at the instant of maximum balance velocity. Stator thickness is made equal to the diameter of the magnet. The circuit is comprised of a tapped winding, resistor, transistor and energy source. The coil-form terminals serve to mount the resistor and transistor as a complete subassembly.

Operation

As the balance magnet enters the stator, the transistor remains in its normal cut-off or open circuit state. During this entry period, the attraction force of the permanent magnet acting upon the stator imparts energy to the balance. As the magnet begins to leave the stator, a signal is generated of the proper polarity to start a regenerative process of triggering the transistor to full conduction. This triggering process occurs in less than 0.0002 second. Flow of current through the transistor and part of the winding then produces a magnetic field which repels the balance magnet away from the stator poles. The driving forces of attraction and repulsion may be designed to deliver equal amounts of energy so that the natural balance rate is not disturbed. Deviation of the hairspring from true isochronal operation may be compensated by unbalancing these forces. The resistor critically damps the circuit to prevent self-oscillation.

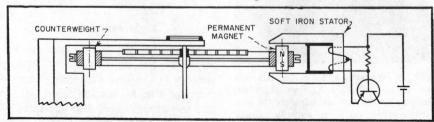

FIG. 1—Arrangement of balance wheel and transistor drive system

Chapter 13

INDUSTRIAL DETECTION CIRCUITS

CREVASSE DETECTOR BLAZES GLACIAL TRAILS

By H. P. VAN ECKHARDT Project Engineer, Pathfinding Section, Mine Detection Branch, Research and Development Laboratories, U. S. Army Corps of Engineers, Fort Belvoir, Virginia

Sled electrodes, in contact with ice-snow surface, set up an electrical field in ice pack and pick up constant readings when ice is solid and safe for travel. As tractor approaches a crevasse, bridged with snow, the flow of electrical current is disrupted and an alarm warns operator of hidden chasm. Transistorized system uses the crevasse walls to simulate capacitor

CREVASSES, hidden pitfalls often wide and deep enough to swallow men and equipment, have haunted Arctic explorers for many years. Bridged over slightly with snow, these chasms in the ice are particularly dangerous, in summer Arctic white outs and snow storms.

Detecting Methods

Until recent years, the only methods used for detecting crevasses were aerial photography and hand-probing with long rods. Aerial observation proved effective only under highly favorable weather conditions and hand-probing was extremely tiring, tedious and slow.

A research program resulted in the development of electronic techniques employing surface-electrodes for the effective detection of crevasses.

The detector employs a double-system of electrodes. A wide system detects crevasses in a path around and in front of a vehicle. A long system detects crevasses with extra thick snow roofs missed by the wide system, and distinguishes between large crevasses and narrow cracks.

System Details

Each system consists of four large dish-pan-shaped sled electrodes in contact with the ice-snow surface. The wide-system electrodes are pushed in front of the vehicle in a fan-wise arrangement on wooden booms. Two pans act as current electrodes which set up an

Crevasse detector has made it possible to explore many hundreds of miles of ice and snow in the Arctic and Antarctic and has never failed to detect a crevasse (see cover)

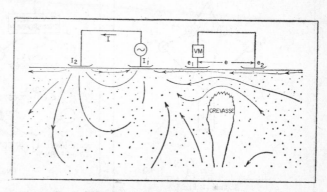

FIG. 1—Sketch shows the electrical field set up by sled electrodes in contact with glacier surface. A three dimensional pattern flows through the ice. Crevasse is detected by electrodes

electrical field in the surrounding ice pack. The remaining two signal electrodes pick up readings from this electrical field. When the ice is solid and safe for travel, the signal is comparatively constant. As the vehicle approaches a crevasse, the flow of the electrical field is disrupted and an alarm signals danger.

The long system operates simultaneously with the wide, but on a separate frequency. Its electrodes are arranged differently. One electrode is pushed ahead by the vehicle, which also acts as an electrode, and the other two are towed behind at 20-ft intervals.

Alarm

As the vehicle travels over the glacial surfaces the detector reports its findings by a two-channel recorder, mounted in the vehicle. An alarm box containing a pair of special relay meters, a red light and a buzzer warns the driver of crevasses. Audio warning is also available through earphones. A light and a buzzer signify component failure in the detector.

Electrode Pattern

The sled electrodes are in contact with the glacier surface. As seen in Fig. 1, a source of alternating current, I, is connected between two current electrodes I_1 and I_2 and a three-dimensional pattern of current flows through the ice. Since ice is a nonconductor, this is displacement current, like that flowing through the dielectric of a capacitor connected to an a-c source.

A potential-difference measuring device is connected between the signal electrodes e_1 and e_2. Any marked distortion of current pattern by an obstruction, such as a crevasse near the electrodes, will cause a change in the voltage reading, e.

Ice Coupling

The electrodes are large enough to provide good coupling to the ice. Since the assembly moves over the surface to determine safe trails, the electrodes' effectiveness in contact with snow and ice is bound to vary. The resulting variations in the electrode-voltage drops should not be allowed to affect either the input current or the output voltage reading appreciably.

Spacings and arrangement of the electrodes determine the ice-sampling depth. In general, the smallest practical spacing between any pair of the four electrodes should exceed the depth of the thickest snow bridge anticipated. Also, the spacing should be several times the dimensions of the electrodes themselves so that the variations in snow contact will not appreciably alter the effective electrode spacings.

Symmetrical electrode patterns are avoided as they place the electrodes on a common equipotential in the current field and would fail to indicate crevasses oriented parallel to that equipotential plane. The most suitable arrangement found is shown in Fig. 2.

A block diagram of the crevasse detector is shown in Fig. 3. The system is operated from a 24-v system that can be readily changed for use entirely from the 12-v vehicle storage battery.

Transmitter

The main transmitter consists of a bridge-T oscillator and a power amplifier. These two assemblies are bolted within the transmitter chassis which is provided with three front-panel controls.

The oscillator, Fig. 4A incorporates heavy degenerative feedback in which a small incandescent lamp is used as a nonlinear compensating resistance. The oscillator provides constant output frequency and voltage for any supply between 12 and 32 volts at temperatures as low as −20 F. Oscillation frequency is governed by capacitors C, in Fig. 4A. Various values of these ca-

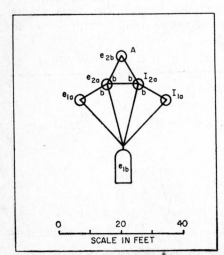

FIG. 2—Electrode arrangement found to produce the best detection results

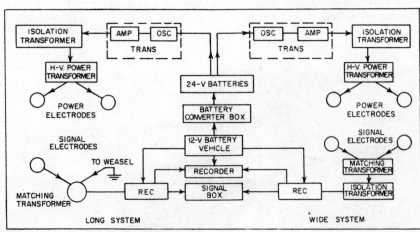

FIG. 3—Block diagram of crevasse detector showing the double-electrode system that distinguishes between large crevasses and narrow cracks The detector operates from a 24-v source that can be converted to operation from the 12-v vehicle storage battery

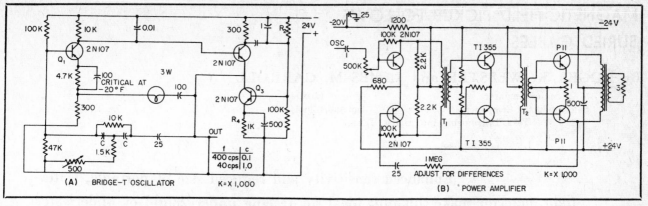

FIG. 4—The main transmitter consists of two distinct parts, a bridge-T oscillator (A) and a power amplifier (B)

pacitors provide frequencies of 100, 150, 230 and 350 cps. These frequencies are adjusted by the 500-ohm trimmer control which varies a shunt resistance in the tuning circuit. The oscillator fine-frequency trimmer is adjusted slowly since there is a frequency-change lag in the nonlinear feedback stabilizing system.

Audio Output

Oscillator a-f output is fed to the power amplifier shown in Fig. 4B, consisting of a phase-inverter and two push-pull stages. The input amplifier and phase-inverter employ two 2N107 transistors, coupled through transformer T_1 to the intermediate stage employing two TI 355 *pnp* transistors which are operating class AB. These, in turn, apply driving power to the low-resistance interstage transformer T_2, manufactured to order.

The interstage transformer,

which drives the final power stage, employs a pair of P11 transistors operating Class B. Types XH-25 or XH-10 may be substituted for these transistors which are no longer manufactured.

Receivers

The receiver input signal, Fig. 5, from the electrodes and isolating or matching transformers is attenuated to a suitable level at a constant impedance of 1,000 ohms by the T-pad and passed to the preamplifier. The supply voltage of the 2N107 preamplifier is stabilized at 5.8 volts by a reversed TI 620 silicon diode shunt, operating at the Zener point. Signal voltage then passes through a band-pass L-C filter employing a 10-henry inductor, is further amplified and applied to the driver circuit employing two 2N185 transistors in push-pull.

The signal is applied to the final output amplifier which uses a 355

transistor operating class A and to one of the large voltage step-up driver transformers mounted on a separate chassis. The signal power is rectified in push-pull and applied to the recorder pen motor and to the relay-meter. The frequency-selector switch shown in Fig. 5 provides for reception on four different bands corresponding to those of the transmitter plus one for the 60-cps emergency vibrator supply.

Test Results

In tests covering a 200-mile unexplored trail in Greenland, the unit never failed to detect a crevasse. The U.S. Navy has also enjoyed complete success with the use of the detector in its Antarctic operations.

Exhaustive tests have shown that: operating frequencies of 200 cps or lower give the largest crevasse indications relative to background fluctuations. The background signal fluctuations are sometimes so complex that crevasse anomalies may be disguised. However, proper electrode spacings and visible recording of the signal over the distance travelled assist in distinguishing the crevasse anomalies.

In addition to crevasses, buried buildings, other large objects are readily detected. Small portable outfits towed by a man are fairly successful, providing adequate electrode spacings are used.

The crevasse detector works best at low temperatures when near surface melt moisture is absent. However, frictional electric noise generated by motion of the potential electrodes is bad at low temperatures.

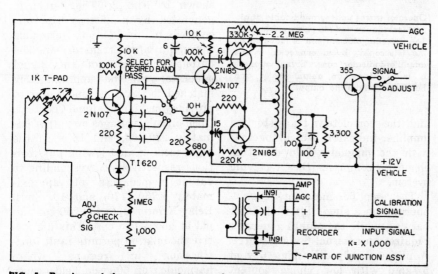

FIG. 5—Receiver of the crevasse detector showing the driver circuit for one recorder channel along with checking and adjusting circuitry

MAGNETIC-FIELD PICKUP FOLLOWS BURIED CABLES

By ROGER R. WEBSTER and JAMES M. CARROLL

Senior Engineer
Semiconductor Components

Project Engineer
Apparatus Division

Texas Instruments, Inc., Dallas, Texas

Enhanced sensitivity and reduced size of portable, water-tight, detector make this unit ideal for tracing underground or underwater cable systems used for airport and harbor lights. Instrument locates open circuits by sudden drop or loss of 250-cps signal generated by vibrator. Ferrite core is used in the detecting element pickup coil. Signal is amplified by a three-stage, transistorized, tuned amplifier

TRACING buried or submerged cables and locating cable faults, is a serious problem in the maintenance of airfield lighting systems. Faults may vary from direct shorts to open-circuited cables, but high-resistance leaks which break down in wet weather probably give the most difficulty.

Insensitivity and inaccuracy of previous detectors have made fault-locating difficult. Incomplete and inaccurate records of the location of underground cables contribute to the difficulty. Thus, when a fault develops, it has been more practical to abandon existing cables rather than spend long fruitless periods attempting to locate the fault.

Faced with this continuing problem, the Navy Bureau of Aeronautics desired a cable-fault detector that would replace the older vacuum-tube model tester.

Detector

The answer, the transistorized AN/TSM-11 cable test detector follows underground cables buried as deep as 12 to 15 feet. Open-circuited cables are followed and the fault located to within a short distance when the cable is not too far from the detecting element. Sufficient charging currents will flow in buried cables with open circuits to permit the cable test set to follow with good accuracy.

The detecting set consists of two major units, the signal generator

Operator traces underground cable at airport installation and pin-points cable fault. The detecting-element coil is mounted on the telescoping boom connected to the amplifier-indicator case. Signal strength is indicated on an output meter and headset monitors the output

and the completely transistorized amplifier-indicator, which together with the magnetic field detecting element, comprise the receiving system.

The amplifier-indicator, Fig. 1, operates on a single internal 22.5-v battery, while the signal generator requires an external 6-v d-c source. A power supply delivering 6 v at 5 amp with low ripple content powers the signal generator. A 6-v auto storage battery may be used.

A 250-cps a-c signal, from the signal-generator is applied, through an impedance matching transformer which accommodates various cable conditions, to the end of the cable to be tested. Signal current in the cable produces a magnetic field which induces voltage in the pickup coil of the magnetic-field detecting element. The induced signal is amplified and applied to an indicating meter and headset. Cable faults are located by a sudden change in the intensity of the received signal, usually by a sudden drop, or by complete loss of the 250-cps signal.

Signal Generator

The signal-generator circuit is shown in Fig. 2. The signal is generated by a 250-cps vibrator, coupled to the output terminals through a tapped transformer. The output impedance of the signal generator can be approximately matched to the cable under test to provide maximum-signal current. Open-circuit voltages are approximately 250, 64, and 16 v in the high, medium and low output positions respectively, corresponding to output impedances of approximately 2,750, 170, and 10 ohms. Relay K interrupts the 250-cps signal at about two cps. This distinctive modulation permits fault identification in regions where harmonics of 60-cycle fields cause high background noise.

A push-pull connection on the

primary of the output transformer avoids d-c saturation of the transformer core and allows the use of a small transformer. Input current requirements are minimized and vibrator life increased because of the decreased contact current.

In the input and output circuits, r-f filtering gives a better wave shape and minimizes high-frequency noise generated by the vibrator.

Amplifier-Indicator

The signal from the magnetic-field surrounding the cable is amplified and rectified. Signal strength

is indicated on an output meter and the headset is used for monitoring the amplifier output.

Ferrite Core

Sensitivity was greatly increased by using a ferrite core in the pick-up coil. Coil and core are mounted on a telescoping boom connected to the amplifier.

The three-stage amplifier is mounted on an etched circuit board. It is tuned to 250 cps by the series L-C circuit in the emitter circuit of the second stage. The output is down 5 db from maximum at 215 cps and 285 cps. Three type

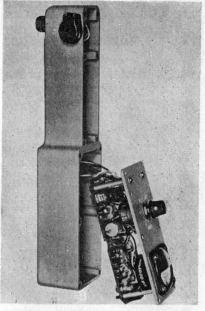

The amplifier unit is mounted on an etched-circuit board. The detecting element may be rotated by knob at top to obtain maximum signal deflection

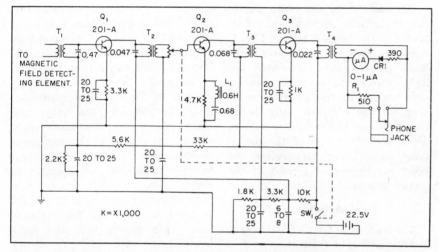

FIG. 1—Amplifier-indicator unit of cable detector is tuned to 250 cps by the series 1-C circuit in the emitter of the second stage. Three transistors give overall gain of 85 db

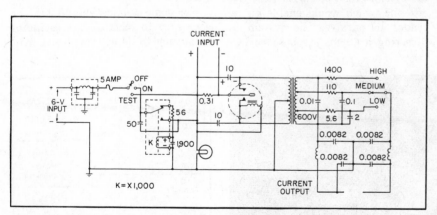

FIG. 2—Signal generator of cable detector. A range of output impedance allows matching the cable under test to provide maximum signal current

201-A triode *npn* grown-junction germanium transistors in the amplifier give minimum overall gain of 85 db. The input impedance of the unit is approx. 10 ohms, designed to match trailing leads in sea water when following submerged cables. Output impedance of approx. 600 ohms matches the headset.

Amplifier output, rectified by a diode, is indicated on a d-c meter. Resistor, (R_1) in Fig. 1, terminates the output when headset is out.

Watertight Case

The amplifier case is watertight to a submerged depth of 3 feet, except for the phone jack which may be plunged when the headset is not in use.

The complete unit is carried easily in two watertight portable cases, which house the major units.

The cable-fault detector can localize practically all types of cable faults so that the cable may be exposed and repaired.

INTRUDER ALARM USES PHASE-SENSITIVE DETECTOR

By S. BAGNO and J. FASAL   Walter Kidde & Company, Inc., Belleville, N. J.

Transistorized burglar alarm has electronically modulated infrared light source and synchronous phase-sensitive demodulator pickup unit. Pulsed light technique overcomes adverse effects of continuous or varying ambient light conditions. Alarm also sounds if power supply or interconnecting lines are tampered with.

INTRUDER DETECTION systems must be virtually foolproof, even under the most difficult conditions of operation, and they must also provide safeguards against false alarms or being rendered inoperative by intruders. Furthermore, the system must not respond to a variation of the light intensity and should not be affected, within practical limits, by such climatic conditions as fog, rain or snow.

Because intruders have developed increasingly more ingenious methods of defeating conventional alarm systems, a phase-sensitive system has been developed. An invisible infrared light beam keeps introuers unaware of its exact location. Any tampering with the power supply or any other interruption of lines produces an alarm.

The basic principles of the photoelectric alarm system are illustrated in the block diagram, Fig. 1. A 55-cps transistorized power oscillator powers and modulates the light source and supplies a reference signal to the phase-detector stage of the receiver. Any deviation between the reference-voltage phase and the phase of the modulated light beam deenergizes the relay in the output circuit of the phase detector.

Modulation

The oscillator frequency is made slightly different from the line frequency to eliminate the possibility of synchronization with the 60-cps line which would enable an intruder to paralyze the system by directing a 60-cps stroboscope into the housing of the phototransistor.

The beam modulation frequency is directly related to the percentage of light modulation. The tungsten-lamp filament reaches its maximum temperature when the collector current of the oscillator transistor goes through its peak value and cools off to its minimum value when its collector current goes through zero. If the time interval between maximum and minimum is small, (frequency is high) the filament does not have time to cool off so much as when the frequency is low. This means that the light modulation with high frequency will be less than with low frequency.

For some applications, it may be necessary to increase the oscillator frequency to detect small and fast-

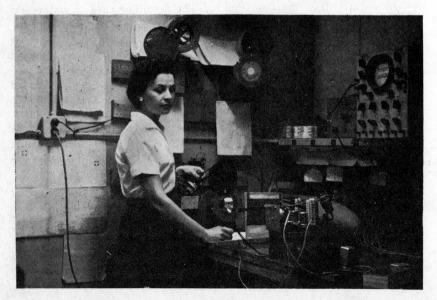

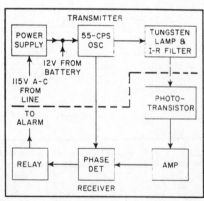

FIG. 1—Basic elements of transmitter and receiver units that make up alarm

Alarm system operation is checked with perforated sheet between receiver and transmitter units

184

moving objects, but the price is always the loss of sensitivity, as a result of a lower percentage of light modulation.

Due to the optical characteristics of the phototransistor and the type of glass used, only the portion of the radiated modulation falling above 2 microns is detected, as illustrated in Fig. 2. Since that portion of the spectrum shorter than 2 microns includes most of the energy radiated by high-temperature light and only the portion of the low-temperature light most sensitive to temperature modulations, the phototransistor response indicates that the received modulation percentage is relatively constant and independent of the average filament temperature over a wide range.

A series-emitter resistance can stabilize a phototransistor against ambient variations in collector current.

This effect can be used to make the steady-state d-c response of the phototransistor and its a-c response independent of each other. Thus, for steady-state light, the phototransistor has the low sensitivity of a photodiode, whereas for modulated light it can approach the full sensitivity of a phototransistor —a gain of more than 40 db over the d-c steady-state light response. In that way the phototransistor can operate in bright ambient light without becoming saturated and insensitive to light modulations.

Phase Detection

The basic schematic of the phase detector is shown in Fig. 3 in simplified form.

The amplified signal from the phototransistor drives the base of the phase detector (Fig. 4B).

The emitter-collector voltage of the phase detector is supplied from the 55-cps transistor oscillator through an unfiltered bridge rectifier and through a d-c relay located in the a-c branch of the rectifier bridge. The emitter-collector voltage is therefore a full-wave rectified signal (Fig. 4D) derived from

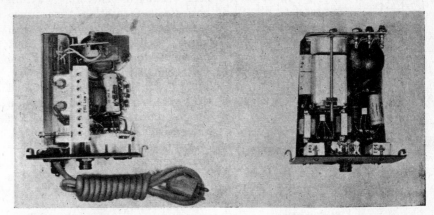

Unhoused transmitter and receiver chassis are shown at left and right respectively

the reference signal in phase with the phototransistor signal (Fig. 4C). The phase shift between phototransistor and reference signals that results from the thermal lag of the filament is corrected in the amplifier.

The resultant collector current (Fig. 4E) of the phase detector depends on the instantaneous voltages on the base and collector and can be considered as the modulation product of both. This current must always be in phase with its driving voltage. Therefore, on the d-c side of the rectifier, where the adjacent half waves are in the same direction, this current will be in the same direction although modified in amplitude by the signal on the base of the transistor.

Similarly, on the a-c side of the rectifier, this current although equally modified in amplitude by the transistor, will have its adjacent half waves alternating in direction in response to the driving alternating potential and be in phase with it (Fig. 4F). This current is therefore a nonsymmetrical alternating current that contains a d-c component by which the relay is energized. The a-c component of the current is bypassed.

There are several cases to consider in the behavior of the phase detector under varying operating conditions.

Suppose first that the phase between the phototransistor signal and reference voltage had been re-

versed (Fig. 4I). This is caused by a reversed primary or secondary winding of one of the coupling transformers.

The considerations leading to the waveform of the collector current and the current through the a-c circuits and through the relay are the same as before. The only difference is that the sequence of the smaller and larger amplitudes of the wave is now reversed. Therefore, the d-c component in the relay circuit or the relay voltage itself has to change its sign as shown in Fig. 4N.

Other Phase Conditions

When no light or nonmodulated light reaches the phototransistor, as shown in Fig. 4O, the phototransistor signal is zero, while the bias reference and full-wave rectified reference signals are the same as before. However, the resultant collector signal will be proportional only to the rectified reference signal because the base signal is the constant bias voltage.

The current through the a-c circuit will be a sinusoidal current without any d-c component so, in this case, the relay is deenergized and drops out.

Another interesting and important condition exists when there is a phase shift of 90 deg between the photo signal and reference signal, as shown in Fig. 4V.

To obtain the resultant collector signal consider adjacent quarter-

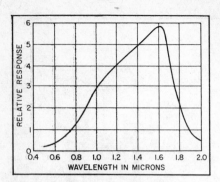

FIG. 2—Typical response of phototransistor to various wave lengths

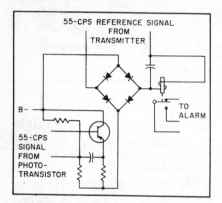

FIG. 3—Simplified representation of basic phase-detector circuit

waves; the first quarter-wave is equal to the fourth, although the mirror image of it. Similarly, the second and third quarter-waves are equal mirror images of each other.

Therefore, the sum of the first and second quarter-waves are equal to the sum of the third and fourth quarter-waves. That they are mirror images of each other does not alter the fact that their areas are equal. Thus, on the a-c side of the rectifier, the positive half-waves will exactly cancel out negative half-waves and give a zero d-c component. The relay is deenergized and the alarm sounds.

Other Cases

For other cases where the phase shift lies between 0 and 90 or 90 and 180 deg, a d-c component will exist in the relay circuit that becomes smaller as the phase shift approaches 90 deg and larger as the phase shift approaches 0 or 180 deg.

From the previous case, it may be seen how the system will behave when the frequencies of the phototransistor signal and reference voltage are slightly different. The two frequencies behave as if the phase between them were continuously changed. Therefore, within periodically repeating time intervals, the two waves will be in phase, 90 deg or 180 deg out of phase, passing through all possible phase conditions. Consequently, the relay is periodically energized and denergized and thus triggers an alarm during the unenergized intervals.

As the same detecting device acts on both half-waves that are being compared, any change in the detecting device acts symmetrically and cannot affect the balance.

Phototransistor Properties

There are some properties of the phototransistor that affect the reliability of the whole system.

The collector current of an uncompensated germanium transistor will double for each 8C temperature rise. This can be somewhat obviated with proper temperature stabilization.

Furthermore, a phototransistor, changes continuously and asymptotically, the phase between light modulation and photocurrent. Under unfavorable conditions, it may slowly reach a phase shift of more than 15 deg during a day of operation. When the phototransistor is disconnected from its power supply the phase immediately returns to its original position.

This phase shift becomes more important the higher the emitter-collector voltage. The effect is based on the formation of capacitance bounded channels, which are nonuniform diffusion *p* and *n* regions penetrating from the surface of the semiconductor into the base. These channels have a certain capacity to the barriers of the semiconductor which increases with time when the static charge builds up under the action of the internal d-c field.

Since the capacitance of a *p-n* junction is a function of voltage and the voltage in turn is a function of the charge that has been built up through resistance with time, the overall capacitance becomes de-

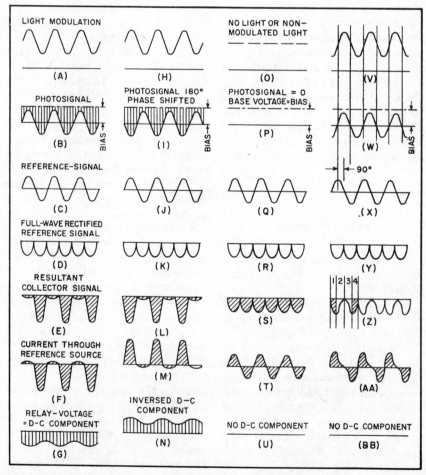

FIG. 4—Typical waveforms in system for different operating conditions

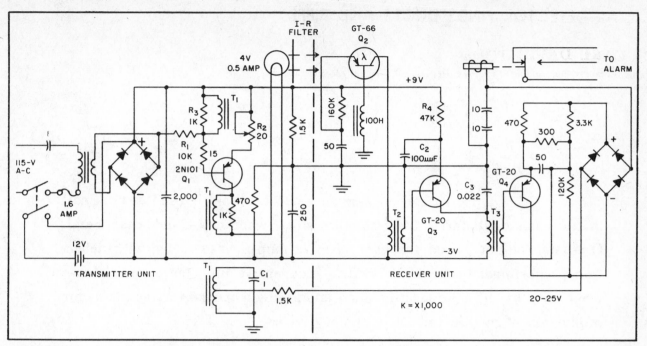

FIG. 5—Complete system normally operates from 115-v a-c power source with storage battery floating; battery takes over load if line supply fails. Capacitor C_1 tunes winding of T_1 to 55 cps, the oscillator frequency

pendent on voltage and time. The entire system corresponds to a rather complicated R-C network in which the capacitance is a function of time and voltage. That time varying network is the origin of the variable phase shift.

Although the network presents large phase shifts to low frequencies (55 cps) the response is hardly affected at higher frequencies.

The phase drift between phototransistor signal and reference voltage must be eliminated as much as possible. By reducing the emitter-collector voltage, the phase shift can be reduced to a nonmeasurable amount. This also reduces the noise level.

Light Dispersion

An important consideration is the dispersion of the light beam energy so a large portion of the light is lost and does not reach the receiver. The narrower the beam angle the more serious such a situation can become.

In general, the loss in light varies inversely as the area that the transmitter beam would tend to cover in the plane of the receiver. For a given distance between transmitter and receiver, the area of the optical image of the light source would increase inversely as the square of the f-num-

ber of the optical system of the transmitter. Thus, going from an optical speed of f2.5 to a speed of f0.25, which can be obtained for a condenser system, decreases the sensitivity to lens fogging 100 to 1. For this reason, the highest speed optical system was used; a sealed beam lamp with a parabolic reflector.

Circuit

The schematic of the photoelectric alarm system is shown in Fig. 5.

The primary of the power transformer, which is a saturable type, forms a voltage regulator, in conjunction with the 1-μf capacitor, that maintains a constant d-c supply of approximately 12 v, independent of line voltage variations of $+$ 20 percent but not independent of load.

Where standby power is required, a 12-v battery is connected to the projector unit. In normal operation the battery is trickle charged by the power supply. In the event of a-c power failure, the battery automatically assumes the load.

Resistor R_1 provides proper base current for oscillator Q_1; R_2 controls degeneration to vary the a-c voltage on the lamp. The winding of T_1 that is in series with the lamp

and collector of Q_1 provides regenerative feedback to the winding that is connected to the base and emitter; R_3 provides loading to prevent parasitic oscillations.

Capacitor C_1 and its associated winding tune the oscillator to approximately 55 cps. The center tap of this winding supplies an a-c voltage of 20 to 25 v which is relatively independent of lamp intensity setting.

Modulated light falling on the photosensitive junction of Q_2 modulates its collector current. This 55-cycle current is coupled through T_2 to amplifier Q_3, the emitter of which is biased through decoupling network R_4, C_2.

Since the thermal time constant of the lamp produces an inductive phase lag in the transmission of the modulated light, it is necessary to correct the phase of the amplified signal with phasing capacitor C_3. The corrected signal is coupled through T_3 to the phase-detector stage.

The phase-detector output results in a filtered d-c across the relay, holding it in the energized position.

The normal sensitivity of the system extends to 700 feet, with a four-to-one safety factor for dust and misalignment factors that affect operation at that distance.

METAL DETECTOR FINDS DUCTS AND PIPES

By CARL DAVID TODD

Design Engineer, Semiconductor Products Department, General Electric Company, Syracuse, New York

Search coil encased in Faraday shield is used as tank coil of transistor oscillator whose frequency shifts when coil comes near metal. Oscillator output is fed through selective amplifier to meter with deflection proportional to frequency shift. Inexpensive unit can speed defense production by aiding in locating ducts and piping when converting plants for production of missile and navigational devices

LOCATING OBJECTS such as conduits, gas and water pipes or heating or ventilating ducts when their exact position is not known requires some method to obtain approximate location before any work may be done. A metal detector is of great assistance in such situations, but the cost of commercial detectors make the purchase of a device of this type impractical except when it is used extensively.

One solution is to build a detector such as described in this article.

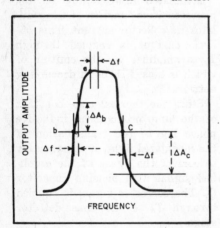

FIG. 1—Selective amplifier response curve shows various possible operating points for amplifier operation

This metal detector is simple in construction and operation, is completely portable yet performs well. Total cost of the device runs approximately $15.

Principle

If a sheet of conducting material is brought near the tank coil of an oscillator, its frequency increases or decreases slightly because of electromagnetic coupling between coil and sheet. This action exists in the use of a tuning wand for the alignment of tuned circuits.

In the metal detector, the search coil is the tank coil of an oscillator. Thus the presence of any metal will cause a shift in the oscillator frequency.

Theory

If the signal from the variable frequency oscillator feeds a selective amplifier, the amplifier output amplitude will be a function of the oscillator frequency. A typical response curve of a selective r-f amplifier is shown in Fig. 1.

If the presence of metal causes a shift in the oscillator frequency by the amount Δf, then the amplifier output will vary by ΔA_b or ΔA_c depending on the operating point on the response curve. If the fixed operating point—that is, the frequency of oscillation when no metal is near the search coil—were at point a the shaft in output would be small depending on the amplifier bandwidth in comparison with the shift in frequency, Δf.

However, if the operating point is at either b or c then the change

in output for the same frequency shift is much greater. The metal detector may be operated at any of the points a, b or c at the discretion of the operator.

To indicate the output level of the selective amplifier a rectifier-amplifier circuit such as shown in Fig. 2 may be used. The r-f signal from the amplifier feeds the base of a transistor. With no signal, the d-c collector current is small. However, when an a-c signal is applied to the base, the transistor will be biased on for the half-cycle in which the base is negative with respect to the emitter.

This causes a collector current to flow that is a function of the amount of a-c signal input.

The meter reads the average of the semisine wave of the collector current, since the meter movement cannot follow the individual r-f pulsation.

Circuit

The complete schematic is given in Fig. 3. The oscillator is a Colpitts type to allow the tank coil to be of the untapped variety, thus simplifying the search coil connections. The tank coil comprises the search coil and is connected to the detector chassis by a shielded coaxial cable to eliminate external capacitance effects.

The output of the oscillator is

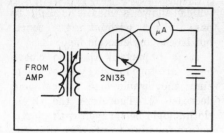

FIG. 2—Circuit indicates r-f output level since collector current is a function of a-c signal input

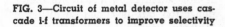

FIG. 3—Circuit of metal detector uses cascade i-f transformers to improve selectivity

taken from the emitter and fed into the selective amplifier by isolating resistor, R_1. This resistor serves to keep the resonant circuit of the input transformer, T_1, from affecting the frequency of oscillation and also decreases the loading effect on the resonant circuit of T_1 thus increasing Q and selectivity.

Only one stage of amplification was used in the model, although a second stage could be used to give greater selectivity and thus greater overall sensitivity of the instrument. If a second amplifying stage were added, the overall gain of the system should not be much more than with one stage so as not to overdrive the meter.

The gain of the amplifiers can be decreased by inserting an unbypassed resistor in the emitter lead thus causing degeneration.

The combination of transformers T_1 and T_2 in cascade serve to increase the selectivity of the selective amplifier. If more transformers are used, it may be necessary to increase the overall gain because of the additional transformer losses.

The value of the supply voltage is not critical, but once set, should not be changed. Collector capacitance is a function of voltage and thus a change in supply voltage would detune the amplifier. There is also the possibility that a change in supply voltage could cause oscillation owing to the shift in value of feedback capacitance required for stable operation.

Construction

The metal detector was constructed in a 5 by 7 by 2 in. aluminum chassis. The oscillator and selective amplifier were built on a small plastic laminate subchassis as may be seen in the photograph. The additional socket, also visible in the photograph, was not used.

The transformers were a type designed as i-f transformers in transistorized receivers. The frequency was chosen at about 435 kc, a frequency low enough to prevent interference in radio receivers using 455-kc i-f amplifiers.

The amplifier is aligned using the level indicator as an output indicator. If a signal generator is not available, the oscillator may be used to align the amplifier. To do this the search coil is brought near a standard a-m radio receiver tuned to any station. The oscillator is tuned until a zero beat is obtained in the radio causing the detector oscillator to operate near 455 kc.

The i-f transformers will usually be tuned to this frequency and some output should be visible on the meter. If not, the oscillator should be tuned back and forth slightly until some output is visible. The i-f

Internal view of detector. Components are mounted on plastic laminate chassis

transformers are then tuned until the output meter reads a maximum.

To decrease the center frequency of the amplifier, decrease the oscillator frequency a small amount by increasing the tuning capacitance. The i-f transformers should be adjusted to give a maximum output at this frequency thus completing the alignment.

To insure that variable capacitor C_1 is at midrange at the aligned frequency capacitor C_2 may be changed slightly. This insures that the operating point on the frequency response curve may be selected on the desired slope.

If the output level is insufficient to give greater than full-scale deflection at the resonant frequency, then resistor R_1 may be decreased slightly.

Neutralizing

While the value of the neutralizing capacitor, C_3, is shown as 95 $\mu\mu$F, the exact value may be somewhat different. This capacitor prevents the amplifying stage from oscillating.

In some configurations, slightly higher or slightly lower values of neutralization capacitance may be needed, depending on the transformers used, the collector capacitance of the transistor, and the supply voltage.

The search coil is an air coil with an inductance of roughly 500 μH. The coil consists of approximately 50 turns of number 30 enameled wire around a 2-in. cardboard tube. A larger diameter coil with the same inductance would give greater sensitivity.

Shield

If precautions to prevent external capacitance effects were not used, the oscillator frequency would be shifted whenever the coil came close to any object, metallic or not. To eliminate this problem, the entire coil is enclosed in a Faraday shield, which allows a magnetic field to pass through without interference, but blocks any electric field.

It is important that no closed loops are produced by the shield since they would load the coil and decrease Q. Therefore the shield cloth should be sprayed with plastic before forming to prevent adjacent wires from shorting to each other.

To operate the completed metal detector the oscillator is tuned by the variable capacitor until a deflection is noted on the meter. For maximum sensitivity the oscillator is not tuned from maximum meter deflection but is tuned to either a slightly higher or slightly lower frequency point.

If the search coil is brought near any metallic conductor, the meter deflection will either increase or decrease a slight amount. The direction of the change in meter deflection depends on the choice of points b or c for an operating point on the frequency response curve.

Chapter 14

AIRCRAFT, MISSILE AND SATELLITE TELEMETRY

PHASE-SHIFT OSCILLATOR FOR STRAIN GAGES

By **WILLIAM H. FOSTER**
Research Engineer, Electronic Engineering Company of California, Los Angeles, California

Completely transistorized strain-gage oscillator for resistive-type gages produces frequency-modulated signal output that is directly proportional to applied force such as stress or pressure. Though intended for aircraft and missile flight testing, unit also has applications in spectroscopy, thermodynamics and mechanics

ONE PROBLEM associated with flight testing of aircraft and missiles is the telemetering and/or recording of stress, pressure and other information. During flight testing such data is sometimes gathered by resistive-type strain bridge transducers that convert stress variations to voltage level changes.

Varying voltage levels are not directly compatible with f-m telemetering or magnetic tape recording. Accordingly, it is necessary to employ a converter for changing such amplitude-varying voltages to frequency-varying signals. The strain-gage oscillator to be de-

scribed accomplishes this conversion with low power consumption and maximum space utilization.

System Description

The strain-gage oscillator is essentially a phase-shift oscillator used in conjunction with a four-arm resistance bridge. Unbalance of the bridge causes the oscillator frequency to vary, thus generating an f-m signal which is directly proportional to bridge unbalance or strain.

Transistors are used exclusively as the amplifying elements throughout the circuit.

Heart of the oscillator, shown in

Fig. 1 and 2 is a stable high-gain a-c amplifier that operates in conjunction with the phase shifter and phase corrector to comprise a form of phase-shift oscillator when the bridge is balanced. The amplifier output signal is applied directly to the phase shifter where it undergoes 90-deg phase shift and attenuation. The phase shifter output signal is then fed through the mixer to the input of the phase corrector, the bridge being balanced and furnishing zero signal for mixing. The phase corrector shifts the signal another 90-deg and attenuates it, making the signal of the proper phase and amplitude to

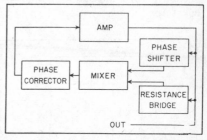

FIG. 1—Block diagram of transistorized strain-gage oscillator

Complete strain gage oscillator weighs only 8 oz and occupies space of 6 cu in. Units have been constructed with bandwidths of 80 percent (± 40 percent frequency and an output of 5-v p-p across a 2,500-ohm load

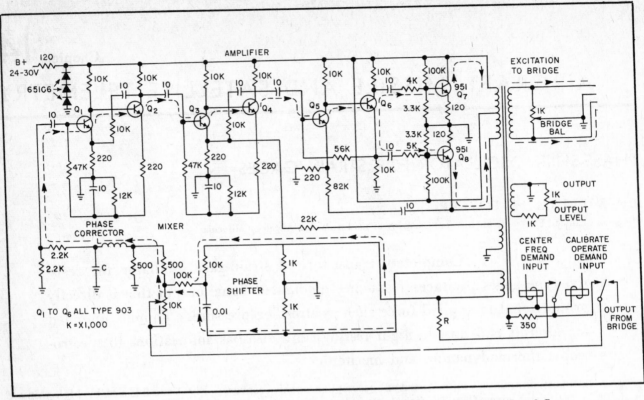

FIG 2—Operating and band-edge frequencies of strain gage oscillator are determined by values of R, L and C

satisfy the Nyquist criterion for oscillation ($K_\beta = 1$).

When the resistance bridge is unbalanced, as a result of applied stress or pressure for instance, the resulting bridge output voltage is resistively mixed with the local feedback signal (from the phase shifter) in the mixer. The sum of the two signals yields a voltage of the same relative amplitude as in the case of the balanced bridge, but of differing phase; this new signal is then applied to the phase corrector.

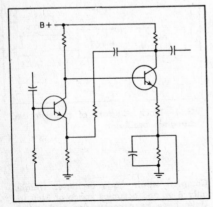

FIG. 3—Basic transistor amplifier pair used in amplifier portion of oscillator for best a-c and d-c stability

Consequently, the Nyquist criterion for oscillation is satisfied at a different frequency which is directly related to the new mixer output signal. The mixer output signal is in turn proportional to the bridge output voltage which is a direct function of the applied excitation phenomenon such as stress or pressure.

Amplifier

Referring to Fig. 2, the amplifier consists of eight transistor stages connected in the common-emitter configuration to provide a minimum open-loop gain of 10^6 and 180-deg phase reversal of the input signal at low frequencies.

The transistor amplifier pair shown in Fig. 3 was employed as the most satisfactory compromise between a-c and d-c stability. It tends to be self-compensating for d-c drift due to ambient temperature variations. Additional d-c temperature stability is obtained by supplying the first transistor bias current from the second transistor emitter circuit.

This configuration has practically no loss in theoretical gain and has

the additional advantages of offering good operation with transistors of widely differing parameters and being practically unaffected by replacement of transistors.

For a-c temperature stability and high input impedance, negative feedback is applied around the common-emitter-pair amplifier.

From Fig. 2, it is seen that the first four transistors comprise two common-emitter-pair amplifiers The fifth and sixth transistors comprise a third common-emitter-pair amplifier with the sixth transistor connected as a paraphase amplifier capable of driving a push-pull pair of class AB common-emitter output transistors.

In addition to the extreme a-c and d-c temperature stability characteristics, the amplifier exhibits specific low and high-frequency roll-off and linear phase response as a function of frequency.

Low- and high-frequency roll-off are essential to avoid undesirable oscillations which normally occur as a result of reactive effects inherent in multiple transistor feedback amplifiers.

Linear phase response is desirable to ensure linear deviation of frequency above and below the nominal or center frequency.

Phase Shifter

When the oscillator is operating at nominal or center frequency, the output signal from the amplifier is phase shifted by 90-deg, attenuated somewhat and applied to one of the mixer input terminals.

The phase shifter shown in Fig. 2 does an efficient job. The secondary of the transformer, together with a resistance and capacitance comprise a bridge network. If a signal is applied through the transformer and the output is taken, as shown in Fig. 2, from the common point of the R-C network to the grounded center tap of the transformer, this output signal may be adjusted for any phase from about 0 to 180 deg with respect to the input, depending upon the ratio of resistance and capacitance. However, once adjusted in a given strain gage oscillator, the phase shift variation is approximately ±5 deg throughout a bandwidth of ±7.5 percent.

Mixer

The mixer circuit receives the phase shifter and bridge output signals and resistively mixes them, yielding a simple vector sum which is applied to the phase corrector.

Both the function and design of the mixer are simple. Presenting a high input impedance to the resistance bridge and the phase shifter, the mixer does not load the bridge unduly and it does not enter into the 90-deg phase shifting adjustment. Presenting a low impedance to the phase corrector circuit, it is not loaded by the phase corrector and the reactance of the phase corrector is not reflected back to the bridge or phase shifter circuit.

The phase corrector circuit is a simple low-pass L-C filter. This filter provides the additional 90-deg phase shift and attenuation necessary to sustain oscillations of the phase-shift oscillator. Since the circuit is correctly terminated and driven from the proper effective generator impedance, it provides the ideal linear phase shift versus frequency characteristic desired

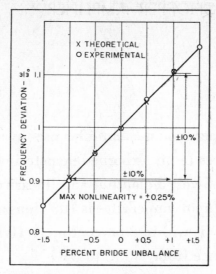

FIG. 4—Comparison of theoretical and experimental performances of oscillator

throughout the operating range of frequencies.

Calibrator

Included within the strain gage oscillator are relays and appropriate calibration resistors such that, upon demand, the oscillator presents a center-frequency signal corresponding to a balanced-bridge condition and then a band-edge frequency corresponding to a full-scale unbalanced bridge. The obvious value of these signals is in the playback system, where the calibration frequencies may be compared to standards and appropriate corrections for center-frequency drift and/or sensitivity change can be applied, if needed, to the actual data obtained from the oscillator.

During the center-frequency calibration period a precision resistor is placed between the mixer bridge input and ground, replacing the bridge output. This causes the amplifier input to be solely a local feedback signal which in turn causes the oscillator to operate at its center frequency.

During the band-edge or full-scale frequency calibration period, a precise fraction of the bridge excitation signal is placed at the mixer bridge input, replacing the bridge output as in the case of center frequency calibration. This is made to correspond to a full scale unbalanced bridge output signal and causes the oscillator to oscillate at a band-edge frequency.

The oscillator is capable of pre-

cise self-calibration because the ratio, not the absolute value, of the local feedback and the bridge output signals causes the frequency of oscillation to deviate. The oscillator offers an advantage over voltage-controlled oscillators in that it requires no external transducer excitation voltages.

Performance

Laboratory models and prototypes of the transistorized oscillator have been constructed and subjected to extensive temperature testing. The test results are shown in Fig. 4. The individual units shown in the block diagram have also been tested independently of the oscillator and are now operating satisfactorily in the field.

Applications

Because of its small size and low power requirement, the transistorized strain-gage oscillator finds major application in the fields of telemetry and remote control, particularly in the flight testing of aircraft and missiles. However, other applications are numerous.

For example, the oscillator may be used in conjunction with thermistors or hot wires to yield a good instrumentation device in the field of thermodynamics. A simple modification of the oscillator, in conjunction with photo diodes or photo transistors, yields a valuable instrument in spectroscopy.

In conjunction with pressure transducers or accelerometers utilizing resistance bridge principles, the oscillator yields simple measurement of mechanical phenomena.

Variations of the oscillator have been constructed for use as low-level voltage-controlled oscillators. Measurements of d-c levels in the millivolt range are readily and accurately obtainable.

The strain-gage oscillator may be used in conjunction with resistance-bridge transducers ranging in resistance from 100 ohms to 1,000 ohms. Total power required is less than 500 mw, including the necessary resistance bridge transducer excitation voltage of 2 v rms.

Units have been constructed that operate efficiently on 50 mw. This unit was developed under USAF contract No. AF04 (611)-683.

MISSILE TELEMETER USES CHOPPER AMPLIFIER

By JOHN H. PORTER

Portronics, Inc., Rochester, New York

Chopper-type d-c amplifier uses available channels to indicate missile temperatures in an airborne telemetering system. Unit has voltage gain of 1,000 with 5-volt d-c output and linearity within 2 percent over the full output range. Input impedance is 100 ohms and response is flat from zero to 10 cps. Stability is within 2 percent up to 10 g vibration at 1,000 cps or over temperature range from —65C to 85C

IN an airborne telemetering system, the thermocouple may be far physically separated from the telemeter and make the thermocouple appear as a generator of finite, but not necessarily constant, internal impedance. Typical impedances may be 10 to 50 ohms and may vary as much as 20 percent due to temperature changes.

Terminal emf of this generator may vary from 0.1 to 50 or more millivolts, depending on the application. While response time of a thermocouple is generally considered rather low, the amplifier should pass faithfully a range from 0 to 10 cps. The reference junction can be considered as a source of bucking voltage in series with the couple.

Most telemetering systems operate from voltage actuated devices, with the possible exception of transistorized subcarrier oscillators, or the ptm or pwm systems and a load of 250,000 ohms is typical. An amplifier with a basic voltage gain of 1,000 is adequate to supply up to 5 volts d-c output. A simple attenuator in the amplifier input or output can be used to adjust the output level for less gain.

Power available in most instances is 115 v at 400 cps, with a

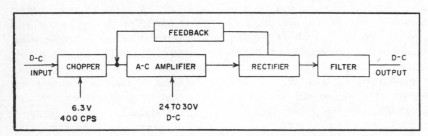

FIG. 1—Block diagram of transistorized chopper system utilizes synchronous modulation of d-c input and demodulation of output from a-c amplifier

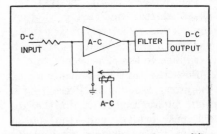

FIG. 2—Basic full-chopper d-c amplifier was adapted to missile instrumentation

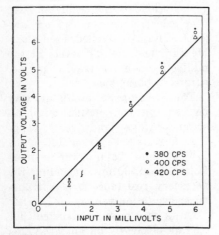

FIG. 3—Line frequency variation for 380, 400 and 420 cps at 6.3 volts rms

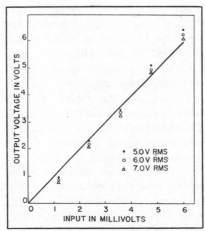

FIG. 4—Chopper drive variation for 5, 6 and 7 volts rms at 400 cps

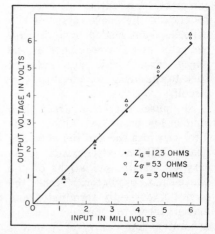

FIG. 5—Generator impedance variation at 3, 53 and 133 ohms

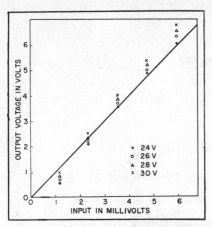

FIG. 6—Power supply voltage variation at 24 to 30 volts

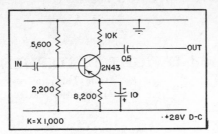

FIG. 7—Typical d-c amplifier circuit. The system employs three iterated common-emitter stages

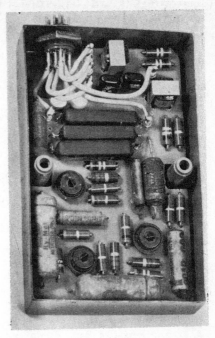

Compact amplifier unit weighs less than eight ounces. All connections are made to nine-pin connector shown at top left

±5 percent tolerance on both parameters and d-c between 24 and 30 from a generator or transformer-rectifier combination.

Transistorized telemeters to generate enough r-f power for successful ground station operation still require vacuum tubes in the output stage. This implies a B+ voltage of 180 to 350 or more, as well as a low voltage filament supply, generally and nominally 6.3 at 400 cps.

Vibration surveys within the airframe necessitate a B+ voltage for associated a-c amplifiers. However rapidly advancing transistor development may soon produce a unit whose input impedance is high enough to be compatible with vibration pickups and obviate the need of a plate power supply.

Dissipation of heat is becoming more of a problem and highest possible efficiencies must be sought in the amplifier. Other major problems are space limitations and vibration isolation facilities.

All of these considerations, resulted in the development of a transistorized adaptation of a full-chopper amplifier having a linearity of 2 percent or less over full output range and stability of less than 2 percent change in characteristics in an environment of up to 10 g vibration at 1,000 cps, or over the temperature range from −65 C. to +85 C.

System Chosen

The block diagram in Fig. 1 shows the transistorized system adapted from the full-chopper d-c amplifier in Fig. 2. Typical performance of the system is shown in Figs. 3 through 6.

The chopper portion employs two surface barrier transistors in the grounded-collector configuration. Matching of chopper units was found necessary; however 20 pairs were obtained from a random lot of 50 transistors, by selecting a pair for equal emitter currents.

Chopper output is a particularly clean square wave with no switching transients, as it alternately connects the amplifier input to the signal source and to ground.

Input impedance appears to be a minor function of generator impedance but for source impedances of up to several hundred ohms is in the vicinity of 2,000 ohms.

Overall open-loop gain of 100 db is realized by the amplifier, which consists of three iterated common-emitter stages.

A typical stage is shown in Fig. 7. Component values produce a stability factor $S = 1.1$ and each stage is designed to operate with 10 volts at 1-ma collector current. To provide partial isolation against variations in power supply impedance, a decoupling filter is used in the supply to the first stage. Output of the third stage is capacitively coupled to a voltage doubling rectifier circuit, using miniature selenium diodes, permissible in this application due to the high load impedance and is smoothed by a capacitive filter. The output return is common to the negative of the 28-v supply.

Efficiency of such a network is nearly as high as that of a synchronous demodulator and is considerably more economical of transistors and other components.

STRAIN-GAGE SYSTEM FOR AIRCRAFT TELEMETERING

By WILLIAM O. BROOKS and DWIGHT L. STEPHENSON

Ramo-Wooldridge Corp.
Los Angeles, California

Los Alamos Laboratories
Los Alamos, New Mexico

High-accuracy transistorized strain-gage system for aircraft and missile applications has one-percent linearity and provides 0 to 5-volt d-c output suitable for subcarrier oscillator modulation in f-m telemetering systems. Bridge excitation supply powers three paralleled gages having total resistance of 40 ohms

HIGH-PERFORMANCE AIRCRAFT and missiles have made the need for high-accuracy telemetered strained-gage systems more critical.

This article describes the design of a prototype strain-gage instrumentation system having a 0 to 5-v d-c output suitable for telemetering. In addition to the strain-gage, the system as shown in Fig. 1 includes a standardized bridge-excitation supply capable of supplying 1 to 3 strain-gage bridges and amplifiers and a strain-gage amplifier-converter to provide a d-c signal from the a-c strain-gage output.

Bridge Excitation Supply

The regulated a-c power supply is shown in block form in Fig. 2. The design approach used was to standardize the input voltage and then amplify it to the required power level with a stable amplifier. The requirement of ±1-percent regulation meant that a stable a-c voltage standard had to be developed first.

The a-c voltage standard was developed around the use of Zener diodes to clip the incoming variable a-c into a fixed amplitude square wave and then filtering out the harmonics to a sine wave by use of a low-pass filter. The 1N429 diodes chosen are of the double-anode type and clip both sides of the waveform symmetrically. When operated at an a-c peak-current rating of 7.5 ma, they have a low temperature co-efficient.

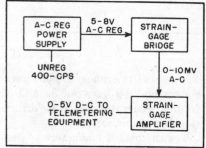

FIG. 1—Strain-gage instrumentation system

With an input line variation of as much as ±10-percent it is necessary to use two clipping stages to limit the output variation to the required minimum.

The output of the clipper circuit goes to a low-pass filter with a flat response from 380 to 420 cps and a rejection of 40 db at the second harmonic of 760 cps to obtain the lowest possible distortion. The best filter obtainable was flat within ±0.5-percent at room temperature over this frequency range.

Variations in output from the filter due to copper losses and so on amounted to as much as 3 percent total over the temperature range concerned. Two deposited-carbon resistors were used for compensation, each having a negative temperature coefficient of approximately 300 parts per million in series-circuit legs.

Transistor Amplifier

Silicon transistors were required in the amplifier shown in Fig. 3 to

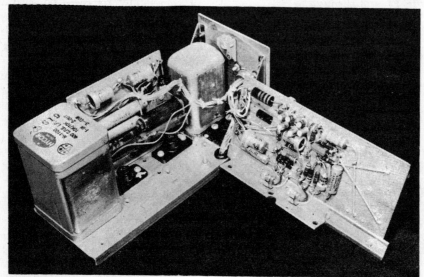

Overall view of prototype version of complete strain-gage instrumentation

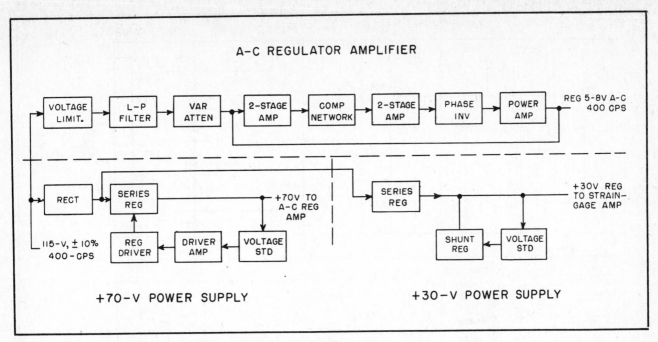

A-C REGULATOR AMPLIFIER

FIG. 2—Block diagram of bridge excitation and strain gage amplifier power supply

cover the wide temperature range. The available high-power transistors of the silicon variety were type 970's having a dissipation rating of 8.5 at 25 C or 3.5 at 100 C. at +100 C, their class B sine-wave output is approximately 2.5 w. Considering 80-percent output transformer efficiency, this permits an output of approximately 2 w.

Since these transistors do not have a military specification, variations in leakage current, current gain and power output are great. For use in class B, with a minimum of distortion, matched pairs must be used.

The emitters of the 970's tie directly to ground. Elimination of the emitter resistances is necessary to obtain the required output power.

Base bias is set at a point giving approximately 2 ma of total no-signal collector current to overcome the switching step of class-B operation. A thermistor network in each 970 base keeps this 2 ma constant by continuously lowering the base bias voltage with increasing temperature, thereby compensating for the increase of collector current brought about by increased leakage current within the transistor.

Driver

Experiments proved that the driver requirements could be met with a standard phase inverter resistance coupled to the output transistors. The output impedance of the pushpull 970's is 2,000 ohms collector to collector and the input impedance is 140 ohms base to base or approximately 35 ohms each.

The output stage operates best with an impedance mismatch by the driver; a high driving impedance is required to obtain the smallest step in the output waveform. Sufficient power output with good wave form is obtained from the type 953 driver transistor with an 820-ohm load resistor in the collector and emitter circuits respectively.

Large values of coupling capacitors prevent loss at low frequencies when working into the low input base resistances of the type 970 output transistors.

The 40 db of negative feedback in a single loop requires special compensation to prevent instability and oscillation. This compensation becomes more critical with variable loading. A value of 35 db feedback at 40-ohms load increased to 45 db with 120 ohms load and to 55 db with open-circuit loading as shown in Fig. 4. To insure stability, a 10-db safety margin is required making a total of 65 db, an exceptional amount in one loop around a five-stage amplifier, when the

output transformer is included.

This quantity of feedback was made possible by special shaping networks. The networks are combined and added in one low-level stage prior to the driver permitting the other stages prior to the driver to be direct coupled, thus giving the full range response with little frequency attenuation.

D-C Power Supplies

The 100-ma, 70-v supply is of the series regulator type and uses a type 970 transistor as an emitter follower. A type 953 transistor, connected as an emitter follower, provides necessary current gain to control the 970 power transistor.

Resistor R_1 is a current limiting protective resistor that prevents burnout of transistor Q_2 if the circuit malfunctions. Transistor Q_4, connected as a grounded emitter, provides the necessary gain and phase reversal for the correct voltage control.

The voltage standard is a series-string of 15 4.9-v Zener diodes in series with the base of Q_4. These low-voltage Zener diodes were used instead of one 70-v diode to increase dissipation limits and to give a low temperature coefficient when used over the required wide temperature range.

Resistor R_2 causes approximately

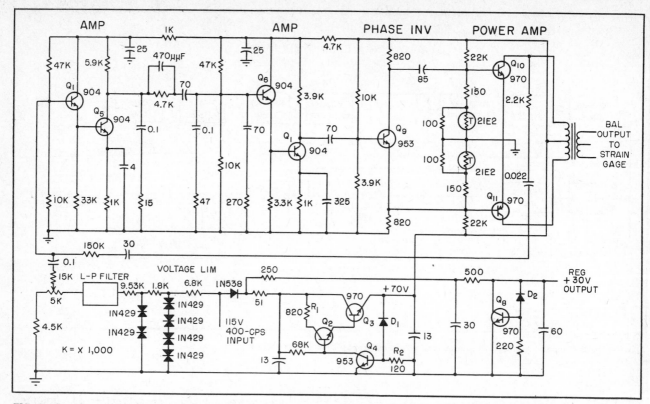

FIG. 3—Complete circuit diagram of excitation and power supply. Diode D_1 represents 15 651C4 diodes connected in series; D_2 is six 651C5 diodes in series

5 ma of continuous current drain through the Zener diodes to place them in the most stable operating part of their temperature characteristic.

The 30-v d-c supply is of the shunt-regulator type and uses the same type of Zener diode voltage

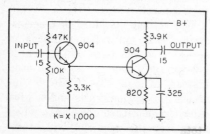

FIG. 4—Response curves for a-c regulated power supply with 40-ohm (A), 120-ohm (B) and open-circuit (C) output loads

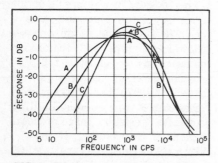

FIG. 5—Two-stage amplifier has 45-db gain

standard; however, additional amplifier stages are not needed because only ±5-percent regulation is required.

It can be seen from the photographs the exitation-supply unit is a preprototype unit and would need repackaging to take vibration tests. It was developed to determine if it would need a complete potted assembly to distribute heat well enough to stay within the ±1-percent amplitude-variation specifications over the specified temperature range. The unit passed the test at +0.6-percent and −0.4-percent variation with a maximum of 2-percent distortion indicative that the unit could be repackaged in the unpotted form.

Strain-Gage Amplifier

The strain-gage amplifier receives its signal input from an a-c excited strain-gage bridge. This input signal is balanced to ground, and is at approximately 0 to 10 mv.

The amplifier output is used to feed a telemetering input and is 0 to 5v d-c into a 1-meg load from a 10,000-ohm output impedance.

Frequency response is 0 to 10 cps and the requirement for overall

linearity an stability of 1 percent from the a-c input to the d-c output dictates a feedback loop encompassing both amplifier and rectifiers so as to eliminate rectifier non-linearity and instability with temperature changes.

Design Approach

Rough calculations using estimated variations in transistor and diode parameters, both production variations and those induced by temperature variations, indicated that approximately 40 db of feedback would be required to achieve the desired linearity and gain stability. The a-c input to d-c output voltage gain required was 54 db; therefore total open-loop amplifier gain was determined to be 94 db. Shaping networks produce an additional loss of 20 db, making the total gain requirement 120 db.

Due to the extreme temperature variations encountered, silicon diodes and transistors are used throughout, transistor bias networks are designed to provide good stability factors.

Figure 5 illustrates a typical two-stage grounded-collector to ground-emitter amplifier and the associated

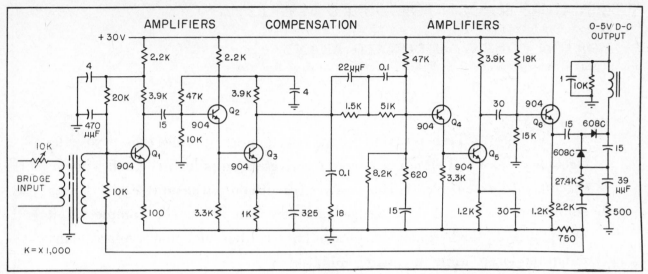

FIG. 6—Transistorized strain-gage signal amplifier converts a-c input into d-c output through voltage doubler and low-pass filter

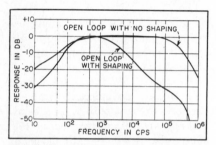

FIG. 7—Strain gage amplifier response

bias stabilization network. This amplifier has no apparent shift of operating point over the required temperature range; typical gain is 45 db and output voltage capability is 3.5v rms. Direct coupling serves the multiple purpose of reducing phase shift and the number of components required while also providing excellent stability.

The stability factor of the grounded-collector stage is calculated to be $S = 3.3$. Stability of the second stage is also excellent.

A-C Amplifier

Two of these basic amplifiers plus a phase-inverting grounded-emitter stage and an emitter-follower output stage make up the a-c amplifier portion (Fig. 6) of the strain gage amplifier.

A silicon-diode voltage-doubler and an L-C filter provide the a-c to d-c conversion. The 500-ohm wire-wound resistor, R_{fb} provides the rectifier current feedback impedance. The feedback loop therefore includes the diodes and the input transformers. Considerable amplifier frequency-response shaping

was required by use of lead-lag networks to achieve a satisfactory phase and gain stability margin.

Amplifier Performance

Figure 7 illustrates the frequency response both open loop and closed loop including the shaping networks. Due to the 120-db open-loop gain, it was necessary to make the plots by holding the output constant at an arbitrary 0 db while varying the input voltage.

Figure 8 illustrates the a-c output waveform as seen at the input to the voltage-doubler circuit. The steep sides indicate essentially zero feedback and extremely high amplifier gain during zero diode current, while the rounded tops represent

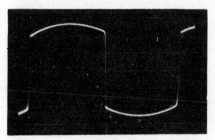

FIG. 8—Waveform at input to voltage-doubler circuit in Fig. 6

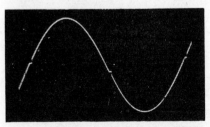

FIG. 9—Niches in feedback voltage waveform indicate diode cutoff points

sent the amplified sine wave with heavy feedback.

Figure 9 shows the feedback voltage; the niches indicate diode cutoff points, and therefore, no feedback.

Environmental tests indicated excellent linearity and stability from approximately −10F to +185 F. Linearity remained good below this point; however, gain stability dropped off somewhat. For this reason the final packaged unit contains a miniature thermostat and two parallel internal heaters.

The metal container is lined with sheets of plastic foam for heat insulation. The entire package is capable of maintaining an 80F temperature differential and therefore keeps the internal temperature well above the low temperature limitations. The entire amplified assembly is potted with a standard potting compound and a 20-percent (by weight) filler is used to enhance the thermal conductivity and thereby minimize hot spots.

Measurements over the entire temperature range indicate that stability and linearity are within 1 percent from 0.1v to 5v d-c. Ripple on the d-c output is less than 50mv under all conditions.

The complete system provides 1-percent linearity and stability over the following environmental conditions: temperature, −55 to 185F; vibration, 20 to 2,000 cps at 10 gs; altitude, 70,000 ft; shock, 15 gs; input voltage variations, 105 to 125v rms; input frequency variations, 380 to 420 cps.

SUBMINIATURE BEACON FOR GUIDED MISSILES

By MORTON COHEN and DONALD ARANY

Senior Engineer, Senior Project Engineer, Radio Receptor Co., Brooklyn, New York

Transistorized S-band transponder provides echo to missile-tracking radars and signal that could activate a missile fuel-cutoff system. Circuits of preselector, video receiver, modulator, transmitter and power supply are given and their design considerations discussed. Complete unit occupies a 2.5-inch diameter, 6-inch long volume and may replace many older and larger units in existing missiles

BEACON TRANSPONDER AN/DPN-43, to be described, normally provides two basic functions. It supplies an amplified echo for missile-tracking radars and an audio-command signal that could be used to activate a missile fuel-cutoff system. It is possible that this unit may replace other beacons of larger size and differing shape.

System

The beacon contains six main subsections: duplexer, preselector, crystal video receiver, modulator, transmitting cavity and power supply. The system shown in Fig. 1 receives and transmits over the frequency range from 2,700 to 2,950 mc. The receiver sensitivity is better than − 35 dbm and the transmitted power output is greater than 1 watt.

Radio-frequency pulse pairs spaced 3-μsec apart are used to interrogate the receiver. Coincidence circuitry in the decoder of the video receiver then operates the modulator. The modulator in turn drives the r-f cavity to transmit a single pulse reply. The power supply operates from a six-volt battery with an estimated operational life of about 30 minutes. Total weight is approximately 2.5 lb and the beacon will operate at ambient temperatures in excess of 70 C and below 0 C and at altitudes exceeding 100,000 yards. It will sustain shocks of at least 100 g.

Duplexer

The duplexer consists of two electrical lines W_1 and W_2. Their lengths are so chosen that W_1 is

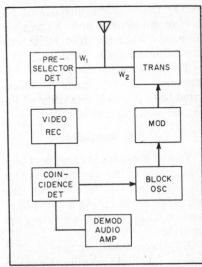

FIG. 1.—Block diagram of compact transponder operating in S band

Beacon video amplifiers are adjusted for proper response characteristic

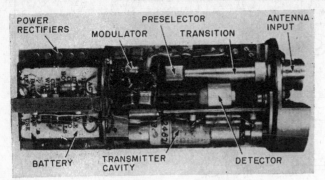

Internal views of beacon show compact construction and parts layout required to keep size and weight down

$\lambda/4$ and W_2 is $n\lambda/2$ at a nominal mid-band frequency of 2,825 mc. Over the ±5-percent nominal frequency swing, the impedance looking into W_2 from the antenna is quite high. Consequently there is little loss of energy from the antenna to the receiver input. Looking into W_1 from the transmitter cavity, the impedance is also quite high. Therefore most of the transmitter's output power is diverted into the antenna where it belongs.

Preselector and Transmitter

The preselector shown in Fig. 2 consists of two line sections that are coupled together through an iris and are capacitance tuned. Nominal bandwidths obtained with this design are > 6 mc at − 3 db, < 60 mc at − 20 db and < 120 mc at − 35 db down.

Crystal CR_1 is a 1N32 used as the crystal detector. The video crystal is also tuned by a fixed length of line.

The transmitting cavity, also shown in Fig. 2, consists of a ruggedized fixed-tuned ultrahigh frequency oscillator triode intended for grid-pulsed oscillator service between 2,700 and 2,950 mc. The cavity is a tunable plate tank coupled back to a fixed-tuned cathode line section. Power is capacitively coupled to the antenna output. Power output is approximately 1 to 10 watts across 50 ohms.

Crystal Video Receiver

The crystal video receiver shown in Fig. 3 consists of 6 transistor stages, Q_1 through Q_6. This receiver

has a 35-db r-f dynamic range which implies a 70-db video dynamic range. This results from the square law operation of the detector.

Three main problems are encountered as a result of this large dynamic range: pulse widening wherein the output pulse width becomes a function of the input signal level, spurious responses that appear at the output because of overshoot at high input levels and the s/n ratio tendency to deteriorate owing to limiting in many of the stages. To insure proper operation of the decoder none of these conditions can be tolerated.

Storage

One cause of pulse widening is that semiconductors exhibit to a varying extent a storage property that discharge after the input energy has ceased. This storage energy varies with input energy level. Hence this effect causes successive widening of the pulse with each succeeding stage. Silicon junction transistors, while superior for their temperature stability, exhibit this storage effect.

To counteract this problem 2N128 surface-barrier germanium transistors were chosen. This transistor has a negligible storage time constant making it suitable for this application. However, they do not exhibit the good thermal stability of silicon transistors.

A second cause of pulse widening results from the fact that the rise and decay time of the pulse is finite. The f_{aco} of the 2N128 is 30

mc, β is approximately 20 for the common emitter configuration and $f_{\beta co}$ equals $1/\beta \times f_{aco}$ resulting in an $f_{\beta co}$ of 1.5 mc. This corresponds to a rise and decay time of 0.24 μsec. Hence the 1-μsec pulse is more of a trapezoid than a square pulse.

Therefore a low-level signal will be degraded only to the above extent. A high-level signal rise time will be improved because of limiting. However, limiting will cause pulse widening, thus the output pulse width will be a function of the input level.

To eliminate this problem, an overshoot is purposely introduced in Q_2 by introducing a short time constant in the emitter circuit. This forces the pulse to rapidly cross the base line with an extremely small decay time. Hence the squareness of the trailing edge is restored and pulse widening is minimized but an overshoot problem is created.

Compromise

If each transistor stage were biased full on or full off most of the overshoot would be clipped off. However these two extremes are far from optimum operating points for a pulse amplifier of this nature. In the region near cutoff, α is low hence the stage gain is low. In the saturation region, the input impedance is low and power consumption is relatively high. Thus to obtain a reasonable gain per stage, a compromise must be employed.

It was necessary to make certain that any overshoot existing in the

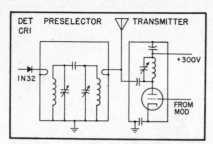

FIG. 2—Preselector and transmitter feed common antenna through duplexer and transition section of coaxial cable

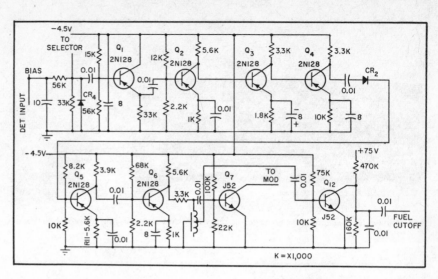

FIG. 3—Video receiver uses direct coupling wherever possible to maintain stability

early stages was not appreciably amplified by the latter stages. Otherwise the overshoot would cross the base line and appear at the output as a spurious response of the same polarity as the desired signal.

To minimize the overshoot, diode CR_2 was placed between Q_4 and Q_5. Also the quiescent operating points of these stages were carefully chosen to minimize the amplification of the overshoot with only a slight loss in gain.

Overloading

Another problem occurs when the transmitter cavity and preselector cavity are separated by a small frequency increment. The receiver overloads for all high-level signals. Diode CR_4, back-biased slightly, limits the incoming video signal to 0.6 volt. As a result the transmitter and preselector may be tuned to the same frequency with only a 2-db loss in receiver sensitivity. The burn-out rating of CR_1 in the detector is just high enough so the transmitter cavity power does not burn out the diode.

Because germanium transistors are employed, each stage had to be temperature stabilized over the operating range of 0 to 70 C. A standard technique in achieving this stabilization is to insert a bypassed resistor in the emitter circuit.

The value of series emitter resistance does not effect the dynamic operation of the stage as it is bypassed but it does control and limit the optimum setting of the quies-

cent operating point.

Direct coupling was used where practical, thus eliminating coupling overshoots and ensuring good temperature stability with minimum loading effects.

Noise

Since the minimum input signal level is well above the noise level, the design allows biasing the input stages to reduce residual noise. The signal to noise ratio has consequently been improved rather than deteriorated through the receiver.

Limiting occurs over the entire dynamic range. Additional helpful results of limiting cause a better pulse rise time characteristic and supply a constant amplitude output pulse to the coincidence circuitry. Overall video gain is about 100 db.

Coincidence circuitry at the input of Q_7 enables the detection of pulse pairs. An open-circuited delay line, 1.5 μsec long, reflects a pulse of the same incident polarity. A succeeding input pulse 3 μsec later will therefore occur in coincidence with the reflected pulse al-

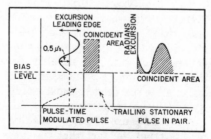

FIG. 4—Pulse-time modulated pulse is compared with stationary second pulse in coincidence circuit to get fuel cutoff signal

lowing Q_7 to conduct and thence initiate the blocking oscillator.

A tap on the delay line drives an audio amplifier that supplies an audio command signal for external use. The leading pulse of the pulse pair is pulse-time-modulated at an audio rate with a nominal excursion of ±0.5 μsec. The delay-line tap converts the pulse-time modulation to a pulse-position modulation due to coincidence action as shown in Fig. 4.

Suitable time constants in the demodulator audio amplifier Q_{12} afford an audio output signal of about 150 millivolts.

Modulator

The modulator shown in Fig. 5 consists of blocking oscillator stage Q_8 and two grounded collector stages Q_9 and Q_{13}.

A negative pulse, impressed upon the collector of Q_8 is reflected as a positive pulse at its base. A positive bias is applied to the emitter through network R_1 and R_2 which biases off the base. If the trigger is of sufficient amplitude with respect to the base bias level the transistor will regenerate and blocking oscillator action will ensue. The overshoot of the blocking oscillator pulse will affect the recovery time since the overshoot drives the base more negative and any trigger impressed during this interval will not allow the blocking oscillator to fire.

A junction diode could short out this overshoot decreasing its amplitude but, because of the long storage time effect in such diodes, recovery time would be excessive.

A point-contact computer diode such as the 1N191 has less of a storage effect, however it is still excessive. By putting two 1N191 in series, the storage effect is greatly reduced and a recovery time of better than 30 μsec is obtained.

Count Down

The beacon is required to operate up to a 4-kc repetition rate without count down and from 4 to 10 kc with a count-down rate that would ensure that the transmitter cavity would operate within its maximum dissipation rating.

One count-down method purposely introduces a large overshoot in the blocking oscillator pulse. However this method results in a recovery time equal to the time constant of the count-down circuit.

The approach assumed incorporates a circuit with a controlled time constant which will not affect the recovery time of the blocking oscillator. The energy is then integrated and fed back as a dc back bias to the base of Q_8. This energy has to be a function of the output repetition rate of the blocking oscillator.

Feedback

Operation of the circuit requires that the negative pulse output of the blocking oscillator be reversed in phase. This positive pulse then drives Q_9 which in turn drives Q_{13}, both grounded-collector stages. These stages are required to obtain power gain to drive the grid of the transmitter cavity. Capacitor C_1 is charged through R_3 to ground when CR_5 conducts.

After the initiating pulse energy has ceased C_1 discharges through R_3, R_4 and R_5 in series. The discharge current through R_4 is negative with respect to ground. The d-c level of this voltage is a function of the output repetition rate of the blocking oscillator. This voltage is fed back through isolating resistor R_6 to the base of Q_8 thereby affording count down.

Power Supply

Two germanium power transistors operating from a 6-volt battery make up the power supply shown

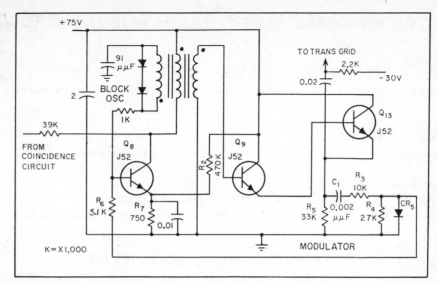

FIG. 5—Modulator provides 4-to-10-kc repetition rate with count down

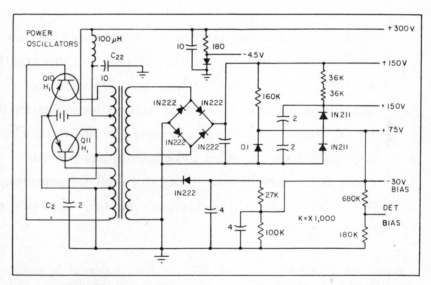

FIG. 6—Oscillator-type power supply converts 6-volt d-c source to circuit potentials

in Fig. 6. Transformer T_1 has a step-up winding and a feedback winding which allows the transistors to oscillate push-pull at a repetition rate of 1 kc.

The transformer input waveform is essentially square wave which is then stepped up to 300-v peak and rectified in a full-wave bridge to furnish the d-c output voltages.

Regulation

Suitable regulation is accomplished by Zener diode operation. The total power drain including the remote control fuel cutoff device is approximately 2 watts. The video amplifier draws only 2 ma at —4.5 volts.

The major problem in a power oscillator design of this nature is to obtain a high transformer input impedance. This is necessary so that the transistor can see a load impedance sufficient to cause a loop gain greater than one, thereby ensuring oscillations.

High-frequency oscillation is helpful since it minimizes filter component circuitry. Here the frequency of oscillation is not limited by the transistor or transformer but by the reflected load impedance.

Starting capacitor C_2 has been incorporated to ensure starting of oscillations under all environmental conditions.

Thanks are due B. Karp of our mechanical engineering section.

The beacon was developed under a Signal Corps contract originated at Ft. Monmouth, N. J.

CYCLOPS CORES SIMPLIFY EARTH-SATELLITE CIRCUITS

Part I
By WHITNEY MATTHEWS
Head, Applications Branch, Solid State Division, Naval Research Laboratory, Washington, D. C.

Part II
By R. W. ROCHELLE
Head, Magnetic Amplifier Section, Solid State Division, N.R.L., Washington, D. C.

Part III
By C. B. HOUSE and R. L. VAN ALLEN
Research Engineers, Applications Branch, Solid State Division, N.R.L., Washington, D. C.

Part IV
By D. H. SCHAEFER and J. C. SCHAFFERT
Research Engineers, Applications Branch, Solid State Division, N.R.L., Washington, D. C.

Requirements for the satellite electronics systems are much the same as for any equivalent aircraft or rocket system, but are greatly magnified in importance. Weight reduction is mandatory. Operating power must be held to an irreducible minimum. And, since many unknowns exist, equipment must be operable over a wide range of ambient conditions. This discussion of satellite electronics is not intended to be a complete discourse on the topic. It serves, rather, as an introduction to the so-called Lyman-alpha environmental satellite of Vanguard with emphasis on the telemeter encoder, memory and meteor counter.

Part I—An Introduction to Scientific Satellites

SATELLITE STRUCTURES will, in general, consist of hollow magnesium-alloy spheres 20 in. in diam with total weight limited to 21.5 lb. About 10 lb is devoted to the shell, internal structures and mechanism for separation of the satellite from the burned out third-stage rocket.

All electronic equipment and associated batteries are installed in the centrally located, pressurized instrumentation compartment, except for instruments and antennas which must be attached to the outer shell. The front cover of this issue shows encapsulated circuitry being inserted in the 5.5-in. diam, 7-5-in. high instrumentation compartment.

Other Instrumented Satellites

Instrumented satellites will be of several types, each containing a different combination of scientific experiments. In addition to the Lyman-alpha satellite discussed in the remaining sections of this article, additional types are being prepared.

A second satellite design will incorporate two experiments. In one, an airborne proton resonance magnetometer will study the earth's magnetic field at high altitudes. Signals will be transmitted only on interrogation.

In the other experiment, air-density measurements will be made by determining the drag on an inflatable sphere. These measurements will be similar to those made by the satellite themselves but will be made faster because of the greatly enhanced drag-to-mass ratio.

A third satellite type will be directed toward a study of radiation balance in space. Four small spheres will be located at the antenna tips. Spheres of differing absorptivities and emissivity will be used, with and without radiation shields. A study of their individual temperatures can reveal much valuable information. Signals will be recorded on tape and transmitted only on interrogation.

A fourth scientific satellite is devoted to a study of the earth's albedo for meteorological purposes. As the satellite spins, photosensitive devices will scan the earth's surface. Detail will not be great, but general distinction will be made between land masses, water and cloud cover for correlation with weather phenomena. Again, information will be recorded for playback only upon interrogation.

All telemetered scientific data will be transmitted by amplitude

---

* Now with Magnetics, Inc., Butler, Pa.

Encoder portion of Lyman-alpha satellite. Memory and counter are similar in appearance. Each unit is encapsulated and is about ¾-in. high. Unit modules are assembled on two rods to form a single package

modulation of the Minitrack radio tracking transmitter operating at 108 mc. Tracking data is obtained by phase measurements at 500 cps. Tracking accuracy demands that any modulation signal during the tracking interval contain no signal components in the range from 500 to 2,500 cps. Signals transmitted only on interrogation can use this band by waiting until tracking data has been obtained.

Spectrum Utilization

To reduce noise by narrowing the passband of the receiver, the signal should contain maximum information in the narrowest possible bandwidth. Maximum signal frequency components have been set at 15 kc for this reason.

Part II—Satellite Telemetry Coding System

SEVENTEEN TRANSDUCERS located on the shell and in the internal package of the Lyman-alpha earth satellite measure such parameters as temperature, collision with micrometeorites and solar Lyman-alpha radiation. Signal inputs from each of these transducers must be encoded for modulation of the Minitrack transmitter. Figure 1 is a block diagram of the complete satellite.

By using transistors and magnetic cores in a system combining both f-m and time-sharing modulation, weight of the encoder was reduced to 3.8 oz and the batteries to 2.8 oz. Expected life was over a month of continuous operation. The

resulting system has a capacity of 48 channels of telemetered information.

Outputs from the satellite transducers are in the form of variable resistances or, as in the case of the Lyman-alpha test, in the form of currents or voltages. The encoder takes these currents or voltages and makes the frequencies of tone bursts proportional to them. The on time of the burst and the time between bursts is proportional to the resistive values of the transducers. Three channels are represented by each tone burst.

The modulator output is a series of tone bursts in the frequency range from 5 to 15 kc.

Timing Multivibrator

Gates, which determine the lengths of the tone bursts, are generated by a timing multivibrator and a transistor matrix. In its simplest form, the timing multivibrator is as shown in Fig. 2. Two transistors, a square-hysteresis-loop magnetic core, and two transducers (R_1 and R_2) are used to produce a square-wave output.

Transistor Q_1 drives the magnetic core towards positive saturation. Transistor Q_2 takes it to negative saturation by regenerative action of the base winding.

Transducers R_1 and R_2, which

ABOUT LYMAN-ALPHA RADIATION

Solar radiation always contains energy in the Lyman-alpha region. Since the earth's atmosphere is quite opaque in this radiation, high-altitude rocket or satellite techniques are mandatory to obtain data. Marked increases in radiation are anticipated during solar flares. Random and infrequent occurrence of the flares, short life of research rockets, and time required for rockets to be placed in position for these measurements, present formidable obstacles to the gathering of data pertaining to this radiation.

A satellite under continuous observation would be an almost ideal vehicle for study of this phenomenon. Since continuous observation is impractical, compromises may be made which permit collection of valuable information. Background radiation from a quiescent sun may be studied by measurement of instantaneous values of this radiation as the satellite passes over the data collection stations. Correlation between visually observed solar flares and Lyman-alpha radiation may be obtained by storing and transmitting information regarding the maximum value attained during each orbit

might be thermistors or pressure gages, drop the battery voltage across the core. This action is accomplished by the magnetizing current flowing through R_1 and R_2 during each half cycle. Flux in the core at any time is the time integral of the voltage across the core. Reduced core voltage increases the time needed before saturation of the core is reached.

Variations in transducer R_1 will cause the length of the positive half-cycles in the output to vary independently of the negative half-cycles. Similarly, variations in R_2 will change the length of the negative half-cycles, independently. As actually applied, the half-cycle lengths can be varied over a dynamic range of 5 to 30 millisec for transducer resistance changes from 0 to 5,000 ohms. Only the positive half-cycle is used to gate on a higher frequency square-wave magnetic-core multivibrator. This multivibrator is termed a tone-burst oscillator.

A system containing only a timing multivibrator which gates on one tone-burst oscillator would be capable of telemetering three channels. These channels would be for frequency of tone bursts, length of the tone burst and time duration between tone bursts.

Extension of the system to more than three channels is accomplished as shown in Fig. 2. As many as six or eight base windings may be added to the magnetic core of the

timing multivibrator. Each center-tap to the base winding is brought out externally and biased so that none of the transistors are turned on. If a negative voltage or gate is applied to centertap A, transistors Q_1 and Q_2 will conduct alternately. Transducers R_1 and R_2 will determine length of the positive and negative half-cycles in the output, respectively. If centertap B is energized with a gate after removing the gate from A, R_3 and R_4 will control the lengths of the positive and negative half-cycles. Any of the pairs of transducers may be switched in alternately by applying a gate at the proper centertap.

Transistor Matrix

The transistor matrix in the encoder supplies the sequential gates. The matrix switch gates at the end of every full cycle of the timing multivibrator so that each transducer controls the length of a half-cycle in sequence.

A flip-flop follower, Fig. 3, and four tandem binary-counters count down the cycles of the timing multivibrator. The transistor matrix samples the states of the binary stages and produces a unique gate for each combination of binary states. There are 2^N states so that with four binary stages there are 16 unique states. Each gate turns on its own pair of transistors through the center-tapped base winding.

The flip-flop follower removes any loading from the timing multivibrator. Output of the follower is used in an AND circuit with the matrix gates to turn on the tone-burst oscillators or multivibrators during the positive half-cycles only. A wider bandwidth is needed for the instantaneous Lyman-alpha channel and the solar cell in the satellite because readings will be modulated by the satellite roll rate. By paralleling matrix output gates, a group of channels is repeated six times each frame to provide more telemetry time. This arrangement increases the channel bandwidth effectively. Normally, four binary stages with 16 base windings on the timing multivibrator core will produce 48 separate channels of information. By paralleling matrix

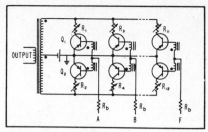

FIG. 2—Timing multivibrator for 49-channel telemetry system

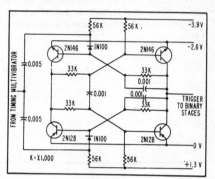

FIG. 3—Circuit diagram of the flip-flop follower used to count down

output and using only six base windings, Fig. 4, some channels are repeated several times during the 48-channel frame.

Binary stages, Fig. 5, are unique in that a steering-circuit transistor replaces the two back-to-back diodes used normally. Current gain of this transistor is utilized in triggering. It drives the bases of the binary transistors through capacitive coupling and is decoupled from the low saturation impedance of the on transistor by the 3,900-ohm resistor.

Half the time, the steering transistor is used in the inverted-alpha condition, since the binary carries the emitter more negative than the collector. Many surface-barrier transistors have betas in the inverted-alpha connection almost as large as in the regular connection. The four-transistor flip-flop or binary connection reduces total quiescent drain on the batteries.

The tone-burst oscillators, gated on by the matrix, may be of several different types. The instantaneous Lyman-alpha and solar aspect cell drive one type, which is a variable-frequency magnetically coupled multivibrator. Two magnetic cores, Fig. 6, are battery-driven by the two transistors. Base windings are regenerative. When one core is driven towards satura-

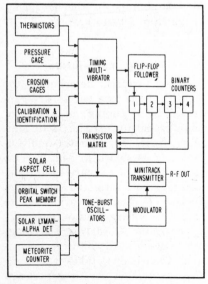

FIG. 1—Block diagram of the complete Lyman-alpha earth satellite

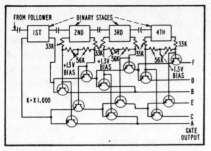

FIG. 4—Binary stages and transistor matrix. All transistor types are 2N146

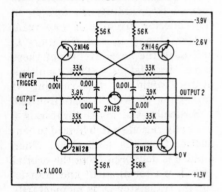

FIG. 5—Schematic diagram of a binary stage. Four are used in system

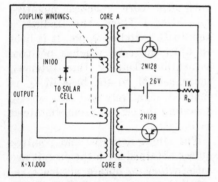

FIG. 6—Variable-frequency magnetically coupled tone-burst oscillator

tion by its transistor, the second core is reset through a coupling winding. At saturation, the circuit switches and the second core begins resetting the first. Injection of a current or voltage in the coupling circuit causes a change in the reset. This change varies the multivibrator or tone-burst-oscillator frequency. Variation from 30 to 700 μa can change the frequency from 5 to 15 kc. With a slight change in the coupling winding circuitry, 0.5 v from a solar cell will cause the same frequency shift.

Another type of tone-burst oscillator, termed a Cyclops will be described in Part III of this article, dealing with the analog magnetic memory of the satellite. There are three Cyclops oscillators in the

meteoritic collision module and two in the peak reader. One operates every other orbit while the other stores information. The Cyclops oscillators plus the variable-frequency magnetically coupled multivibrator make a total of six tone-burst oscillators in the satellite. Each is turned on by the sum of the gates from the matrix and the follower.

Outputs of the tone-burst multivibrators are parallel-added to drive the modulation stage, Fig. 7. Multivibrator outputs are decoupled through diodes. Since only one multivibrator is gated on at a time, there is no interaction between multivibrators. Modulation transistors serve the dual purpose of amplifying and clipping. Clipping insures 100-percent modulation of the transmitter.

Synchronization

Because tone-burst lengths and spaces are functions of transducer values, the frame rate is variable. If the average resistance of the transducers is low, the frame rate will be fast. More information can be sent per unit time by this system than in conventional ones which allot a fixed time duration for each channel. Since the frame rate is not constant, the signal must be unique so that individual channels may be identified. By using a few fixed values of resistance in place of some transducers, a key is formed and calibration is provided.

Part III—Analog Magnetic Memory

STORAGE and readout requirements encountered in certain Lyman-alpha experiments demand special equipment. In this experiment, the Lyman-alpha line of hydrogen in the solar spectrum must be measured and correlated with optical observations of solar flares. One reason for the measurements—suspicion that variations in Lyman-alpha, due to solar flares, may be a contributory cause of radio fadeouts.

In the Lyman-alpha satellite, information transmission will be continuous. This operating method requires two storage elements. One stores peak intensity information during one orbit. The other trans-

mits information gathered during the previous orbit. Since the storage element must transmit continuously the information it received previously, nondestructive readout must be used. In addition, functions of the storage elements must be switched automatically once each orbit as the satellite passes from darkness into sunlight.

The developed system accepts information in the form of d-c values from a few μa to one ma. It remembers the maximum current value that has been applied to it and retains this information until it is erased deliberately. This informa-

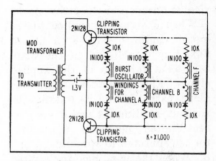

FIG. 7—Clipper and modulator stage

tion is presented directly in the form of alternating current with frequency a function of the stored information level.

Measurement is accomplished with a technique originated at the Naval Research Laboratory. Radiation impinges upon an ion chamber designed to respond only to the desired spectral line. The resulting minute current is amplified by an electrometer and fed to the memory unit. Information stored in the form of flux level in a square-hysteresis-loop magnetic material will remain until removed by external means. This method for storing the peak Lyman-alpha information is used.

The magnetic core must assume a certain flux level for a given value of current flow. The core must remain unaffected by any subsequent currents unless they are larger than any previous currents. In such a case, the core must assume a new flux level. In other words, the core must not integrate in the steady state.

Once the method of storing the information was established, the next problem was to translate the information into a usable form. To make the system compatible with the requirements of the encoding system, generation of an alternating voltage waveform with frequency proportional to information was necessary. It was felt desirable to obtain the readout with a single magnetic core and appropriate circuits.

During this experimentation, a two-aperture memory core named the Cyclops, Fig. 8, was suggested. It could be fabricated by drilling a hole through one wall of a tape-wound core. Outer convolutions of a tape-wound core shield the inner convolutions. As a result, the device is insensitive to stray fields.

When a signal is applied to the input windings, Fig. 8, a stored flux ϕ is established in the main core as a function of the applied signal. Flux ϕ divides at the hole along paths ϕ_A and ϕ_B linking the readout windings A and B. A current-limited magnetic multivbrator is connected to windings A and B. When the windings are excited alternately, there are flux changes in opposite directions around the small hole but in the same direction as ϕ in the main core.

Nondestructive Readout

Assume that flux ϕ has been set by a signal from the input winding. Assume, further, that flux ϕ saturates the iron on each side of the hole at ϕ_A and ϕ_B in the direction indicated in Fig. 8. The input signal is removed and the multivibrator is turned on. Because of the difference in B_r and B_{max}, a small clockwise flux change about the hole will be caused by current in winding A when the upper transistor gates. A similar small flux change in a counterclockwise direction will be produced in winding B when the lower transistor gates. Since, in this case, the amount of flux changed can only be small and the time each transistor is switched on is short, the multivibrator operates at a high frequency.

Next, assume that a reverse current is passed through the input windings on the main core so that flux level ϕ is reduced to zero and the main core is demagnetized. When the multivibrator is ener-

gized, there will be a large amount of flux reversed each half-cycle. This results in a low operating frequency for the multivibrator. For intermediate levels of ϕ, there will be corresponding intermediate frequencies.

Memory-Core Circuitry

Use of two Cyclops memory cores to record peak current from the Lyman-alpha detector leads to other requirements. A means is needed for wiping out the memory and resetting one core. A commutative switch is necessary to connect the other core containing stored information to the telemetry encoder.

In Fig. 9, the left side of the curve is used to store current information from the electrometer tube. It is best suited to store currents in the required range of 20 to 700 μa. Reset must be provided to restore the core saturation each time the memory is erased. For precise measurement, it is unsatisfactory to simply reverse current in the winding. It is also poor practice to apply a single reverse voltage pulse and attempt to return to the exact frequency each time.

To maintain high accuracy, a resetting circuit was devised consisting of one negative and one positive pulse. First, the negative pulse drives the core to negative saturation in the same direction that negative control current would normally set the core. Next, the positive pulse restores the core to positive saturation or to the point of origin at a high frequency. By this technique, repeated resets with or without previous input currents restores the frequency to within one percent of its original value.

Figure 10 includes the transis-

torized resetting circuit used with the peak memory cores. The circuit as shown is for two cores. One half is essentially a duplication of the other except for connections to the orbital switch and frequency range. One multivibrator has a frequency range from 4 to 8.5 kc; the other, 8.5 to 14 kc. The two different ranges permit core identification.

A magnetic multivibrator is formed by windings L_3 and L_5 in the collector circuits of pnp transistors Q_1 and Q_3 and windings L_7 and L_9 in the base circuits of these same transistors. This multivibrator would be free-running if transistor Q_5 were short-circuited.

Basically, the multivibrator is a normally closed switch biased to conduction from the +1.3-v bus. When a pulse is delivered from the orbital switch via terminal C' and transistor Q_7, transistor Q_1 is momentarily switched on and L_3 is energized. A voltage induced in L_7 clamps Q_1 on until the core reaches negative saturation. Then, the inductive kick in L_9 switches Q_3 on. It is held on until the core reaches positive saturation. During this interval, the capacitor to the base of Q_5 is charged negatively and cuts off Q_5, disabling the circuit from sustained oscillations.

The single negative and positive pulses obtained with the circuit are satisfactory to obtain accurate reset of the core in about one millisec. Switches at A and A' are caused to lag C, C' to prevent current flow in the memory winding during the one-millisec reset interval.

Temperature characteristics of the resetting circuit were improved by use of thermistors in the bias circuits of the main switching transistors Q_1, Q_2, Q_3, and Q_4. At elevated temperatures, where I_{co} normally increases, thermistor resistance is lowered. This action increases the bias current from the +1.3-v bus which passes through the diodes in series with the base windings. The bases of the switching transistors are thus kept at a constant positive bias voltage.

Core Input and Readout

Windings L_1 and L_2 are the input or set windings on the main section of the cyclops cores.

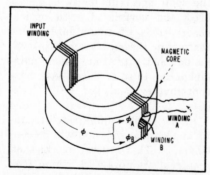

FIG. 8—Basic Cyclops core and winding configurations

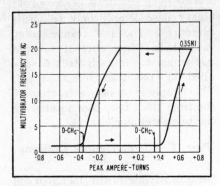

FIG. 9—Multivibrator frequency of a wound toroidal core as a function of current. Slope of the curve is influenced by the inner diam—outer diam ratio of the cores as well as the material

These windings carry the cathode current from the electrometer tube and are controlled by switches at A and A'.

Windings L_{11} and L_{12} are the multivibrator windings in the holes of the cores. Switch terminals labeled B and B', Fig. 10, connect only one multivibrator for readout to the negative bus depending upon the state of the orbital switch. Connection to ground or the positive bus is made through a 2N128 transistor which is switched on and off at the command of the encoder.

When both positive and negative bus connections are made, the appropriate multivibrator is turned on. Next, either output winding L_{13} or L_{14} delivers a tone burst to the modulator of the Minitrack transmitter. A full-wave diode bridge couples the output windings of each Cyclops multivibrator to the modulator. This bridge provides the maximum available driving voltage to the bases of the push-pull class-B modulator transistors. Even with a temperature-stable resetting pulse on the core, there is an output frequency decrease with an increase in temperature caused by thermal effects on the core alone. These thermal effects are compensated by an inverse thermal characteristic of the multivibrator transistors. With use of the thermistors mentioned previously, plus the self-compensating circuit components, the complete circuit maintains ±5-percent accuracy over a temperature range from −50 to +70 C.

Orbital Commutation

Design of the commutator which interchanges the two memory cores once each orbit called for minimum power dissipation since the switch circuit would be in continuous operation. A rough approximation, taking into account the contemplated battery type, indicated 10 mv-amp drain would cost one oz of weight for two to four weeks of operation.

One channel of information presented to the encoder is the instantaneous aspect or angle of the satellite axis with respect to the sun. This information is gathered by a silicon-junction photocell mounted on the equator of the sphere. A regular $1\frac{3}{8}$-in. diam round solar cell was secured and a $\frac{3}{8}$-in. diam hole cut in the center. Back of this, three $\frac{1}{2}$ by $\frac{3}{32}$-in. strips of the silicon junction material were mounted and connected electrically in series to give a higher voltage for the orbital switch.

The orbital switch is designed to trip on the first increase in incident light level occurring after any five-minute absence of solar radiation. It then locks out against any recurrent variations which may occur at a rate of more than once per five-minute period. When the switch trips, it initiates a pulse. This pulse wipes out the memory of the element that has been transmitting information and connects it into the electrometer circuit prepared to read and store peak-intensity information. Simultaneously, the pulse connects the element which has been storing information into the readout circuit.

A circuit for obtaining these ac-

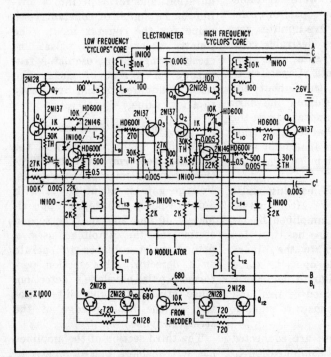

FIG. 10—Resetting circuitry used with the low- and high-frequency Cyclops memory cores

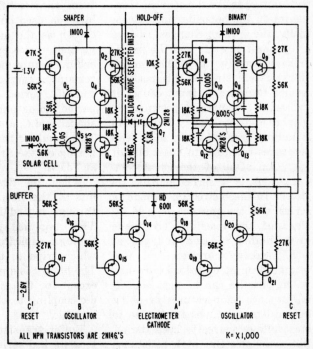

FIG. 11—Orbital switch circuit. Shaper is essentially a monostable multivibrator

tions is shown in Fig. 11. The multivibrator stages employ so-called complimentary multivibrator circuitry. Ordinary multivibrator flip-flop connections are used but npn and pnp transistors are connected in series and substituted for the load resistors. The common base voltage which turns one transistor off will turn the other on. Any load tied to the common collectors will be supplied through a low-impedance source when on but a high impedance is presented to the battery when off.

Shaper Circuit

Operation can be clarified by considering operation of the input shaper circuit. For the present, transistors Q_1 and Q_2 will be assumed open. Transistors Q_3, Q_4, Q_5, and Q_6 constitute the basic monostable multivibrator. In the normal state (no signal), input transistor Q_5 is off and Q_3, in series with Q_5, is saturated on. Corresponding transistors on the other side are in the opposite condition.

As the signal rises, increased current begins to flow through the series npn transistor Q_3. When this current becomes large enough so that Q_3 is no longer operating in the saturated region, a voltage appears across the emitter-collector terminals of Q_3. This voltage is transmitted to the d-c coupled transistors on the opposite side. Eventually, the voltage changes produce a regenerative state and the multivibrator rapidly flips. The multivibrator will stay in this position until a decrease in input causes it to resume its normal position.

Increased leakage currents in switching transistors at high temperatures can tend to upset circuit operation. One of the usual compensations is a series emitter resistor. In these circuits, a diode in series with the emitters is used instead of the resistor. The diode gives higher biasing voltages at small leakage currents and less voltage drop at high load currents.

The circuits can be kept operating at high temperatures by using low resistance values from base to bias voltage source to keep the off transistor off. But high resistance values are required at low temperatures to turn the on transistor on.

Connection of transistors Q_1 and Q_2 across the base and emitter terminals of the multivibrator transistors Q_3 and Q_4 offsets these conflicting requirements.

Transistors Q_1 and Q_2 act as switches driven from the collectors of their respective multivibrator transistors. A low impedance connects the base and emitter of the multivibrator transistor when it is in the off position. When a multivibrator transistor is on, base-to-emitter resistance is high (open switch) to prevent bypassing of the base current. This compensation is required on the npn transistors only since the type of pnp transistor chosen is relatively insensitive to high temperatures.

Hold-off Capacitor

When the monostable multivibrator first trips on, it generates a fast-rising, flat-topped pulse. This pulse charges the hold-off tantalum capacitor, C_1, in Fig. 11. Any subsequent pulses due to high-frequency spin rate which appear before the end of the designed hold-off period, will not be able to pass enough current through C_1 to affect Q_7. They will merely recharge C_1.

The discharge resistor in parallel with C_1 may be chosen for the desired time constant. Hold-off times of several hours were available from the rectifier-capacitor combination used but five minutes was chosen as the most desirable compromise between spin rate and orbital requirements. Capacitor C_1 remains charged throughout the sunlight period. During the dark period, no output appears from the shaper and C_1 discharges gradually. The circuit is then armed for triggering on the first pulse generated at "dawn".

When the first charging pulse hits C_1, the steep wavefront passes through it and is amplified by Q_7. This amplified pulse is then fed through capacitors to the binary which trips to the opposite state. The binary is similar to the shaper except for the method of feed and d-c coupling to all bases to make it bistable.

Buffer transistors are connected in a stacked series configuration to minimize current drain from the binary. With this connection, the

binary need supply only the base currents to the first transistor of each stack. Base current for the last transistor in each stack is supplied by the load-circuit supply. This supply is on only a fraction of the time since it is commutated by the telemetry encoder.

Part IV—Micrometeorite Collision Counter

SMALL MICROPHONES mounted on the satellite skin will be activated by collision of the satellite with meteoritic particles of about 10^{-8} grams traveling at speeds of about 60 km/sec with respect to the satellite. Counting rates will probably vary from one particle every ten minutes up to ten particles per second.

Signals transmitted by the satellite will consist of a number of tone bursts separated by blank spaces. Frequencies of three of these tone bursts will be indicative of the cumulated number of impacts detected by the microphones. Frequency of one burst will indicate the unit digit, another the tens digit and a third the hundreds digit of the cumulated count.

A block diagram of the equipment is shown in Fig. 12. The pulse former takes the transistor-amplifier output, a pulse of a few μsec duration, and forms it into a low-impedance signal of constant v-sec value for the counter input. The counter accepts these pulses increasing the output oscillator frequency on receipt of each. The tenth count resets the frequency to a low value.

Amplifier Sections

The first four transistors of the amplifier perform linear amplification upon the microphone outputs. Voltage gain is about 60 db with 3-db response points at approximately 40 and 200 kc. The second section of the amplifier uses a fifth transistor biased to operate as a detector. This section performs pulse-envelope detection, pulse stretching, and differentiation of the leading edge of the stretched pulse.

The third section of the amplifier has a transistor biased almost to cutoff. It amplifies the differentiated stretched pulse to an ampli-

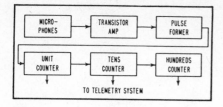

FIG. 12—Block diagram of micrometeorite experiment

tude suitable for driving the pulse former of the counter. This output pulse corresponds in time to the leading edge of the ringing-pulse output from the microphones.

The pulse-former circuit is similar to the resetting circuit described previously in the section on the magnetic-memory readout. The pulse former converts the pulse from the amplifier into a square pulse of 200-μsec length with an output impedance of a few ohms.

Counter Circuitry

Each decimal counter has as its most basic elements, two square-loop-type magnetic cores. These cores remember the last flux level to which they were set. One is a Cyclops core, as described in the previous section on the satellite memory.

One complete stage of counter circuitry is shown in Fig. 13. Core I is a standard square-loop core while core II is a Cyclops. All transistors are used only as switches. The transistors are either fully conducting or fully cut off at all times. The input circuitry forms nine small pulses followed by a tenth large one. At the end of ten counts, the reset circuitry resets the Cyclops, partially completely resets core I, and provides input voltages for the next stage.

Input Circuitry

A schematic diagram of the input circuitry is shown in Fig. 13. Voltages A and D are the square-wave inputs from the pulse former. In actuality, A and D are windings on the pulse-former core or the Cyclops of the preceding stage. Voltage C' is a voltage from the preceding stage that is transformed through a saturable square-loop core (core I of Fig. 13). For counts one through nine, voltages A, D and C' are all present and in the polarities shown. Voltage C' has a

magnitude somewhat greater than A. Under these conditions, transistor Q_3 conducts fully while Q_2 is cut off. The 1.3-v battery voltage is applied across the Cyclops after undergoing a drop across silicon rectifier D_1 and resistor R_2.

Upon receipt of the tenth voltage input pulse, the core that has been transforming the voltage C' is saturated. No transformer action takes place making C' zero, while A and D are still present. Both transistors become fully conducting and the 1.3-v battery appears across the winding with only negligible drop. This is the mechanism that allows nine small pulses followed by a large tenth one to appear across the Cyclops winding.

Reset Circuitry

Reset circuitry is shown in the center section of Fig. 13. A large negative voltage spike is developed across the core windings following the tenth input pulse. This negative spike is so large that the 1.3-v battery in the base circuit of Q_4 is over-ridden. A negative voltage appears upon the base of Q_4 and the collector-circuit 4-v battery appears across the windings in the collector circuit. This voltage, in turn, is transformed to the base windings which puts Q_4 even more fully into the conducting state and holds it there even after the initiating kickback voltage has disappeared.

The 4-v battery voltage takes both cores back towards positive saturation—core II directly, core I

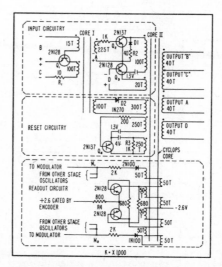

FIG. 13—Complete counter stage

by coupling through D_2. Core I is designed to have much less volt-sec capacity than core II. Core I goes into saturation when core II has only moved a small way up its hysteresis loop.

When core I goes into staturation, a short-circuit is thrown across both cores due to the coupling. Voltage on the base windings of Q_4 falls below 1.3 v and Q_4 cuts off. The circuit is now ready to receive more counts. During reset, a large negative voltage pulse is developed across the windings. This reset voltage pulse becomes the voltages A, B, C and D for the following stage from the output windings as shown in Fig. 13.

During the interval that a particular oscillator frequency is to appear in the transmitted signal, a gating signal is used to apply oscillator power. This signal turns the appropriate oscillator on for the proper interval. The output terminals go to a modulator that reshapes the somewhat triangular waves of the oscillator into square waves. The square waves then modulate the transmitter. The oscillators of the three stages are tied together in parallel at the points M_1 and M_2 in Fig. 13.

Units have been built that will consistently count to ten and reset over the range from 0 to +80 C. Changes in calibration with temperature do occur, but can be accounted for since the instrument-compartment temperature will be a telemetered quantity. Gain of the amplifier is constant to within 20 percent from −20 to +70 C. Weight of a module which includes amplifier, pulse former, and three counter stages is 3.1 oz before potting; 5 oz after.

Power Requirements

The amplifier requires 1 ma current drain at 4 v. The counter itself has zero current drain when no counts are being received. Approximately 15 or 20 ma flow for each count. This flow lasts for only 200μsec per pulse. A 1,000 ma-hr battery is able to keep the counter counting to over 10^9 counts.

Readout oscillators require about 3 ma of current at 2.6 v but are on only about one-sixth of the total time.

MAGNETIC-CORE MEMORY MONITORS
EARTH SATELLITE

By C. S. WARREN, W. G. RUMBLE and W. A. HELBIG
Defense Electronic Products, Radio Corporation of America, Camden, New Jersey

Telemetered data from U.S. earth satellite will be decoded by transistor-operated magnetic-core memory. Circuits required to numerically translate input information and present modified output information use alloy-junction transistors as current drivers, gated-pulse amplifiers, voltage amplifiers, high-speed switches and flip-flops. Memory storage capacity is 6,400 bits arranged as 256 characters of 25 bits each

ONE OF THE FIRST transistor-operated magnetic-core memories used in the field for the U.S. earth satellite program *Vanguard* uses a unit, known as a linearizer memory. This memory is essentially a data converter which operates on input information according to a predetermined numerical transformation and presents it in modified form as an output. Input data is used to address the memory and the desired output data is stored in corresponding memory locations. Provision is also made for monitoring the outcoming data with an automatic plotter or similar device.

Data telemetered from the earth satellite to a receiving unit may be coded in any desired manner since

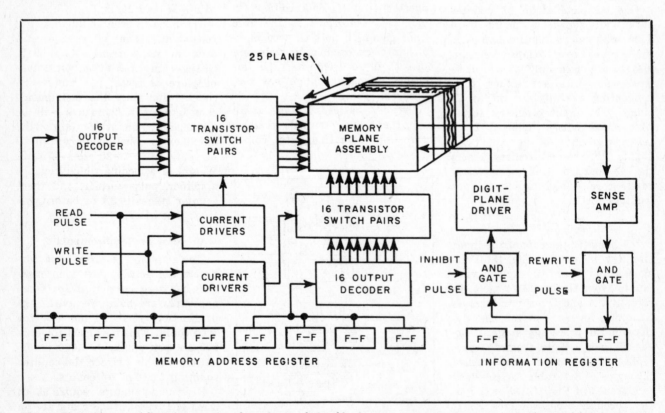

FIG. 1—Block diagram of linearizer memory shows basic relationships between one magnetic-core memory plane and transistorized input and output circuits. Each of the other 24 planes have identical arrangements to provide a total capacity of 6,400 volts. Driving and sense circuits are completely transistorized

Paper tape reader preloads linearizer memory with calibration information. Coded input signals telemetered from the earth satellite are digitally converted by the calibrated memory

Memory plane assembly, shown with cover removed, has thermostatic control to provide stabilized temperature for ferrite cores. Transistorized circuits facilitate high-speed random access

the linearizer memory can be precalibrated to decode the signals. For example, if the information received originated from a temperature sensing instrument in the satellite, the linearizer memory can convert the coded telemetered data into a directly usable output which is fed into other types of computers. The entire data from a satellite can be recorded on magnetic tape and read into the linearizer at a later time.

Storage Capacity

The memory system, shown in the block diagram in Fig. 1 uses 25 memory planes, each having 256 memory cores, which provide a 6,400-bit storage capacity. Each memory matrix is square and has 16 cores along each axis. Since the X-axis planes are connected in series as are the Y-axis planes, an excitation voltage applied to one X winding and one Y winding selects 25 cores which are identically placed in each of the memory planes. Information stored, therefore, is arranged into 256 characters of 25 bits each. Each plane has inhibit and sense windings.

At the start of the memory cycle,

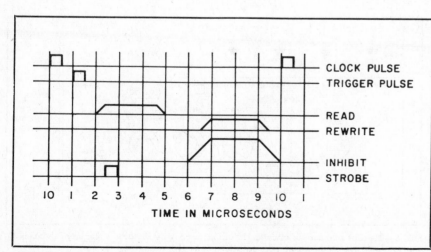

FIG. 2—Timing diagram shows sequence of pulses controlling memory cycle

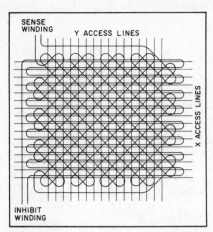

FIG. 3—Simplified diagram of one memory plane matrix. Sixteen access lines on both the X and Y axes provide 256-bit storage in each plane

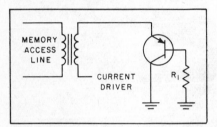

FIG. 4—Simplified basic transistor switch used in access driver circuits

the memory location to be interrogated is set into an address register by the 0.5-μsec trigger pulse shown on the timing diagram in Fig. 2. The address register consists of eight flip-flops; four address the X axis and four address the Y axis.

Outputs from the two groups of transistor switch pairs feed the access windings along the X and Y coordinates of the memory matrices. Current drivers associated with each group of switches supply the current pulses which excite the memory. Each driver consists of two pulse amplifiers; one for read polarity and one for rewrite polarity. These circuits control amplitude and rise time of the driving pulses.

Read Pulse

When the address flip-flop and decoder settle out, a read pulse of the proper polarity to drive the selected memory core to the ZERO state is applied. A voltage then appears on the sensing wire output if on ONE was previously stored.

Following the read pulse, a rewrite pulse is applied, driving the selected core to the ONE state. To allow writing of a ONE, each of the 25 planes is provided with an inhibit winding and driver. These apply a half excitation-current pulse having the same polarity as the read pulse to all the cores in a plane when ZERO is to be stored or regenerated.

Regeneration

Regeneration circuits consist of two parts: the sensing amplifier with its associated output gate and the digit-plane driver with its associated input gate. Sense windings series-link the cores in each plane and connect to amplifiers which rectify and amplify all signals above a predetermined threshold value. The sense amplifier output sets an information register flip-flop. The information output is obtained from the sense output gate 3 μsec after the start of the cycle.

During interrogation of the memory, the sense amplifier output is gated into the information register by the digit-plane drive. When new information is to be stored the sensing output gate is blocked and information from the computer is supplied to the register.

Memory Plane

Memory plane construction and winding arrangements are shown in Fig. 3. Equal numbers of cores along any one access line are linked in opposite senses to cancel a large percentage of noise resulting from

half-excited cores in the memory.

Three characteristics were considered in selecting the memory core: switching time, drive current and noise voltage. Since the switching time of the core is inversely related to the current and voltage drive requirements, it was considered desirable to select a memory core having a long switching time which is consistent with reasonable access time.

To allow the core itself to participate in the switching required for its selection, the hysteresis loop generated by the drive current must be square. Although good rectangularity improves the signal-to-noise ratio, noise voltage is reflected back to the drivers when the induced voltage peaks. This is caused by half excitation of the memory cores and appears as reverse bias across the transistor driver.

Memory cores selected for application in this memory have an outside diameter of 0.08 in. and a full excitation drive current of 500 ma. With a 0.5 μsec rise time for the driving current pulse, the core requires 2 μsec (measured from the 10 percent point of the drive pulse) to reverse its state of magnetization. The maximum voltage readout for half excitation is 10 mv; the voltage output for full excitation of a core storing a ONE is approximately 100 mv.

Temperature Effects

Since the ferrite cores are the most temperature sensitive elements in the memory, stabilization

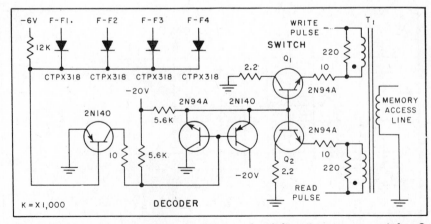

FIG. 5—Decoder and switch circuits. In circuit used with linearizer memory, switches Q₁ and Q₂ each are made up of two parallel connected transistors

of operating temperatures is reqired to assure that the signal and disturb or noise outputs of the memory do not vary. Present memories maintain an adequate signal-to-noise ratio over a limited temperature range of 10C. The range can be extended by the following methods: selecting improved ferrite material; providing automatic temperature compensation in sensing circuits; providing automatic temperature compensation in drive circuits: or maintaining the memory core matrices at the maximum required operational temperature.

Only the last method was found both feasible and presently attainable. Using this technique, the memory core matrices are maintained at 45 C ± 5 C. Stable operation was accomplished by enclosing each core in an insulated box and thermostatically controlling the temperature.

Access Drivers and Switch

A fast, efficient switch capable of handling large current pulses is required between the single source input and the appropriate line of the memory matrix. Transistors are ideally suited for this because their low saturation impedance permits relatively large currents to pass with low power dissipation. Since current gain and speed are also desirable, the transistor must not be operated too far into saturation. Operation at the knee of the grounded-base collector characteristic curve assures low storage and low dissipation at full current gain.

A circuit which operates at this point without the use of an additional collector voltage supply is shown in Fig. 4. If a low-value resistor is used for R_1 the transistor will present a low input impedance.[1] In the switching circuit described here, base resistor R_1 is replaced with an emitter-follower which provides extra current gain and low base resistance.

Drive Pulses

The 250-ma read and rewrite current pulses that drive the memory cores are generated by a

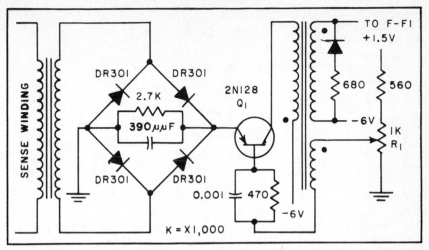

FIG. 6—Sense amplifier. Reverse bias on base of transistor amplifier Q_1 prevents false triggering

constant-current pulse driver and directed into the proper access line by the voltage-selected transistor switch. Figure 5 shows the switch circuit for a single line of the matrix. Two switch transistors are used for each memory access line through the plane; one for the read pulse and the other for the rewrite pulse.

Recovery Time

Since the switch transistors are operated in saturation, a symmetrical emitter follower is required to insure fast recovery between memory cycles. During each memory cycle, emitter circuits of all read and rewrite transistors are pulsed from the read constant-current drive. By transformer-coupling the current pulses to the memory plane with T_1, only one access wire is required for both the read and rewrite pulses.

To prevent additional reverse voltages from appearing across the emitter-base diode of the switch transistor, the output pulse transformer is specially designed to give low leakage inductance. By using a toroidal core made from high-permeability ferrite wound with trifilar windings having a one-to-one turns ratio the low leakage inductance is obtained.

Decoder

The decoder consists of a 64-diode matrix having 16 outputs, each of which feeds one transistor

amplifier. The circuit for one decoder output is shown in Fig. 5.

Each decoder output feeds the base of an emitter follower associated with the address switch.

Input current requirements of the decoder are low enough that amplifiers are not required between the address register and the decoding matrix. Decoding is accomplished in less than 1.5 μsec after the arrival of the information pulse at the address register.

Sensing Amplifier

The sensing amplifier shown in Fig. 6 contains a blocking-oscillator type transistor amplifier which is triggered by the output from a diode bridge network. Since the memory sensing wire exhibits a low output impedance, sufficient amplification of the readout signal input is obtained by transformer coupling.

To assure that readout voltages of both polarities are sensed, the signal is rectified by the diode bridge network. Because semiconductor rectifiers respond nonlinearly to voltage signals, small signals are greatly attenuated relative to large signals. This factor increases the s/n ratio at the rectifier output to about twenty to one.

Complete elimination of spurious signals and the standardization of all readout signals from the memory is accomplished with transistor amplifier Q_1. All signals below a specified level are pre-

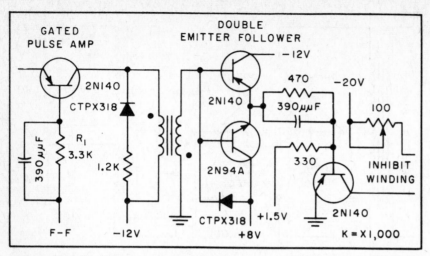

FIG. 7—Digit-plane driver. Sensing amplifier output is gated through information register during interrogation but is blocked out during storage of new information. Double emitter follower reduces input requirements of circuit that drives inhibit winding

vented from triggering the amplifier by a small adjustable reverse bias applied to the emitter-base of Q_1 by potentiometer R_1. A signal of at least 12 mv above the controlled threshold level is required to obtain a full output from the sense amplifier. Since the noise impulses appearing on the sense wires are relatively few, a minimum s/n ratio of ten to one is obtained, therefore, the problem of false triggering is eliminated.

Digit Plane Driver

The digit plane driver shown in Fig. 7 contains a high-current pulse amplifier capable of supplying half excitation-current pulses to all of the cores within a given plane. Voltage and current drive requirements are dictated by the size of the memory matrix.

A conventional grounded-emitter amplifier with the memory load in the collector circuit makes up the output stage of the amplifier. The transistor is operated in saturation; therefore, the collector-current amplitude is a function only of the collector voltage and the 100-ohm variable resistance.

When a high-frequency transistor is used, the rise time of the current pulse is controlled by the time constant L/R of the collector circuit. In this instance, L represents the inductance of the memory matrix plus a small added inductance. To reduce the input requirements, an emitter follower drives the output stage.

Gate requirements for the inhibit function are provided by a gated pulse amplifier. The gating function of the circuit is accomplished in the base-emitter diode of the transistor. When the control level from the information register flip-flop is at ground, the emitter diode remains reverse biased under the maximum excursion of the positive 3-v input pulse. However, when the control level is at a negative six volts, the pulse forward biases the emitter-base diode and an output pulse results. Thus the inhibit pulse is gated through only if there is no readout from the sense wire.

A power gain of approximately ten is obtained from this circuit using currently available *pnp* transistors. The inhibit-pulse gate can be designed to handle pulses having a 5-μsec duration.

Logic Circuits

Flip-flops throughout the memory use a complementary symmetry circuit with two *pnp* and two *npn* transistors.[2] At any one time, one *pnp* and one *npn* are conducting in saturation while the other two are held in the cutoff region. Address register flip-flops are conservatively designed to supply a maximum current of 10 ma with a voltage drop of 6 v.

Because of their many applications and variations, both *pnp* and *npn* transistor gated pulse-amplifier circuits are used in the logic associated with the memory.[3] A

basic circuit of this type is part of the inhibit gate shown in Fig. 7.

Future Memories

Transistor operated memories are an answer to the need for high-speed random-access storage in cases where power requirements, size and weight must be minimized. Application of transistor circuits is not limited to the size and type of memory described here. With the exception of the sensing amplifier and decoder, these circuits could be extended to a 64 by 64 memory plane without much change.

In extrapolating the above circuits as building blocks for larger memories, three problems have arisen. First, increased voltage is required for driving and switching circuits. Second, better discrimination in the sensing amplifier is necessary to overcome the decreasing signal-to-noise ratio. Third, larger address decoders are needed to retain speed of operation and output current requirements.

Solutions

The first problem results from larger inductive loads presented by larger memory planes. Its solution is a function of the availability of transistors with high inverse voltage breakdown characteristics.

Since the s/n ratio of the readout from a 64 by 64 plane is low, the sensing amplifier must incorporate an additional stage of amplification. A strobe gate would be required at first output to discriminate between the read-out signal and the noise generated from half-excited cores.

A 64-output decoder using 80 transistors and 180 diodes is now being tested. It is expected that decoding will be accomplished in less than 1 μsec with a maximum output current of 10 ma with this unit.

REFERENCES

(1) W. A. Helbig and W. G. Rumble, A High Current Switch for a Transistor Operated Memory, *Proc NEC*, 1956.

(2) T. P. Bothwell and L. Kolodin, A Bistable Symmetrical Switching Circuit *Proc NEC*, 1956.

(3) G. W. Booth and T. P. Bothwell, Logic Circuits for a Transistor Digital Computer, to be published in *Proc PGEC*.

Chapter 15

SCIENTIFIC AND MEDICAL INSTRUMENTS

PILL TELEMETERS FROM DIGESTIVE TRACT

By STUART MacKAY and BERTIL JACOBSON
Karolinska Institutet, Stockholm, Sweden

Radio sounding device, small enough to swallow and pass through the gastro-intestinal tract, generates 400-kc signals that transmit internal temperature and pressure information. Powdered iron core is pressure sensor while transistor base-collector resistance is temperature sensor. Receiver uses nonlinear capacitors to sweep frequency band

SUCCESSFUL passage and operation of a radio telemetering device through the gastro-intestinal tract makes it possible to transmit internal temperature and pressure information, useful in medical diagnosis and physiological studies.

Transmitter

Called an endoradiosonde, the transmitter, modulator-transducer that is swallowed is a capsule that measures 0.9 cm in diameter and 2.8 cm long. Since much smaller components are becoming available, specific details will change. The typical circuit shown in Fig. 1, generates a sufficiently powerful signal so that it does not require the use of a shielded room. The common-emitter connection is used in a Hartley circuit with a tapped coil.

Components

The coil contains 600 turns on a hollow 1⁄16 inch diameter form that is 1⁄16 inch long. Into this moves a piece of powdered iron of 1⁄4-inch length and turned down to 1⁄8 inch in a lathe. The resonating capacitance is in the range of 100 $\mu\mu$f.

Best blocking action is obtained with Telefunken transistor OC 612 though the smaller CK 784 also was satisfactory. In some of the units, to save space, a hollow double capacitor was wound. Thin metal foil combined with 10-micron polystyrene tape, 5-mm wide, gives two

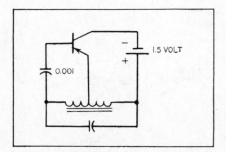

FIG. 1—Transistor quenching oscillator. The quench frequency depends on temperature and the r-f frequency on pressure. Adequate signal is generated to penetrate the body's attenuation

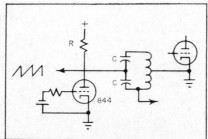

FIG. 2—Sweeping frequency oscillator based on properties of nonlinear capacitors. These act in series to tune the tank circuit and in parallel to generate the sawtooth to produce panoramic effect

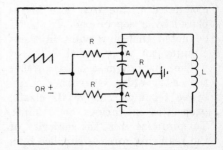

FIG. 3—Connection for many saturating capacitors under control of a small amplitude input sawtooth (or switches and a small direct voltage) and yet having small nonlinearity to the a-c of the tank

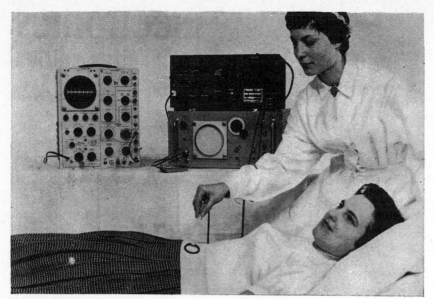

Nurse determining position of endoradiosonde. Signal is picked up by antenna loop connected to receiver. Oscilloscope at left gives temperature, right-hand scope gives pressure while drugs are tested as they progress through patient

cylindrical capacitors: a center piece with one metal piece outside and one inside, when the sandwich is wound up. The two capacitors of this unit required metal lengths of 5.5 centimeters and 0.55 cm.

The coil itself, is the transmitter antenna, its field being adequate for detection. Circuit components are sealed with Araldite.

Problems

The radio frequency, relatively insensitive to voltage changes, is affected by the shunt capacitance from end-to-end on the capsule. To eliminate the possibility of an air bubble giving an erroneous frequency shift, electrostatic shielding is desirable. Coating the inside with a layer of silver paint does not stop oscillation or radiation. An axial scratch eliminates any shorted turn effect.

These circuits are amplitude modulated if the iron core is replaced by a copper one. It should be noted that any point in these circuits can be grounded.

The diaphragm can be another source of error. A rubber membrane tends to change its elastic properties in various body fluids. To minimize this, the majority of the restoring force is supplied by a spring. An outer changeable limp diaphragm, employed for sanitary reasons, will also help.

It is desirable to eliminate the effects of orientation due to gravity acting on the core. A balanced pivoted armature is the obvious solution though a more compact method is desirable. The solution has not been found, but some methods for consideration are the use of a second internal weighted diaphragm at the other end of a U tube from the first, a second spring-suspended core at the other end of the coil, a spring suspending both core and coil and the use of a neutral buoyancy core in a liquid-filled chamber.

In some experiments, the battery consisted of an iron and a gold electrode with the subject's internal fluids acting as an electrolyte. For better stability the batteries are now constructed from the materials of a dismantled flashlight cell. The original batteries were formed around the lead from a pencil but the present ones sandwich the chemicals between flat sheets of zinc and carbon. Most of their volume is depolarizer.

During any normal experiment, 2-4 days, the battery voltage does not change. Radioactive batteries[2, 3] might be employed to give a longer life in a smaller space. If a radioactive transistor could be developed this would also help.

Frequency

Skin depth considerations and transistor performance limitations led to the choice of a 400-kc signal. This frequency is modulated by the motion of the iron core caused by pressure changes. The transistor is temperature sensitive and the pulse repetition rate of blocking transmits this temperature information. The latter reading, somewhat dependent on pressure, can always be corrected by the pressure reading which is relatively unambiguous.

Blocking Action

The blocking action depends on the charging of the base capacitor during oscillation because of rectification at the emitter junction. The base becomes positive. Due to oscillation hysteresis more charge is collected than is needed to keep the oscillation cut off and the transistor is biased off for a finite period. During this period the capacitor discharges through the temperature-sensitive resistances of the emitter and collector, both back-biased, in parallel.

Oscillation resumes after the base becomes a fraction of a volt negative with respect to the emit-

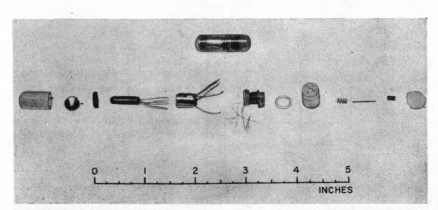

Assembled endoradiosonde above and disassembled unit below showing exploded view of components. Scale indicates relative size

ter. Since the collector resistance is usually lower than that of the emitter, the collector tends to dominate as a thermometer unless prevented from doing so. The period of this relaxation oscillation is proportional to the average of this resistance over the blocked part of the cycle, multiplied by the base capacitance.

If a ten-to-one turns ratio is used rather than approximately one-to-one then blocking is not observed, but instead amplitude modulation with temperature results due to the varying equillibrium voltage on the base capacitor.

The blocking frequency is sensitive to voltage as a first-order effect, decreasing voltage giving increasing frequency.

Receiver

The transmitted signal is received by a 100-turn loop, 4.5 cm in dia, connected to the input of a U. S. Army BC-348-P receiver. This size loop indicates the approximate location of the transmitter within the subject.

The receiver is tuned to indicate the pressure and the signal tone carries the temperature information. Transmitter coil and blocking capacitor are adjusted so that the radio-frequency bursts have an approximately flat envelope, which leads to relatively sharp tuning.

Scanning

In developing a scanning method for the radio frequency, as in a panoramic receiver, the circuit shown in Fig. 2 was evolved. The nonlinear capacitors C decrease their capacitance in response to the increasing direct voltage across each.[4] In series, they act as the tuning capacitor of the tank circuit. In parallel, they respond to the slowly increasing direct voltage. Thus nonlinearity to the radio frequency is small and sine waves are produced, but high control voltage is not required.

The scheme can be extended to four or more nonlinear capacitors as shown in Fig. 3. Here, one can either inject a sawtooth or apply steady d-c and periodically ground points A, in which case the circuit generates its own sawtooth.

High R lowers Q little but the replacement of parts of each R by inductance will make Q maximum. Some point on L is assumed to have a d-c ground. Because of their discharge in parallel, even small capacitors can trigger the thyratron, though a blocking oscillator is a better discharge device.

The oscillator tube is tapped down on the coil for somewhat increased circuit stability. The out-

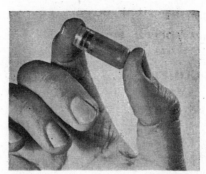

Endoradiosonde, the transmitter-modulator-transducer that is swallowed, is a capsule 2.8 cm long and 0.9 cm in diameter

put sawtooth of the free-running circuit is for a horizontal sweep voltage while the a-c shown represents a local oscillator voltage.

If the output of the radio is then used as a vertical oscilloscope signal, the frequency, in general, will not be a linear function of the signal position, but calibration is possible since there is a reproducible correlation between the two. Sweep in both directions gives a double signal due to one form of hysteresis observed in most nonlinear oscillatory systems.[3] Rather than cycling through the whole frequency band it is possible to use a related feedback arrangement to track the radio frequency either by maintaining a fixed frequency difference between the transmitted and local frequency or by cycling through a small range about the transmitted frequency. Feedback voltage then indicates pressure.

The transmitter is calibrated just before an experiment by applying known temperatures and pressures. Pressure sensitivity of the device can be checked within the subject by applying changing atmospheric pressure. Feedback, in which the output pressure reading is always returned to a fixed

value by altering the surrounding pressure on the subject, is feasible for special observations. Calibration is then not necessary since linearity and sensitivity do not enter (to first orders). Observation would start at greater than atmospheric pressure in the capsule and on the subject. The subject in this case should be accustomed to sudden pressure changes (such as a skin diver). The required surrounding pressure then measures the internal pressure.

Further Applications

Experiments are in progress to incorporate chemical analysis into such devices.[5] Any chemical reaction which is accompanied by reversible mechanical expansion and contraction could be employed in conjunction with the pressure-sensitive device. Certain ion-exchangers swell and shrink, while other macromolecular compounds change their osmotic pressure, for changes in pH. Although the accuracy is low for such chemico-mechanical systems, their simplicity and reliability may make them useful for medical diagnosis.

The use of an antimony electrode as a low-impedance pH sensor should receive further attention, particularly if one can be used without becoming coated and still work while biased to prevent dissolution under all body conditions. A radioactive light source might allow the introduction of optical methods, and be much more effective than a phosphorescent one. An exposed transister will telemeter light intensity and with the above could be used as a photometric pH transmitter.

Invaluable help in construction by Lars Nordberg is acknowledged. This work was done while R. S. Mackey was on leave from the University of California on a Guggenheim Fellowship.

REFERENCES

(1) R. S. Mackay and B. Jacobson, Endoradiosonde, *Nature*, June 15, 1957.
(2) *Radio and TV News*, p 160, May 1957.
(3) A. Thomas, Nuclear Batteries, *Nucleonics*, 13, p 129, Nov. 1955.
(4) R. S. Mackay, Sinusoidal and Relaxation Oscillations Sustained by Nonlinear Reactances, *Jour Appl Phys* 24, p 1,164, Sept. 1953.
(5) B. Jacobson and R. S. Mackay, A pH-Endoradiosonde, *Lancet*, p 1,224, June 15, 1957.

PEN-RECORDER AMPLIFIER FOR MEDICAL DIAGNOSIS

By D. W. R. McKINLEY and R. S. RICHARDS
National Research Council, Ottawa, Canada

Pen recorder amplifier provides transformerless system for recording 3-cps heart signals. Modification of feedback circuit gives audio amplifier with up to 5-watt output flat within 0.2 db from 20 cps to 20 kc

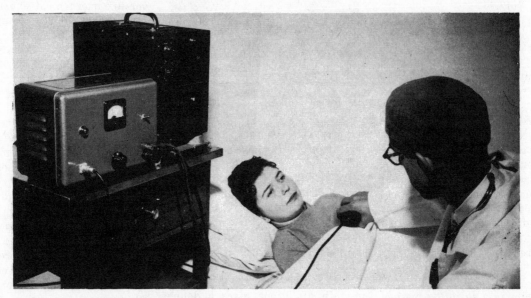

Compact transistorized heartbeat recording system can be operated from either battery or a-c line

RESEARCH WORK in medical electronics often requires an amplifier to couple a heart-beat microphone to a pen recorder. Microphone output is 1 mv at 10,000 ohms and required frequency response of the amplifier is 3 db down at 3 cps and 2 kc. Recorder input impedance is 17 ohms, with 4 volts needed for full-scale deflection.

These impedances, power levels and frequencies suggest the application of transistors and a complementary-symmetry in the output stage to avoid an output transformer. The amplifier shown in Fig. 1 met these performance specifications satisfactorily. With slight modifications, it has also been used as an audio amplifier to feed a loud-speaker at a level of 1 or 2 watts, with low distortion and flat response to 20 kc.

Recorder Amplifier

The overall voltage gain of the amplifier is about 4,000 with the feedback loop. The minimum input impedance is about 10,000 ohms and 1 mv produces 4 volts rms output across a 10-ohm load, hence the overall power gain is approximately 100 db. Direct-current feedback loops in the preamplifier stages ensure a high degree of temperature stabilization. Emitter degeneration is employed in the first stage to raise the input impedance to the desired level.

The driver-output stages have 100-percent internal voltage feedback, with slightly less than unity voltage gain, but there is still a trace of crossover distortion unless the feedback loop to the base of the last preamplifier transistor is in place. Degeneration introduced by this loop is 16 db. The 220-$\mu\mu$f bypass capacitor across the 22,000-ohm resistor in this loop is essential when working with a purely resistive load, otherwise high-frequency transients can cause the power transistors to run away.

The solid curve in Fig. 2, shows the frequency response curve of the amplifier working into a 15-ohm resistive load, with feedback network A. The 3-db points are at 2 cps and 9 kc. The slight increase in

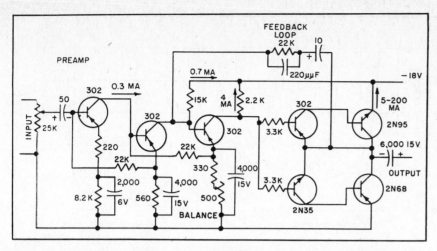

FIG. 1—Transformerless amplifier uses complementary symmetry output stage to match 15-ohm load input impedance of cardiograph pen recorder

response at 5 cps can be removed by selection of the series feedback capacitor and the emitter bypass capacitor, but in practice this bass boost compensates for the fall-off in microphone response at these frequencies.

Figure 3A shows the performance of the amplifier as the load resistance is varied. The maximum undistorted output (1-kc sine wave just below clipping level) is about 1 watt for a 30-ohm load and 2.5 watts for a 6-ohm load.

The efficiency is defined as the ratio of a-c output power to the product of the d-c supply voltage and current to the combined driver-output stage. Curves of total rms harmonic distortion are shown in Fig. 3B for constant output levels of 0.5, 1 and 2 watts.

In Fig. 3C, the solid curve shows the total rms harmonic distortion; and the dashed curve shows the intermodulation distortion, as the input power level is varied. The output signal power, measured here across a 15-ohm load, is directly proportional to input power up to the clipping level. Once this level is reached the distortion increases rapidly.

Clipping

Operating level of the amplifier should be set so clipping does not occur on the peaks when using a loudspeaker load. The pen recorder is an effective peak limiter. Since the undistorted voltage swing of the amplifier is well in excess of the recorder range, harmonic distortion is no problem with the re-

corder. At reasonable operating levels the harmonic distortion is less than 0.5 percent and the intermodulation distortion still smaller, which meets the requirements of a fairly good audio amplifier.

Noise

With a source impedance of 10,-000 ohms and an assumed effective bandwidth of 10 kc (using network A of Fig. 2) the measured noise output at full gain is 12 millivolts, which is 52 db below the maximum rms signal voltage of 5.2 volts across a 15-ohm load. This is quite satisfactory since the dynamic range of the recorder is less than 40 db. Noise factor of the amplifier is about 7 db. If low-frequency response is not required the noise factor might be improved by redesigning the first stage to eliminate the electrolytic condensers, and by using a low-noise transistor.

The 500-ohm balance control in Fig. 1 is adjusted initially to bring the collector potential of the last preamplifier stage to half the supply voltage. Alternatively, it may be set so a strong sine-wave signal

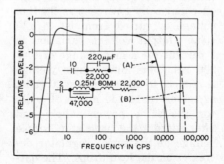

FIG. 2—Response curves of transistor amplifier using two feedback networks

input is clipped equally on positive and negative peaks at the output. Once set, it needs no further attention since the d-c feedback maintains a constant current in this stage over a wide temperature range.

No special selection of transistors is necessary, even for the push-pull driver and output stages, provided that the current gains are within the range of tolerance specified by the manufacturers. The 6,000-μf electrolytic capacitor coupling the output to the load may be omitted by returning the load to a center tap on the power supply. This center tap has to be a low impedance point, which can be obtained easily when using batteries. If a-c operation is desired, two power supplies of opposite polarity are needed to obtain the necessary low impedance.

Direct coupling of the load requires careful selection of the output transistors to ensure equal collector currents, with zero d-c in the load. Even with the best available pair of transistors a small d-c unbalance was observed with increasing output amplitude, resulting in a few milliamperes shift in the load current.

Loudspeaker Amplifier

The only essential change required in the circuit of Fig. 1, to make it useful for driving a loudspeaker, is to modify the feedback loop as shown at (B) in Fig. 2. In this network the distributed capacitance of the 0.25-henry choke was excessive, so the 80-millihenry air-core choke was added in series to reduce feedback at the highest frequencies.

The frequency response of the amplifier driving a 15-ohm loudspeaker is shown as the dashed curve in Fig. 2. It is flat to ± 0.2 db from 20 cps to 20 kc, falling off 3 db at 30 kc. Low-frequency response is less important in this application therefore the values of coupling and bypass capacitors may be reduced by a factor of 10 or so, to bring the lower 3 db point up to about 20 cps. The amplifier will also drive a 7-ohm speaker quite satisfactorily, at levels up to 2.5 watts.

The amplifier is well suited to

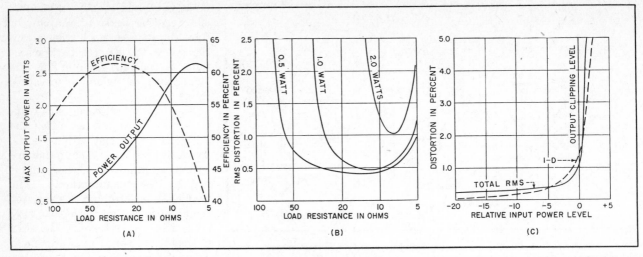

FIG. 3—Power and efficiency (A) rms distortion (B), and rms and intermodulation distortion compared to input power level (C) for transformerless transistor amplifier used to record heart signals at frequencies down to 3 cps

portable operation, using three 6-volt batteries. Over one watt output with 0.5-percent distortion can be supplied to a 15-ohm speaker, with excellent battery economy. Under these conditions the normal overall idling current to all stages is 10 to 15 ma, which increases to 165 ma on peaks. With a 7-ohm loudspeaker the peak current is about 280 ma for 2.5-watts output.

The output transistors work well below their normal ratings and power outputs up to 4 or 5 watts can be obtained by increasing the supply voltage. At still higher levels, the driver stage is the limiting factor, as these transistors can no longer supply adequate base currents to the output transistors.

The 302-2N35 drivers can be replaced by a pair with higher ratings, such as a 2N68-TN95 pair, to yield 10 watts or more of output power on peaks. However, the input impedance of the 2N68-2N95 drivers is considerably lower than that of the 302-2N35 pair, which necessitates a redesign of the previous stage. Furthermore, with the same supply voltage, the idling current of the modified power stages is increased four or five times over the idling current of the original circuit, which may be a serious drawback with battery operation.

Power Supplies

Because of the class-B operation, batteries will give good service in portable applications. Where 110-volt 60-cycle power is readily available, the simplest forms of tran-

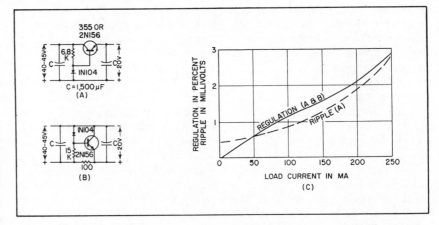

FIG. 4—Series (A) and shunt (B) regulators provide regulation (C) for a-c operation

sistor-regulated power supply are very satisfactory, since the main function of the regulating action is to suppress the ripple voltage. A conventional transformer-rectifier supply, with silicon diodes, is used to furnish full-wave rectified power at 40 to 45 volts to the regulators shown in Fig. 4.

Series Regulation

With constant load the output voltage of the series regulator, Fig. 4A changes by 1 percent as the line voltage is varied from 90 to 130 volts. Output voltage variation is 2 percent for a load change of 0 to 200 ma. Maximum ripple voltage is reduced to about 2 mv in 20 volts, as compared to 150 mv in 20 volts for the same power supply and filter without the transistor. Ripple voltage of the series regulator decreases as the load drops. The output voltage is fixed at slightly less than the breakdown voltage of the

Zener diode, 1N104, which is preselected in the desired voltage range.

The shunt arrangement of the same transistor and diode, shown in Fig. 4B, has a load regulation curve that is almost identical to that shown for the series regulator. The output voltage now is slightly greater than the voltage across the Zener diode. The ripple voltage remains constant at about 2 mv, independent of load. The shunt-regulated supply is a constant-power device, and the total power consumed, including the loss in the 100-ohm series resistor, is greater than for the series regulator, except when the latter is operated continually at maximum load current. On the other hand, an accidental short-circuit across the output of the shunt regulator does no harm, whereas a short across a series regulator usually destroys the transistor.

BUOY TELEMETERS OCEAN TEMPERATURE DATA

By ROBERT G. WALDEN, DAVID D. KETCHUM
and DAVID N. FRANTZ, JR.

Woods Hole Oceanographic Institution, Woods Hole, Massachusetts

Temperature measuring telemetering system in buoy is triggered by audio modulated r-f signal received by transistor receiver. Cycle of transmission includes two standard tones and thermistor temperature data together with identifying code signal. Data can be obtained at distances up to 600 miles by triggering at favorable propagation times

INCREASING cost of operating oceanographic vessels emphasizes the need for a method of making semicontinuous measurements at many different locations independently of the ship. Telemetering anchored buoys can provide measurement of surface or deep temperatures, or when allowed to drift can provide ocean current and circulation data.

The buoy shown in the photographs can be triggered remotely by a radio signal modulated with a distinctive tone or may be triggered by an internal clock at regular intervals. A signal from the buoy, modulated by a thermistor-controlled R-C oscillator, transmits water temperature data in terms of the modulation frequency.

The buoy has been reliably triggered and read at ranges up to 300 miles at frequencies between 2 and 3 mc. This range has been increased to as much as 600 miles by triggering at times of optimum propagation.

Operation

In operation, a number of buoys are anchored at locations determined by the physical problem to be investigated. The buoys transmit and receive on the same frequency, but each buoy responds to an interrogating signal only if the signal is modulated with a tone of particular frequency.

Vibrating-reed relays are used in the output of the buoy receivers to obtain a narrow audio-frequency response that almost entirely eliminates accidental transmissions caused by noise and interference. The interrogating transmitter is modulated by tones derived from a resonant reed-controlled oscillator with a separate resonant reed for each tone.

The reception of the proper tone by the buoy receiver actuates a time-delay relay which allows quick-heating transmitter filaments five seconds warmup time. The transmitter is then keyed with the call sign followed by three consecu-

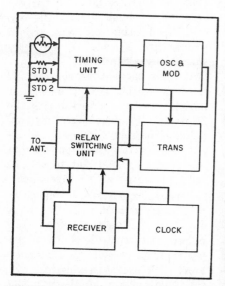

FIG. 1—All sections of telemetering buoy are controlled by relay switching unit

Transistor receiver is given final check before buoy is sealed for launching. Receiver is used to turn on transmitter when propagation conditions are favorable

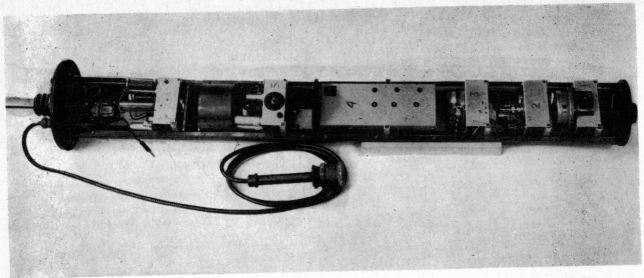

Electronic components for buoy removed from housing. Six chassis from left to right are transmitter, modulator, receiver, timing unit, relay switching unit and clock. All units are powered by battery pack

tive tones. The frequencies of the three tones are controlled by: a standard resistor representing a point at the low end of the temperature — audio-frequency curve, a standard resistor representing a point at the high end of this curve and the thermistor, which measures the water temperature.

The total transmission lasts 15 seconds. At two transmissions per day the estimated battery life is about two months. This f-m/a-m signal is received on a standard communications receiver and recorded on magnetic tape. The tone frequencies are then measured with an eput meter. In the event of garbled reception the buoy may be triggered again at a more favorable time.

The electronic equipment is contained in six aluminum chassis which plug into a cylindrical aluminum and bakelite rack. A two-inch aluminum channel, which forms the backbone of the rack, carries the receptacles and interconnecting wiring. The entire rack fits snugly into the tubing that forms the buoy. A glass fiber center-loaded marine antenna, supported by the end cap, effectively withstands the salt atmosphere and the force of moderate seas that occasionally break over the buoy. The cylindrical battery stack consists of 18 45-volt B batteries and 17 No. 6 1.5-volt cells.

Timing

Figure 1 shows a block diagram of the electronic circuits. The electrically wound clock and a transistor receiver operate continuously with relatively low battery drain. The clock may be set to operate a switch once or twice daily at any desired time. The clock switch actuates a sequential circuit in the relay switching unit that applies filament power to the modulator and transmitter, allows about five seconds warm-up time and then operates the antenna relay and timing unit. The timing unit then keys the modulator to transmit call signal and temperature data.

At the end of the thermistor tone, power is removed from the transmitter and modulator, the antenna is switched back to the receiver, and the device is ready for another cycle. The same sequence of events may also be initiated through the receiver by reception of the proper signal from a remote ship or shore station.

Since the receiver must run continuously during the life of the buoy, the most important design consideration is low battery drain. The receiver, shown in Fig. 2, employs nine transistors in a conventional superheterodyne circuit. To achieve the high sensitivity required for long-range operation, a stage of r-f preselection and two

i-f amplifiers are included.

All stages through the second detector are surface-barrier transistors, chosen for their high-frequency response and low power consumption. The local oscillator is crystal controlled. Stable operation of the r-f and i-f stages is effected by not bypassing a portion of the emitter resistor in each stage.

The push-pull, class-B output stage drives a resonant-reed relay. No forward emitter bias is used in this amplifier, since distortion of the output signal does not materially affect operation of the relay. Power consumption of the output stage is thus negligible in the absence of an input signal. The resonant-reed relay contacts are in series with a sensitive d-c relay shunted by a capacitor.

The discharge time constant of the combination is long enough to hold the contacts closed during the intervals when the vibrating contacts of the relay are open. An input signal with a duration of about one second is sufficient to set the relay switching unit in operation.

Receiver sensitivity is defined as the minimum value of r-f voltage appearing across the antenna terminals required to hold the d-c relay contacts closed. An r-f generator, 100-percent modulated with the appropriate audio-frequency signal, provides the test voltage.

Complete assembly consists of (l to r) battery pack, electronic unit and buoy case

The measured generator impedance and the calculated receiver input impedance are used to arrive at the sensitivity figure. A typical value is 2 to 5 microvolts.

Overall bandwidth of the r-f and i-f sections was found to be about five kilocycles between half-power points. The actual bandwidth of the receiver, however, is determined by the resonant-reed relay, whose bandwidth increases somewhat with the amplitude of the driving current.

Relay Response

The usable response of the relay is at most only a few cycles wide. The result is an excellent overall s/n ratio and almost complete elimination of response to unwanted signals.

Total receiver power consumption with no signal present is 40 milliwatts. With full audio output the batteries must supply about 150 milliwatts.

Because an efficient avc system is not easy to achieve in a transistor receiver, this feature, for simplicity, was omitted. This de-fect, particularly in the high-gain circuit described, results in overloading with a moderately strong input signal. If the signal is not 100-percent modulated, the entire modulation envelope may be clipped off in the i-f stages, causing loss of audio output. This disadvantage is easily overcome by reducing the power of the interrogating transmitter when the buoy is nearby.

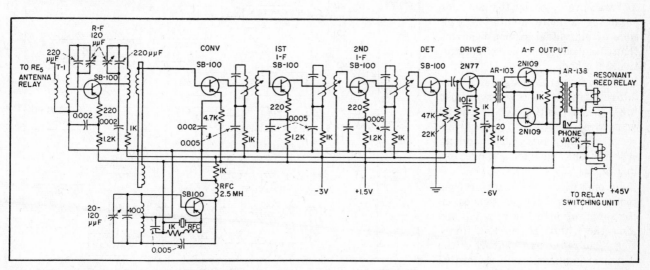

FIG. 2—Transistor superheterodyne receiver turns on transmitter when interrogated with signal modulated with proper audio tone

ULTRASONIC TRACER FOLLOWS TAGGED FISH

By PARKER S. TREFETHEN, JOHN W. DUDLEY and MYRON R. SMITH

Fishery Research Biologist
U. S. Fish and Wildlife Service
Seattle, Wash.

Senior Engineer   Senior Staff Engineer
Seattle Development Laboratory, Ordnance Division
Minneapolis-Honeywell Regulator Co., Seattle, Wash.

Single-transistor pulsed ultrasonic oscillator with 7-hour battery life is clipped to adult salmon on spawning run. After fish is released, movement is tracked with sonar-type equipment on boat to obtain detailed fish behavior patterns in relation to dams. Servo-controlled four-transducer receiving array tracks tagged fish automatically to give elevation and azimuth. Sonar ranging transducer in center of array is thus aimed directly at fish

DEVELOPMENT of automatic ultrasonic tracking and ranging equipment for the Fish and Wildlife Service provides a method for obtaining information on adult salmon behavior in relation to dams in the Northwest. The design and production of this new research tool was prompted by the need for obtaining a more detailed knowledge of individual salmon behavior than has been possible by conventional methods.

The equipment consists of a miniature ultrasonic source (sonic fish tag), a self-positioning directional transducer array with a receiver-servo system (auto-track) and an echo-ranging system arranged as in Fig. 1.

In operation, the tracking system positions itself or homes on the sonic tag attached to a fish and aims the transducer in that direction. The range and bearing of the fish are displayed on a calibrated cathode-ray tube. The depth of the fish is calculated from the range and the angle at which the ultrasonic impulses from the equipment are projected into the water. By plotting the position of the fish and the position of the boat simultaneously, the operator is able to obtain detailed fish behavior patterns.

Ultrasonic Fish Tag

To permit attachment to relatively small fish with minimum

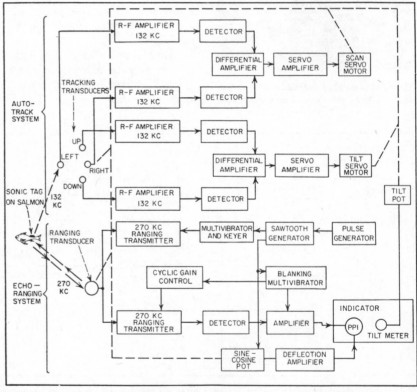

FIG. 1—Complete automatic tracking and ranging system for tagged fish

modification of their behavior, the sonic tag must be of miniature size. Opposed to this requirement is the need for battery life of 8 to 12 hours. Accordingly, circuitry in the sonic tag was chosen to hold current requirements to a minimum.

The additional requirement of approximately neutral buoyancy

modifies the package to some extent. The resultant capsule is 2⅜ inches long, 0.9 inch in diameter, weighs 29 grams in air and displaces 27 grams of fresh water.

The capsule shell, spun from 7-mil aluminum, is provided with a soldered-on hog ring for attachment to the fish. The tag closes with a slip fit and is sealed with

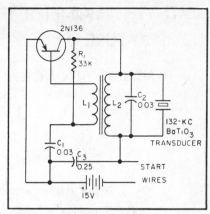

Sonic tag on silver salmon has carefully adjusted buoyancy so as not to affect natural movements of fish. Pulse repetition rate of transistor oscillator in tag is adjustable; with appropriate receiver tuning facilities, up to ten different fish can be tracked simultaneously

FIG. 2—Pulsed 132-kc ultrasonic oscillator used in sonic fish tag can be detected at distances up to 2,000 feet

vacuum grease, a wrap of plastic tape and an over-all plastic dip to ensure water tightness to a depth of 30 feet.

Economy of battery life is improved by pulse operation of the sound source. The common-emitter transistor oscillator circuit used in the sonic tag is shown in Fig. 2. Here R_1 and C_1 establish an interruption rate of 2,000 pps and a pulse length of 200 microseconds. The values of L_2 and C_2 and the clamped capacitance of the transducer determine the 132-kc operating frequency.

The transducer used in the sonic tag is a barium titanate disk 0.040 inch thick and 0.86 inch in diameter. The disk is one-half wavelength in diameter, resonant in the radial mode at 132 kc. It has silver fired on both surfaces and small lead wires soldered to the nodal point or center of each side. The transducer mounts inside one end of the spun aluminum capsule, at the break point of the end radius. Epoxy resin cement secures it in place and gives acoustic coupling

to the aluminum shell. This construction, which provides a marked benefit in matching the acoustic impedance of the disk to the water load, gives the desirable nearly circular radiation pattern shown in Fig. 3.

With this circuit arrangement an r-f voltage of 25 volts peak-to-peak is developed across the transducer, providing a signal level of 5 microvolts at the receiver grid at 400 feet. The 15-gram 15-volt battery has a useful life of 7 hours at a current drain of 3 ma.

Contents of ultrasonic fish tag. Crystal transducer (top) fits into left half of capsule, which is shaped and dipped in waterproofing plastic solution after assembly. Two leads at lower left come out of capsule through joint, for connecting together to complete circuit just before fish is released. Smaller tag is now being developed for use on small fish

BATHYTHERMOMETER TELEMETERS OCEAN TEMPERATURE DATA

By JAMES M. SNODGRASS and JOHN H. CAWLEY, JR.

Special Development Division, Scripps Institute of Oceanography, University of California, La Jolla, California

Two-unit transistorized system lowered from ship gives plot of temperature against depth. Absolute accuracy in depth is better than ±0.25 percent and temperature sensitivity of 0.05 degree C can be obtained. Vibrating wire transducer and thermistor Wien-bridge oscillator provide depth and temperature data, respectively

OCEAN temperature as a function of depth provides valuable information for oceanographic studies. Previous methods of measurement by mechanical means provided an accuracy of ± 0.5 C and ± 5 ft in depth.

Resolution of the order of 1 ft in 1,000 in depth and 0.1 C in temperature are required in present-day oceanographic work. In addition, continuous profile recording is desirable. The system described here is one approach to this problem.

In the block diagram of Fig. 1, frequency variation of a thermistor Wien-bridge oscillator is used to measure temperature. Depth information is provided by monitoring the frequency change of a vibrating wire connected to a flexible diaphragm. Both signals are sent over a single conductor to the ship where they are separated by high-pass and low-pass filters for recording.

Depth Measurement

A Vibrotron is a vibrating-wire transducer in which a mechanical displacement is converted into a frequency change of a vibrating wire. The wire is connected at one end to a pressure-sensing diaphragm. A displacement of the diaphragm changes the tension in the wire thus changing the frequency of vibration. Frequency range of the unit used in this equipment is approximately 9,600 to 11,240 cps.

Electrically, the vibrating wire is similar to an electrically driven tuning fork. The wire is of non-magnetic material and is placed in a fixed magnetic field at right angles to the axis of the wire. When the wire vibrates at its natural frequency, it becomes an a-c generator, generating a voltage that can be amplified by conventional means. If some of the amplified voltage is fed back to the ends of the wire in phase with the generated voltage, vibration is sustained.

An agc circuit is added to the amplifier to control the amplitude of vibration. The output frequency of the system is a pure sine wave, controlled by the axial displacement of the wire. The output frequency is thus nearly a linear function of the pressure applied to the pressure-sensing diaphragm.

Research ship equipped with boom for raising and lowering bathythermometer to measure temperature gradient of ocean water

Depth sensing unit with cover removed to show vibrating-wire transducer (center)

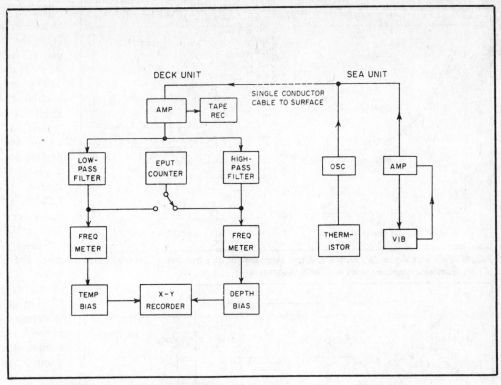

FIG. 1—Outputs of depth and temperature sensing systems are separated in high and low-pass filters for X-Y chart recorder

Figure 2 illustrates the circuitry for the three-stage transistor amplifier used with the vibrating wire transducer. Because the vibrating wire and static wire impedances are relatively low, a step-up transformer is utilized. The grounded-emitter first stage is resistance-capacitance coupled to the grounded-emitter second stage, which in turn, is directly coupled to the grounded-collector output stage. A properly phased positive-feedback loop is fed from the low-impedance emitter of Q_3 to the center-tap of two resistances bridged across the primary of the input transformer.

An automatic-gain-control network is necessary to drive the Vibratron at constant amplitude. Because the relationship between amplitude and frequency of vibration is exponential, it is necessary to maintain the amplitude constant for a given stability. A voltage of between one and two millivolts measured across the vibrating wire is considered satisfactory.

Under the operating conditions chosen, gain of the first grounded-emitter stage is essentially dependent upon the value of r_e. If I_e is varied over a range of 75 to 20μa, the graph of Fig. 3A shows the variation in voltage gain of the first stage versus input voltage.

The measured value of r_e at one milliampere for the type 202 transistor used is 33 ohms. The variation in the measured and computed values results in part from assumptions made in deriving the equation for voltage gain.

The graph of Fig. 3B plots output voltage as a function of input voltage for the three-stage amplifier. Also plotted is emitter current of the first stage as a function of input voltage. The overall result of the automatic gain control action is less than $\frac{1}{2}$-db change in output voltage for a 3-db change in the input voltage at values over one millivolt input.

Temperature Stability

The temperature stability of the amplifier is considered adequate for this application. Sea-water temperature generally is found in the range of 0 to 28 C. Bias stabilization is generally established by the large emitter resistors and the voltage dividers in the base circuits of the first two stages. The degree of temperature stability was measured in actual tests from 5 to 55 C. Fig. 3C shows the results of out-

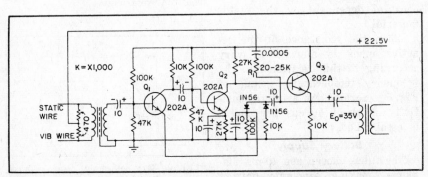

FIG. 2—Transistor amplifier for vibrating wire transducer uses agc to maintain output level at constant amplitude

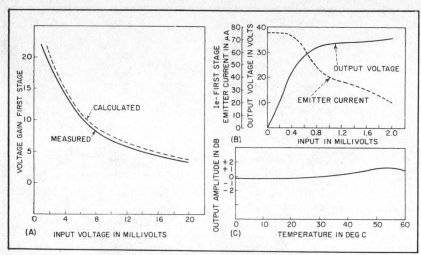

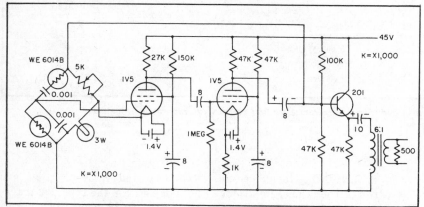

FIG. 3—First stage gain (A), output and agc response (B) and frequency response curve (C) for transistor amplifier used for depth measurement

FIG. 4—Wien-bridge oscillator circuit used to measure water temperature. Units encapsulated in epoxy resin and sealed in pressure tight housing.

put voltage against temperature. Only minor distortion was noticed over this range.

The output voltage is approximately 0.5 volt when working into a loaded 500-ohm line.

Temperature Sensing Circuits

The temperature sensitive oscillator is of the Wien bridge type, with 6014A thermistors as resistive elements in the reactive arms of the bridge. This type oscillator was chosen because of its simplicity and dependable performance. Its one disadvantage is that separate filament batteries are necessary when using filament type tubes. The circuit is shown in Fig. 4.

The bridge balance and oscillator frequency is set by

$$R_1 = 2r$$

$$f_o = \frac{1.59 \times 10^1}{\sqrt{C^2 R^2}}$$

where f_o = frequency of oscillation, R_1 = negative feedback resistor,

r = lamp resistance, R = thermistor resistance and C = capacitance in each half of positive feedback arm.

The temperature coefficient of resistance for the thermistor chosen is −3.9 percent per deg C. Unfortunately, the resistance versus temperature characteristic is nonlinear as shown by the following equation:

$$R = R_o e \exp B[(1/T)-(1/T_o)]$$

where R_o is resistance at reference temperature in deg K, and B is a constant dependent on thermistor material.

When the resistance of the reactive arms is allowed to vary according to this relationship, the frequency - temperature graph of Fig. 5 is formed. An average sensitivity of about 40 cps per deg C is established.

Battery Supply

The tubes chosen use 40-ma filaments, making economic battery operation feasible. The grounded-

collector transistor output stage serves to isolate the output circuit from the oscillator section. An output voltage of 0.5 volt is available when loaded by a 500-ohm line.

The temperature stability of this oscillator is satisfactory for this application. When properly adjusted, a frequency variation of one to two cps over an eight-hour period at constant temperature is normal.

Over a temperature range of 5 to 50 C, a frequency change of four to five cps was observed in laboratory tests.

Speed of response is important since the instrument is lowered and raised through the water at a rate of 2.5 feet per second. A thermal time constant of one second or less is necessary to resolve all the detail present. A step-function change in temperature will result in a frequency equilibrium within two seconds in sea water. This time has been cut to the order of one second on occasion by grinding down the glass surface of the thermistor beads.

Instrument Housings

Both the pressure sensitive transistor amplifier and the temperature sensitive oscillator are encapsulated in a silica-filled epoxy resin. Plug-in construction is used to simplify

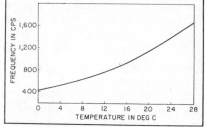

FIG. 5—Plot of oscillator frequency variation with temperature

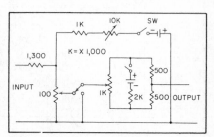

FIG. 6—Zero set circuit used with recorder. Variable resistor and potentiometer are helical type

maintenance and construction.

The instrument is divided into pressure and temperature units and individually housed so that they might be used separately if desired.

The cases are of 5 in. inside diameter, 0.312 in. stainless steel tube. The end plates are of 1-in. brass with O-ring pressure seals. Packing glands are used pass the instrument leads to the surface.

The two instrument housings are pressure tight and designed for a maximum pressure of 5,000 lb per sq in., with an adequate safety factor.

The outputs of the pressure and temperature units are coupled on one side to the instrument case and the other to a single-wire polyethylene-insulated cable. The cable is a 19-strand steel wire with a tensile strength of about 2,800 pounds. The polyethylene jacket brings the overall diameter of the cable to approximately 0.32 inch. The polyethylene jacket also performs the extremely important job of giving the cable buoyancy. This allows the cable to support much greater instrument loads at greater depths, since the cable itself is almost weightless in sea-water.

At the surface the signal is taken off the research vessel's winch via slip-rings. A sea return to the instrument is used with a zinc plate serving as the ground-return connection.

Surface Instrumentation

Surface instrumentation, as shown by the block diagram, consists of a broad-band amplifier capable of handling frequencies up to 15 kc. The frequencies are then separated by a high-pass and a low-pass filter and passed to a demodulator. The resulting d-c outputs are functions of pressure and temperature.

To make these parameters suitable for recording on an X-Y function plotter, a circuit to set zero bias and scale factor is a necessity. It is shown in Fig. 6. This zero offset method is used to set at center scale on the recorder the midrange of both the temperature and pressure functions.

Pressure signal is applied to the X-axis and temperature signal to the Y-axis. This makes it possible to roll the chart manually after each successive run, thus completing a family of bathythermometer curves.

The two channels are also recorded on magnetic tape. A dual channel recorder is used with the combined temperature and pressure information recorded on one channel and a constant 5,000-cps signal on the other. The advantages of this type data storage and playback are self-evident and work is being continued to perfect this phase of the instrumentation problem.

Test Results

This instrument has been used at sea with good results. Fig. 7 is a continuous bathythermometer plot taken at sea off San Diego, California. The research vessel was drifting very slowly. Over the narrow range covered, the temperature scale is nearly linear. Small positive and negative temperature gradients are readily apparent and possible sound channels are indicated.

The vibrating-wire transducer has an absolute resolution of approximately 0.003 inch at the 2.5 ft lowering rate generally used. The overall system is limited by the time constant of the frequency meter. The resulting resolution of the overall depth-measuring system is therefore of the order of 0.3 inch. Absolute accuracy in depth is better than ±0.25 percent or ±2.5 ft in 1,000.

The temperature sensitivity is approximately 40 cps per deg. C. When these parameters are converted to direct current and applied to the function plotter with a ten-inch chart width, a depth variation of ±2.5 ft per 1,000 can easily be read. If ten degrees of temperature are applied to ten inches of chart, a temperature variation of 0.050 deg C can be accurately plotted.

Reproduction of pressure and temperature plots takes place without significant hysteresis effects. This was checked while at sea, where surface temperature and pressure can be accurately measured.

This work was carried out at the Special Developments Division of the Scripps Institution of Oceanography and was supported by funds from the Office of Naval Research and the Bureau of Ships, United States Navy. The authors extend their gratitude to all who took part in this problem, and in particularly to G. T. Barlow and R. M. Blei for their help and assistance in the construction of the equipment.

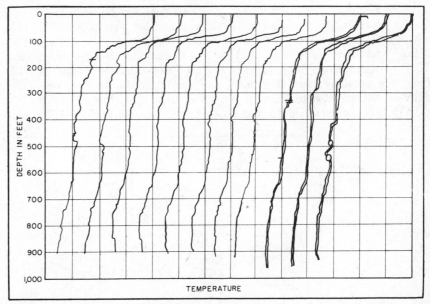

FIG. 7—Typical family of temperature-depth curves obtained with bathythermometer off the coast of San Diego, California

BATHYTHERMOMETER AMPLIFIER

By JOHN H. CAWLEY, JR.

Research Engineer, General Atomic Division of General Dynamics Corporation,
San Diego, California

IN THE article "Bathythermometer Telemeters Ocean Data" (ELECTRONICS, May 1, 1957, p 142) a vibrating wire transducer and transistor amplifier are described as part of the pressure measuring system.

In an effort to adapt the amplifier to more recent transistors, a new circuit using 4 close-tolerance 202-A transistors was developed. This circuit, shown here, differs from the previous circuits principally in the addition of transistor Q_2. It has been added to more

nearly match the output impedance of Q_1 to the input impedance of Q_3. Transistor Q_2 is a simple direct-coupled common-collector amplifier which improves the voltage gain of the first stage by almost 6 db.

This not only improves the waveform at the output but has the added advantage of making the entire amplifier less dependent on h_{fb} of the individual units as well as more independent of temperature variations.

Of 12 transistors selected at ran-

dom from stock, 11 performed interchangeably with no observed variation. Changing the amplifier input voltage from -4 to $+10$ db (0 db = 1.4 mv) caused an output amplitude variation of 0.5 db or less.

Distortion in the output is negligible over this range. This constant amplitude output is important in the operation of the vibrating wire transducer as the frequency is dependent on a constant amplitude drive.

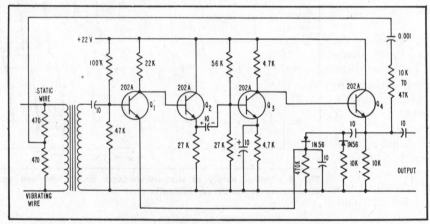

Circuit of improved version of amplifier used in bathythermometer

Chapter 16

COMPUTER CIRCUIT DESIGN

COUNTERS SELECT MAGNETIC-DRUM SECTORS

By A. J. STRASSMAN and R. E. KING
Ground Systems Laboratory, Hughes Weapon Systems Development Laboratories,
Hughes Aircraft Company, Culver City, California

Large-scale data-processing systems require a means of checking the arithmetic and memory systems employed. A 16-digit and 100-digit counter combined with logic circuits provide an automatic writing unit to produce a predetermined binary pattern in any selected word or sector on a magnetic-drum memory. The equipment can also be integrated with arithmetic units to aid or check computation procedures

MAGNETIC-DRUM MEMORIES in digital data-processing equipment require that the stored information be sometimes entered in specified locations on the periphery of the drum cylinder. To accomplish this, some method of determining location is required as well as system for entering the desired information. The equipment to be described can be used as a piece of testing apparatus to check these magnetic drum operating characteristics and also for computation procedures in a computer arithmetic unit.

The ability to select any particular sector of a magnetic-drum memory can be achieved if there is a signal written on the drum to indicate the beginning of each word or sector. These signals are then applied to a counter whose output is selected by switching means to display only a selected number. This number is then co-incidence gated with the desired action of either writing into or reading from the drum at the particular sector selected. Figure 1 shows the block diagram of a system that accomplishes this task.

Words

A pulse indicates the beginning of every word written previously on the drum. This pulse is detected by the read amplifier which triggers the read flip-flop. The flip-flop has the two outputs shown in Fig. 2A; a pulse that is true for only one bit time and its complement which is false for that same bit time and true for the rest of the word.

The true signal P_w is applied to a modulo-X counter where X is the total number words on the drum and for this case is equal to 100. The complement is applied to the modulo-16 digit counter and is used for a reset signal.

The selected outputs of the mod-100 word counter and the mod-16

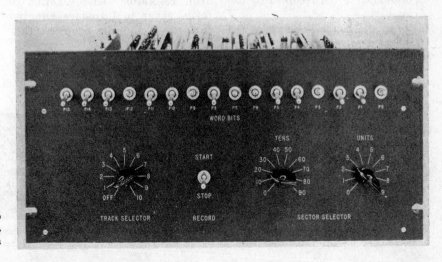

Front view of sector selector shows setup for selecting the word bits, their sequence, the sector and the track to be recorded on

counter are *anded* in logical diode network G_1 and applied to single-input flip-flop 1. Figure 2B shows the output of both the mod-16 and the mod-100 counters.

The output of flip-flop 1 is the selected word time with the selected bits inserted within that word time. Figures 2C and 2D show the output of flip-flop 1 with the coded data in all but one and only one sector respectively. This information is applied to a writing amplifier whose output is fed to a writing head on the magnetic drum. One of several heads can be selected by switch S_1.

The writing amplifier is gated on and off automatically by the logical circuitry shown in Fig. 1 consisting of the gating applied to the input of flip-flop 2. Schematic of the transistor flip-flop is shown in Fig. 3 and a system time diagram is shown in Fig. 4.

Automatic Writing

The binary-coded information is recorded on the drum by the writing amplifier illustrated schematically in Fig. 5.

Manually switching the writing amplifier on and off will normally produce transients in the recording waveform which may alter or destroy any information fed through the writing amplifier or recorded on the drum under the recording head at the instant of switching.

The writing amplifier used to record the binary pattern in the selected word or sector is switched on and off by the action of flip-flop 2. When this flip-flop is on or in the true state the writing amplifier is gated on and allows the output of flip-flop 1 to be recorded on the drum.

When flip-flop 2 is off or in the false state no recording is done and the drum carries stored information. To avoid losing information

Internal view shows how etched-circuit card containing logic circuits and flip-flops are plugged into standard rack-mounted chassis

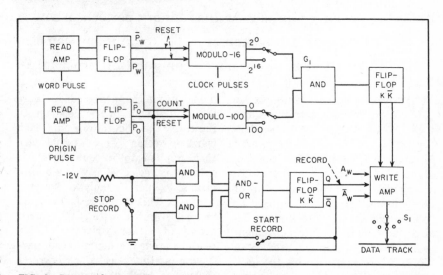

FIG. 1—Sector selector writes preselected word containing up to 16 bits when start-record switch is pressed

the switching time is chosen to coincide with the occurance of the origin pulse beginning at each drum revolution and precaution is taken never to introduce any information onto the drum in this area.

The circuitry used to avoid this area consists of an *and-or* logical net whose output triggers a one-input flip-flop and a normally-closed pushbutton, used to start the operation. Stop action is accomplished by a normally-open switch.

The writing amplifier is normally in the off condition, therefore the Q output of flip-flop 2 is high or on. Flip-flop 2 is held in this

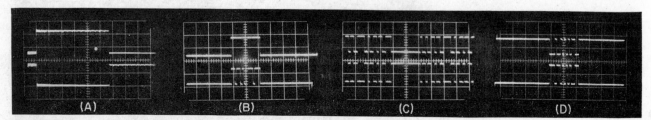

FIG. 2—Waveforms show read amplifier flip-flop output (A) for one bit time, mod-100 (top) and mod-16 (bottom) output as fed to flip-flop 1 (B), output of flip-flop 1 with coded data in all sectors but one (C) and output of flip-flop with data in only one sector (D)

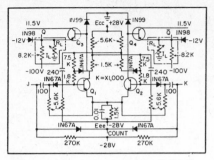

FIG. 3—Output of transistorized flip-flop gates writing amplifier

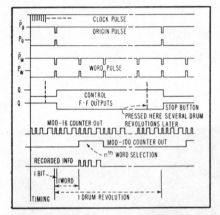

FIG. 4—System time diagram shows operating sequence from time start switch is pressed to time stop switch is pressed

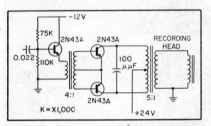

FIG. 5—Transistorized writing amplifier drives recording head over selected drum track in proper sector of rotating drum

condition by feedback through the start switch. When the start switch is pressed this signal is removed, however the gate is still held false by the complement of the origin pulse.

When the origin pulse is received from the drum, all the inputs to flip-flop 2 are removed and the Q output goes high or on. Thus the writing amplifier records when the origin pulse is received.

Recording will continue with the preset pattern injected in the selected word until the stop button is depressed. This triggers flip-flop 2 to its original state when the origin pulse occurs, consequently switching off the writing amplifier. The writing amplifier is held in

this state until the start button is depressed again.

Thus the writing amplifier has been switched on and off at the beginning of each drum revolution only and any transients that may occur will not affect the rest of the magnetic drum.

Sector Counter

Two modulo-10 counters in series form a sector counter that produces a count of 100 before recycling. Standard transistor two-input flip-flops are used as the basic element and diode gating is used to accomplish proper timing.

Since the counter is divided into two identical counters, the logical equations for the tens counter are the same as for the units counter with the addition of the final count from the unit counter logic. Circuitry and logic are shown in Fig. 6.

The single selected word output is decided by the position of the unit-sector number switch and the ten-sector number switch. The outputs of all flip-flops are connected to the input of a 10-position 4-pole switch whose outputs are connected to *and* gate G_1. This produces an on pulse one word long at the selected time only.

An alternate output is the complement of this signal which would cause the output to be on for 99 words and off for the selected word. This allows the same binary pattern to be written in all sectors of the drum except for a guard sector which contains all zeros.

The modulo-16 counter, triggered by logic-circut output, is used

Table I—Flip-flop Truth Table

J	K	Q^{n+1}
O	O	Q^n
O	1	O
1	O	1
1	1	Q^{-n}

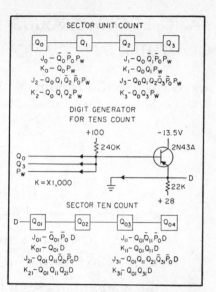

FIG. 6—Logic equations for unit count are shown with circuit of digit generator providing output D, indicating the number 10, to trigger sector ten count that uses lower set of logic equations

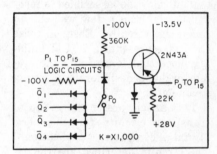

FIG. 7—Modulo-16 digit counter output circuit is driven from P_1 to P_{15} logic circuits

for the digit counter. Figure 7 shows the logic and diode circuitry necessary to accomplish the generation of a binary pattern. Table I shows the truth table associated with the flip-flops.

The output of this counter is *anded* with the output of the sector counter G_1 whose output is applied to flip-flop 1. The output of this flip-flop is a composite of the sector number and binary count (see Fig. 2C and 2D) that is applied to the input of the writing amplifier.

The writing amplifier output can now be switched to any writing head on the selector switch.

This equipment was designed as a piece of test equipment for evaluating certain functional equipments used and developed in the Ground Systems Laboratory of Hughes Aircraft under a contract with the U. S. Army Signal Corps Engineering Laboratory, Fort Monmouth, New Jersey.

BASIC LOGIC CIRCUITS FOR COMPUTER APPLICATIONS

By G. W. BOOTH and T. P. BOTHWELL

Radio Corporation of America, Camden, New Jersey

> Digital computer circuits, including flip-flop, gated pulse amplifier, d-c amplifier, power amplifier and indicator, use high-frequency junction transistors to obtain high reliability and performance characteristics. Circuits operate over temperature range of −30 to +60 C; their low dissipation imposes minimum requirements on power supplies and cooling

ONE of the important properties of a digital computer is that it may be assembled simply and easily from a few well-chosen functional circuits. Each of these circuits represents a logical building block that is useful to the system or logic designer in planning a computer.

This article presents a group of transistor circuits for a general digital computer application.[1-4] The circuits described are the result of a conservative design approach which takes advantage of the high switching efficiencies obtainable with the alloy junction transistor. Precision resistors are used throughout, and are treated as five-percent resistors in the design. Power dissipation, current, and voltage levels are kept low in

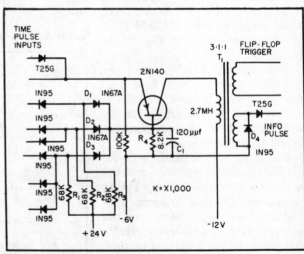

Operator reads out punched tape from NIFTE computer console. Transistor circuits are in rack seen in back of operator's head

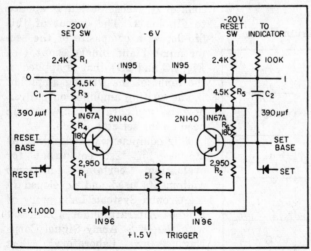

FIG. 1—Flip-flop stage has saturation operation of transistors

FIG. 2—Gated pulse amplifier has 250-kc repetition frequency

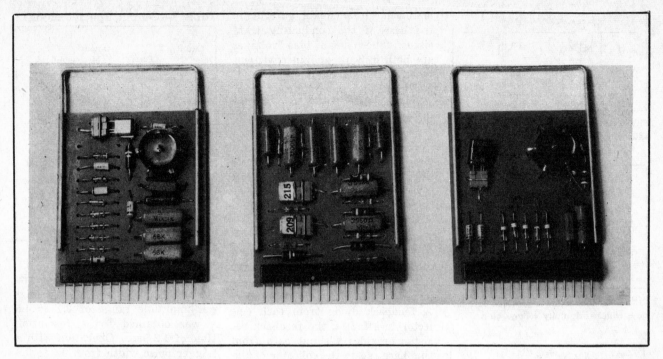

Gated pulse amplifier, flip-flop and power pulse amplifier circuits (left to right) are constructed on plug-in wiring boards

diodes and transistors, yet are large enough to avoid serious problems resulting from noise pickup. Circuit dependence on individual parameters of the transistors is minimized. Commercially available high-frequency alloy-junction transistors are used and a considerable slump of transistor gain from specified values can be tolerated without sacrificing performance.

The two basic circuits are a flip-flop and a gated pulse amplifier. All information flows in the form of 0.5-μsec pulses through chains of one or more gated pulse amplifiers and is ultimately stored in a flip-flop. Gate control information is in the form of d-c levels from flip-flops, in most cases without intermediate amplification being necessary.

Flip-Flop

The first of the basic circuits is a flip-flop. The Eccles-Jordan circuit shown in Fig. 1 was chosen for complementary outputs, its designability and the efficiency obtained by saturation operation of the transistors.

The limited frequency response available in alloy junction tran-

sistors requires clamping of the turn-off transient for fast rise times under load. Collector current in the flip-flop transistors is limited to 10 ma. A collector output swing of six volts was chosen as large compared to diode forward voltage drops, yet small compared to breakdown voltages of diodes and transistors.

A d-c stability analysis determined the circuit parameters to assure a two-ma load specification, stability under conditions of large I_{co} and discrimination against small noise pulses that would tend to trigger the circuit.[5] Reverse bias on the base-emitter diode of the nonconducting transistor is guaranteed with allowances for I_{co} and leakage in diodes of at least 120 μa, which corresponds to the I_{co} that might be encountered at +60 C. Since this figure is based on tolerance extremes of all components, a considerably greater I_{co} can generally be tolerated.

The design assures saturation in the on transistor for worst cases of all resistor and supply voltage tolerances for transistors with large-signal current gains greater than 15 at the 10-ma collector current required for flip-flop operation. No

maximum gain figure need be specified.

Switching Speed

At the instant the conducting transistor turns off, its collector rises toward some potential which is less than the supply voltage. In a flip-flop of this type, the aiming potential should be at least as large as the clamp voltage since the d-c stability analysis is based on a collector voltage equal to the clamp voltage.

Aiming Potential

The aiming potential is defined as the resultant potential at the cut off collector due to the divider action of the collector resistor, the feedback resistor and the load resistor when the clamp diode is disconnected. However, to assure that worthwhile advantage in speed is obtained, the aiming potential-to-clamp voltage ratio must be somewhat greater than unity. For Fig. 1, with a two-ma load current, this ratio is 1:5.

Feedback capacitors C_1 and C_2 were chosen empirically to perform speedup and memory functions for a typical pair of transistors. Too large a capacitance results in slow

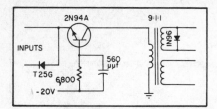

FIG. 3—Power pulse amplifier provides 180-mw output for 30-mw input

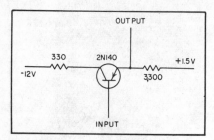

FIG. 4—Emitter follower provides current gain with nearly unity voltage gain

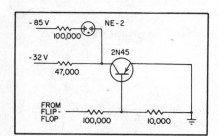

FIG. 5—Biased neon-lamp indicator

switching at the cut-off collector. Too small a capacitance results in tight frequency response requirements on the transistors for a satisfactory range of trigger pulse voltage and width. A compromise in the vicinity of 400 $\mu\mu$f provides a feedback coupling which is relatively independent of transistor characteristics.

To achieve a high repetition rate for the flip-flop, the discharge time constant τ of the feedback capacitors should be as small as possible. This time constant is given by

$$\tau = C_f[R_bR_f/(R_b + R_f)]$$

where C_f is C_1 or C_2, R_b is R_1 or R_2, and R_f is $(R_3 + R_4)$ $(R_5 + R_6)$.

Capacitor C_f must be increased as R_f is reduced; R_f is largely determined by current levels in the circuit. Thus, for minimum resolution time, it is desirable to have R_b as small as practicable.

Common emitter resistor R_e is included to obviate the need for a small bias supply voltage. With R_b small, the bias potential to which R_b is returned must be small to maintain an adequate current in the base of the conducting transistor. Such small bias voltages are both difficult and inconvenient to generate and regulate.

For this purpose, a bias is generated by the saturation current of the on transistor through R_e. The emitter return is set at $+1.5$ v so the collector of the conducting transistor is assured to be positive with respect to ground, providing a reverse bias on ground-returned loads.

Saturation Effects

Pulse steering is incorporated into the circuit to reduce saturation effects of the turn-on trigger pulse at large trigger amplitudes. A feedback diode from each collector to a tap on the feedback resistor provides a shunt path from this base tap to the collector of the conducting transistor.[6] Hence, a large amplitude trigger does not cause a high degree of saturation in the conducting transistor.

On the cutoff transistor, the feedback diode is reverse biased by six volts and has no effect on circuit operation. This feedback diode does not assure nonsaturating operation of the transistor, since some fixed current will flow through the resistor between feedback diode and base for the essentially constant forward bias drop of the diode.

For a high-gain transistor, this fixed current may represent saturation; for a low gain transistor, it may not. However, if the gain of the conducting transistor is sufficiently low as to cause it to be out of saturation by more than a few tenths of a volt, the feedback diode cuts off, providing maximum base current to the conducting transistor. In any case, d-c stability is not sacrificed by the use of the feedback diodes in the circuit.

For the circuit of Fig. 1, a trigger-amplitude range of 2.1 to 4.5 v was obtained for a resolution time of 1.2 μsec. Operation with a trigger pulse width from 0.4 to 0.9 μsec at the 2.6-v nominal pulse amplitude was achieved.

Gated Pulse Amplifier

The second basic logic circuit is the gated pulse amplifier shown in Fig. 2. The gating function of the circuit is controlled by the or-and diode gate in the base lead of the transistor.

Each or input of this gate is derived from a flip-flop whose output is 0 or -6 v. A 2.6-v positive pulse, biased at -6 v, is applied at the emitter. A pulse output will appear only if the base potential is more negative than -3.4 v since, if this condition is not met, the base emitter diode will never be forward biased. This condition can only exist when all and inputs have at least one or input at -6 v.

If one and input is at ground, the voltage division between its associated resistor R_1, R_2 or R_3, and R_4 will raise the base to -3 v. More than one and input at ground will make the base even more positive.

When the gated pulse amplifier is primed D_1, D_2 and D_3 are cut off. Hence the diode gate may be neglected and the pulse amplifying qualities of the transistor amplifier alone considered. The transistor is driven from zero bias to saturation by the 0.5-μsec pulse applied at the emitter.

Capacitor C_1 offers an increase in effective frequency response by

Table I—Output Specifications

Circuit	Output Form	Maximum Load
Gated Pulse Amp	2.6-v pulse 1/2 μsec (10 ma)	2—F-F or 3—GPA or 1—PPA
Power Pulse Amp	2.6-v pulse 1/2 μsec (55 ma)	11 F-F or 17 or 7 PPA
Flip-Flop	6-v level (2 ma)	4 GPA or 2 EF } +1 Ind
Emitter Follower	6-v level (18 ma)	36 GPA

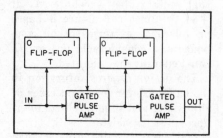

FIG. 6—Two stages of binary counter with high-speed carry

allowing the circuit to operate as grounded base during the 0.1 μsec turn-on and turn-off periods of the transistor. During the rise of the pulse, the transient current in the base lead is determined by the size of C_1 and the low pulse source impedance in series with the transistor base and emitter resistances. During the fall, the capacitor provides a low impedance in the base lead to diminish the effects of storage.

Each of these transient effects are settled in approximately 0.1 μsec and have little effect on the circuit during the flat-top portion of the pulse. When the transistor is operating in saturation the voltage drop from emitter to collector is negligible and $\beta I_B > I_c$. In this condition, a constant voltage appears across the load and the magnetizing current in the output transformer increases linearly.

The pulse output will collapse if βI_B becomes less than I_c or if the pulse falls at the input. The primary inductance of T_1 is chosen to assure that the second of these two conditions determines pulse width. Either case will cause a high collector impedance to be presented to the load.

Diode D_4 provides a low resistance path to damp the overshoot of the transformer at the end of the pulse.

The circuit design allows for a 250-kc pulse repetition frequency.

Amplifiers

Two amplifiers complete the system, one a power pulse amplifier and the other type is a d-c amplifier.

The power pulse amplifier is basically similar to the gated pulse amplifier, with the exception that no gating bias is provided and a 2N94A *npn* transistor is used to obtain a high power output. For a given emitter current, power output is proportional to collector voltage and maximum collector voltage on the available *npn* was higher than on the available *pnp*'s. This amplifier, shown in Fig. 3 is capable of providing a 180-mw pulse output for a 30-mw pulse input.

The d-c amplifier, shown in Fig. 4, is an emitter follower, providing current gain with nearly unity

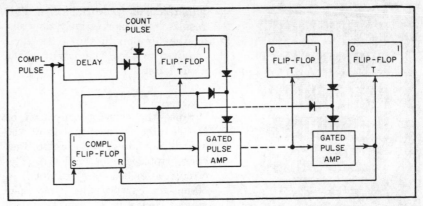

FIG. 7—Binary counter modification allows complementation and reversible counting

voltage gain. To prevent excessive dissipation and to avoid saturation in the transistor, a collector load resistor was introduced. Maximum transistor dissipation was calculated to be 40 mw and maximum load current 18 ma.

Indicator

Although not necessary to the logic, an indicator is desirable for any test of the system and for trouble shooting. The indicator shown in Fig. 5 employs a biased neon tube which is switched off and on by an a-f transistor. The transistor requirements are modest—collector breakdown greater than 35 v and current gain greater than 10.

Dissipation in the transistors is negligible since only the on current of the neon lamp need be supplied. Neon tube requirements are more strict because extinguishing potential is about 55 v and firing potential about 85 v.

Transistor requirements for flip-flop and gated pulse amplifier circuits are modest enough that 85 percent of the high-frequency transistors purchased in 1955 from one manufacturer were acceptable in all respects.

Rejects were primarily due to high leakage current at 20 v. Although the transistors were operable in the circuits, the low output impedance was taken as an indication of a poor junction, with adverse implications on long-term transistor life.

A suitable, currently available *pnp* is the 2N140; a *npn* is the 2N94A.

Complete specifications, formulated along with the test circuits to determine the specifications, are too lengthy to cover here. Three basic requirements are: 1) grounded-base frequency response > 4 mc; 2) large signal current gain (β) > 20 at I_e = 10 ma for flip-flops, > 30 at I_e = 10 ma for gated pulse amplifiers; 3) collector leakage and saturation currents (grounded base) I_{co} at 6 v $\leqq$ 3 μa, I_{co} at 20 v $\leqq$ 6 μa.

Application Rules

Table 1 summarizes the minimum output specifications for the circuits described.

Note that the gated pulse amplifier has two types of input, one requiring a 2.6-v, 3-ma, 0.5-μsec pulse and the other requiring a 6-v, 0.5-ma level. Hence, the outputs of the gated pulse amplifier and power pulse amplifier can drive only the pulse inputs of a gated pulse amplifier.

The outputs of the flip-flop and emitter follower can drive only the level (diode gate) inputs of the gated pulse amplifier. The emitter

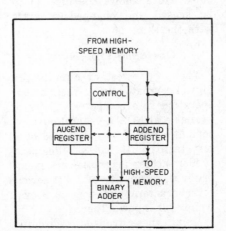

FIG. 8—Adder, excess-three-converter

Rack-mounted transistor adder-excess three converter uses plug-in boards

follower was designed to drive a maximum of 36 gated pulse amplifier inputs. This restriction is a matter of design convenience and not fundamental to the circuit.

Maximum pulse repetition frequency for the gated pulse amplifier is 250 kc. The flip-flop may be set and reset at a 500-kc rate; maximum triggering rate is 400 kilocycles.

Logic Techniques

The binary counter shown in Fig. 6 illustrates the type of logical structure to which the circuitry lends itself.

When a count is added to a binary counter, each bit is complemented according to the following two rules. Complement if: 1) the next least significant bit was one before addition of the count and 2) all less significant bits have been complemented under rule 1. The least significant bit is always complemented.

Hence, all bits of a counter are complemented starting from the least significant bit and continuing through the first bit in which a zero is found to be stored. This function is performed by the chain of gated pulse amplifiers in Fig. 6. The trigger pulse will pass from the first to the second stage and complement the second flip-flop only if the first flip-flop is in the one state before the trigger occurs.

Although the state of the flip-flop may change as a consequence of the pulse, the gated pulse amplifier acts on its initial state. This results from the relatively slow rise of the flip-flop to the -3-v level required to affect the output of the diode gate.

Carry Time

Since the carry propagation of the counter is independent of the switching time of the flip-flop, fast carry ripple can be achieved. The circuit described here results in a 30-mμsec carry-ripple time per stage. Hence in a 12-bit counter, the longest carry time will be 0.36 μsec. Time required for the result to be available is longer than this by the settling time of the flip-flop stages.

With very little additional equipment, the counter of Fig. 6 becomes a reversible counter. Figure 7 shows the logic for a simple version. Here the chain-gate logic performs both complementation and binary count. When subtraction from the original count is desired, the counter is complemented by the complement pulse and flip-flop, triggered with the count pulse, and recomplemented to be read.

Three characteristics of the counters shown might be highlighted for general computer logic applications: the pulse delay encountered through the gated pulse amplifier is small enough to allow a rather lengthy chain of amplifiers to be used with a minimum delay in passage through them; the gated pulse amplifier allows a wide variety of logic because of the or-and cascade and the ability to use these circuits in an interative connection; the relationship between the flip-flop resolution time and the pulse width allows the flip-flop to be sensed and changed in state with the same pulse.

Operating Experience

The proper test of circuits for digital computer use is their actual operation in a computer. There is available at RCA a general purpose computer specifically designed as a test facility. It has a high-speed random-access magnetic-core memory with a capacity of 1,024 seven-bit characters and uses a Flexowriter for input-output.

Using these transistor circuits, two components have been constructed for test in this machine. The first of these was a reversible counter and associated logic for keeping track of iterations during the multiply instruction and was built with a simple breadboard construction. Twenty-three transistors and 81 diodes were used.

Marriage between machine and transistor circuits was performed by a group of vacuum tube amplifiers, pulse shorteners, etc. The unit has operated without error for over 200 hours of computer operation. Routine testing of transistors during shutdown did indicate progressive deterioration of a group of transistors, all of one manufacturer. Since these have been replaced with another type no further deterioration has been encountered so far.

Adder Converter

An adder, excess-three-converter unit for the test computer was constructed as the second test component. Figure 8 is a block diagram of the adder converter. This equipment was paralleled with the existing vacuum tube arithmetic unit. Results computed by the transistor equipment are substituted for those computed by the vacuum tube adder and a parity check is made between the two results.

The basic circuits for this equipment were laid out on five types of individual plug-in units. A total of 81 such plug-ins are used, containing 110 logical and 20 indicator transistors and 450 diodes.

To date more than 300 hours of operation (39,000 transistor hours) have been logged on this unit since debugging, with no transistor or diode failures.

REFERENCES

(1) T. P. Bothwell, G. W. Booth and E. P. English, A Junction Transistor Counter with High Speed Carry, "Transistors I", RCA Laboratories, Princeton, N. J. p 646, Mar. 1956.

(2) D. E. Deutch, A Novel Ring Counter Using Junction Transistors, "Transistors I", RCA Laboratories, Princeton, N. J., p 640, Mar. 1956.

(3) E. W. Sard, Junction Transistor Multivibrators and Flip-Flops, 1954 *IRE Conv Rec*, Part II, p 119.

(4) C. L. Wanless, Transistor Circuitry for Digital Computers, *IRE Trans*, EC-4, p 11, Mar. 1955.

(5) T. P. Bothwell, Design of Nonsaturating Junction Transistor Flip-Flop, AIEE Winter General Meeting, Jan. 1955.

(6) J. Warnock, Junction Transistor Switching Circuits, AIEE-IRE Joint Transistor Conference, Philadelphia, Feb. 1954.

COMPUTER DELAY UNIT USING MULTIVIBRATORS

By WILLIAM A. SCISM
San Diego, California

Three point-contact-transistor one-shot multivibrators are cascaded to provide delay of 40 microseconds per stage. Delay line permits sequential read-in to parallel-form input of digital computer adding register. Rise times are better than 0.1 microsecond per stage

IN an airborne digital computer, it was desired to read data into the register sequentially although the data was presented in parallel form. A delay line was used to allow the carries to propagate before the next bit was read in.

Three conditions were necessary to close the and gates—voltage from the data leads, voltage from a time-sharing circuit and a pulse. The voltage from the data leads were read-in parallel to all gates simultaneously, applied to a block of gates for a time longer than that necessary for the pulses to traverse the delay line and pulses were applied to each gate sequentially from taps on the delay line.

In this manner parallel data was read into the register sequentially. Use of this delay line saved circuitry and weight that would have been necessary had other storage means been used.

Circuit

The delay line consists of three cascaded point-contact one-shot multivibrators as shown in Fig. 1. A negative current pulse through the base emitter circuit of the first stage initiates one-shot operation. The positive pulse from the first stage is ignored by the second stage because of the diode in the coupling circuit.

The trailing negative-going edge of the first stage output passes through the diode triggering the second stage. Hence a positive-going pulse appears at the output of the second stage later than at the first stage by a time determined by the on time of the first stage. The same action is carried through the third stage.

If diodes with a higher back-resistance than the 1N126 are used, a return should be provided for the coupling capacitors.

This circuit provides delays of about 40 microseconds per stage with rise times of 0.1 microsecond. The differentiated output pulse was used to trigger the computer adding register.

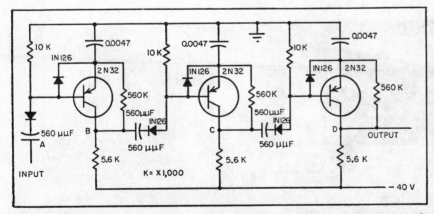

FIG. 1—Transistor delay line for airborne digital computer has delays of 40 microseconds per section for total delay of 120 microseconds

Input pulse A and outputs B, C, D at various points marked on Fig. 1 are shown in the illustration at right

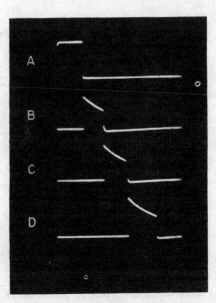

HIGH-SPEED COMPUTER STORES 2.5 MEGABITS

By WILLIAM N. PAPIAN

Staff Member, Lincoln Laboratory, Massachusetts Institute of Technology,
Cambridge, Massachusetts

Development of a transistorized arithmetic and control unit operating at a 5-megapulse-a-second clock rate, a 2.5-million-bit, 6-microsecond internal memory and a 64-register storage unit of index registers and program counters provides the basic units that make up the largest-memory computer built to date. Discussion of characteristics of Lincoln TX-2 computer reveals design details

DISTINGUISHING FEATURES of the Lincoln TX-2 computer that answer the clamor for greater capability in today's digital-computer applications, a clamor usually accompanied by reservations on size, cost, and complexity, are a transistor arithmetic and control unit operating at a 5-megapulse/sec. clock rate, a 2.5-million bit, six-microsecond internal memory unit and a 64-register storage unit of index registers and program counters.

Logic

Two developments in the logic or organization structure of digital machines complement these equipment innovations. They are: an arithmetic element which can be fractured (under program control) from a single 36-bit unit into com-binations of 9-bit units and an input-output system which provides for the concurrent programmed operation of a variety of devices on a priority basis.

A control console and terminal equipments coupled with the features outlined above, combine to

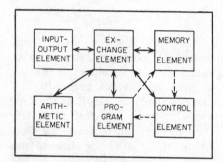

FIG. 1—Block diagram shows major elements of the computer and how information flows between them. Dashed lines indicate internal loop

make the Lincoln TX-2 an extremely capable and flexible digital computer.

Characteristics

The TX-2 is a general-purpose large-scale digital computer that uses the binary code, is a parallel machine using a single-address word structure. The memory word length is 36 bits, facturable into sub-words that are multiples of 9 bits. A number of 18-bit indexing registers are available.

Coincident-current magnetic-core arrays are used as primary storage and have a 6-microsecond cycle time. Bank one contains 65,536 37-bit words; bank two, 4,096 37-bit words and further banks can be accommodated up to a total of 262,144 words (approximately ten million bits). The storage banks

Layout of Lincoln TX-2 computer system. The 2.5-megabit memory is at left and the manipulative elements are behind the console. Engineers at right check out part of the TX-2 manipulative elements

Plug-in package contains complete flip-flop network plus two gate circuits and two additional buffer amplifiers

Receiving sockets for flip-flops used in memory are mounted in frames and their interconnecting wiring soldered by hand

can be operated simultaneously and independently.

Peak operating rates for typical instructions are: 150,000 additions per second and 80,000 multiplications per second for 36-bit words, to 600,000 additions or multiplications per second for 9-bit words.

Transistors numbering about 22,000 and several hundred diodes are used. There are 625 vacuum tubes of which 608 are dual triodes used in the large memory.

Input-Output

The TX-2 communicates with the outside world through its input-output element. Whenever an incoming word (binary-coded number representing data or in-

structions) is to be received it passes from the input-output element through an exchange element, into the memory. Output words travel the other way.

Figure 1 shows how words travel through the exchange element on their way to and from the major elements of the machine. The dashed lines at the left indicate an important loop used in the computer's basic internal cycle.

Typically, an address is supplied to the memory element from the program element. The word from the addressed memory register is sent to the control element which retains and acts out the command portion of the word, and sends the address portion back to the pro-

gram element where it may be modified. The address, which may or may not have been modified, then brings an information word out of the memory element, through the exchange element and into the arithmetic element for an arithmetic operation.

Not shown are the control lines from the control to the other elements to determine and synchronize the internal operations and transfers of information in and between the elements.

Transistor Elements

Manipulations consisting of transfers, shifts, additions, multiplications and so forth, are performed by surface-barrier tran-

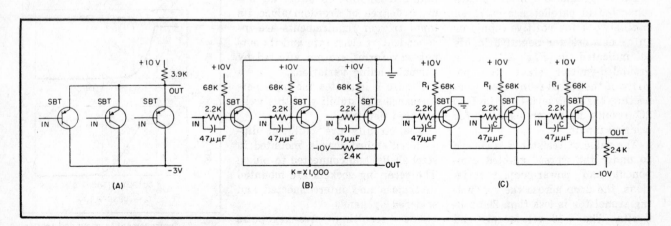

FIG. 2—Typical basic transistor circuits used in the computer are: three emitter-follower stages in parallel (A); three grounded-emitter stages in parallel (B) and three grounded-emitter stages in series (C)

External view of memory shows digit-plane drivers and sensing circuits at left, memory stall with memory-core matrix switches and memory planes in the center and the matrix switching drive circuits at right

sistor circuits operating at rates up to five megapulses a second. Two basic transistor circuits predominate: a saturating grounded-emitter stage and a saturating emitter-follower stage.

Three emitter-follower stages are shown in Fig. 2A combined in parallel to form an *or* network for negative input signals. One or more negative input signals will clamp the output terminal to −3 volts. Ground-level on all of the inputs will raise the output terminal to ground level, thus the same configuration is used as a three-input *and* network.

Three grounded-emitter stages connected in parallel also make an *and* network for positive inputs or an *or* network for negative inputs as indicated in Fig. 2B. In a grounded-emitter stage, the polarity of the signal is inverted. Connecting them in series, as in Fig. 2C results in an *or* network for positive inputs.

The value of resistor R is chosen so under the worst expected component and power-supply variations, the drop across the conducting transistor is less than 200 millivolts. The +10 volt supply and R_1 bias the transistor deeply into the off condition to provide in-

creased tolerance to noise. For the on condition, the input drives the transistor into saturation to provide a solid output level that is independent of variations in input peak amplitudes. Capacitance C shortens turnoff time by speeding the removal of minority carriers from the base, as shown in the experimental circuit of Fig. 3.

Ten of these basic circuits are combined into the standard TX-2 flip-flop detailed in Fig. 4. The flip-flop is basically an Eccles-Jordan trigger circuit with a three-transistor amplifier on each output and a single-transistor amplifier on each input. This flip-flop offers an unusual degree of freedom since its input trigger requirements are independent of clock rate and its output waveforms are not affected by normal loading variations.

Figure 5 indicates the high performance and stability of this well-buffered network.

The circuits are built on dip-soldered etched boards, mounted in steel shells, and connected to plugs. The receiving sockets are mounted in frames and interconnected and soldered by hand.

The resulting five-megapulse system of manipulative elements occupies less than 400 sq ft of floor

space and dissipates less than 800 watts of power.

The Big Memory

The TX-2 six-microsecond, random access, 2.5 megabit memory is a coincident-current unit using two-to-one selection-current ratios for two-coordinate read and three-coordinate write operations. The core array is 256-by-256 with each register holding 36 bits plus a parity checking bit and a spare bit. Vacuum tubes are required only in the high-level driving circuits. The low-level circuits use transistors. Magnetic-core matrix switches supply current pulses to the 512 selection lines.

The 2.5-million cores were made and tested at Lincoln Laboratory. Each ring-shaped core is 80 mils in outside diameter, 50 mils inside, and 22 mils high. A core is switched in one microsecond by a total current of 820 ma and pro-

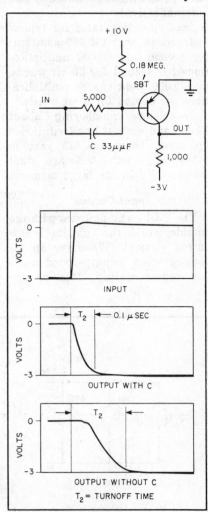

FIG. 3—Improvement in grounded-emitter stage turnoff time by adding C is shown in curves

duces a peak output of 100 mv. A total of 4,096 of these cores is wired into a 64-by-64 plane module; sixteen of these modules are tested and then assembled to form a full 256 by 256 digit plane; 38 digit planes are stacked and interconnected to form the completed array.

The sense, or output, windings of four subassemblies are connected in series-parallel. Each large digit plane has, therefore, four output terminal pairs as shown typically in Fig. 6. Sense amplifier circuits such as shown in Fig. 7 incorporate a balanced amplifier, full-wave rectifier using two emitter followers, pulse amplifier, and line driver. These sense amplifiers feed the memory buffer register in which the signal is sampled at the proper time by an 80-millimicrosecond strobe pulse.

Digit-plane windings are also interconnected in four sections, each of which is driven by the circuit shown in Fig. 8A. Here transistors bring the 3-volt signal levels up to about 9 volts to drive vacuum tubes which provide for further amplification and delivery of a 410-ma pulse to the digit windings. Similar circuits are used to drive the two core-matrix switches on the selection lines.

Each core-matrix switch is made up of 256 tape-wound molybdenum-permalloy cores which are essentially saturable-core pulse transformers with four separate windings arranged and interconnected to form a 16 x 16 two-coordinate switch. These switches operate such that the interval between read and write may be extended several microseconds under computer program control so the memory address register need not hold the given address after the read half of the cycle.

Fast-Access Memory

A small, economical and fast storage unit makes available a reasonable number of intermediate-speed registers to aid in the manipulation of program element words.

This small memory contains sixty-four 19-bit registers and requires a total of 434 transistors, 8 diodes, and 1 vacuum tube. Only 0.6 microsecond is required for ac-

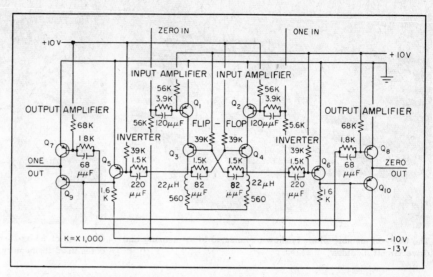

FIG. 4—Basic surface-barrier transistor flip-flop with input and output buffers as used in computer to perform arithmetic operations

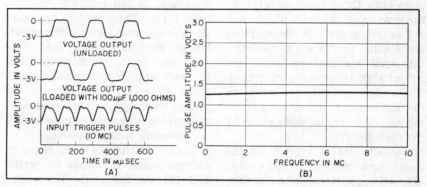

FIG. 5—Curves show output waveforms of flip-flop (A) and trigger sensitivity of flip-flop (B) in dictating high performance and stability of circuit

cess to this memory which completes a full cycle in 4 microseconds.

Each information digit of each register is represented by a pair of ferrite cores. A write operation always leaves the pair with like magnetization in magnitude and direction. A memory register line is selected on a single-coordinate basis using a relatively large cur-

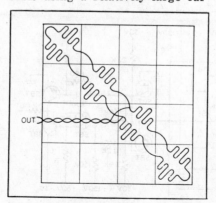

FIG. 6—Connection schematic for one sense winding of a 256-by-256 memory plane in the big memory

rent pulse for the read operation. This register line passes through each one of the pair of digit cores in opposite directions. Consequently, only one core of each pair switches; its partner merely takes a reversible excursion into saturation.

The switched core develops an emf considerably larger than its partner's and of opposite polarity. The digit winding passes through each member of the pair in the same direction so that it picks up the algebraic sum of a large voltage of one polarity and a small voltage of opposite polarity. The polarity of the resultant signal is, therefore, a function of the polarity of the net current during the previous write operation. Since the register is the only line receiving any significant drive current during read, it may be driven as heavily as is required for fast access time.

The write cycle is a two-coordi-

The 64-by-19 memory plane is shown connected to the sensing and drive circuit modules

Close-up of 64-by-19 array showing winding details and ferrite cores used in memory plane at left

nate operation and the net switching current is constrained to be no more than three times as large as the maximum current which will not switch a core. Write currents are applied to the digit windings, the same windings on which the output signals appear during read. These digit currents are one-third the amplitude required for switching and of polarities determined by the digit flip-flops.

No influence is exerted by the small digit current except during the write operation when a two-thirds amplitude current is applied to the selected register line in a direction opposite to read current. This register current adds to the digit current through one and subtracts from the digit current through the other member of each pair of cores on the selected register.

The polarity of the digit current determines at which core, of each pair, the two currents will add and thus, whether the stored

information is a one or a zero.

The cores are 47 mils in outside diameter, 27 mils inside and 12 mils high. They switch at relatively low current with a somewhat longer switching time than the cores used in the big memory. The low switching currents make transistor drive easier and access time is kept short by the large read current which single-coordinate register selection makes possible.

Digit currents are 8 ma, register current pulses are 18 ma for write and 117 ma for read. The open-circuit signal induced on a digit winding by a switching core during read is ±0.5 volt and lasts for about 0.3 microsecond. Surface barrier and 2N123 transistors are used for signal amplification and for all the current drives except the 117-ma read pulse which is supplied from a 6197 pentode.

Figure 8B shows the way in which three surface-barrier transistors are arranged in series to deliver a large, 15-volt pulse to

the 6197 without exceeding their rated 6-volt limits.

Logic Features

The intermediate-speed registers made available by the described memory are to be used to store index registers and program counters. The availability of a relatively large number of program counters makes possible and economical, a multiple-sequence program technique whereby a number of input-output sequences and the internal computer sequences may operate concurrently.

The input-output devices may include paper tapes, magnetic tapes, cathode-ray displays, analog-digital converters, other computers and the like. A multiple-sequence computer is much like a number of logically separate computers which time-share the same memory, arithmetic, and program elements.

When an input-output device needs attention, its program counter is selected and enough computer instructions from its sequence are performed to meet its needs. Control then reverts to some other sequence. The main computational sequence is treated as just another user of the machine and is given attention when the more pressing needs of lightly buffered input-output devices have been met.

Efficient operation and a minimum of programming restrictions result when a priority system is used to rank the program sequences. In TX-2, high-speed free-running devices rank high and get attention first. Electric typewriters have a low priority since they can

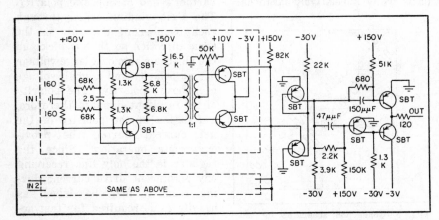

FIG. 7—Sense amplifier circuits drive the memory buffer register where output signal is sampled at proper time

wait indefinitely. The lowest priorities are assigned to main computational sequences.

The final feature is the flexible structure of the arithmetic element which can be altered under program control. In effect, each computer instruction specifies a particular form of machine, ranging from a full 36-bit computer to various combinations of 27-bit, 18-bit, and 9-bit configurations. Figure 9A graphically illustrates four of the possible machine configurations, including two in which parts of the memory are cross-coupled to different parts of the arithmetic element.

Not only does this scheme make more efficient use of the memory for storing data of various word lengths, but it also results in greater over-all machine speed because of the increased parallelism of operation. For example, two 18-bit multiplications may go on simultaneously. This feature is particularly valuable in control and equipment design applications, where incoming data is seldom available to a precision greater than 18 bits, and often only to 9 bits.

The ability to fracture the arithmetic element on an instruction-by-instruction basis requires switching of some of the interconnecting paths between digit stages. For example, parts (5) and (6) of Fig. 9B show the respective basic paths followed by information in the A and B registers of the arithmetic element during a shift right operation in the cases of 36-bit and 18-bit configurations. Part (7) of the figure shows the switching required to instrument just those two configurations for the simple operation.

Experience

The TX-2 computer embodies significant experience gained from components and systems investigations. An eight-stage double-rank shift register containing 99 surface barrier transistors has been on life test since April 1955. It circulated a fixed pattern for 10,789 hours with no errors and no transistor failures.

An eight-bit error-detecting multiplier using 600 SBT's was de-

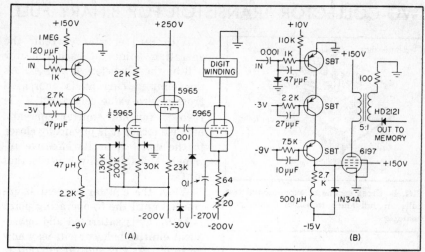

FIG. 8—Digit plane driver circuit (A) delivers a 410-ma pulse to the digit windings. Driver circuit for read windings (B) delivers a 117-ma read pulse

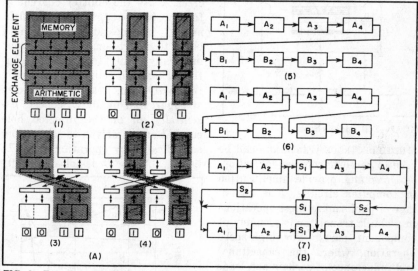

FIG. 9—Four possible configurations that can be used with the TX-2 (A) are the 36-bit configuration (1), four 9-bit, (2), two 18-bit (3) and four 9-bit cross-coupled (4) configurations. Possible paths between digit stages in arithmetic element are also shown (B). Blocks show shift-right operation for 36 bits (5), two 18-bit segments (6) and an 18-bit breakup (7). Switch positions for the 36-bit shift are S_1 closed and S_2 open. Switch positions reverse for the 18-bit shift

signed and completed in August 1955 and has been in nearly continuous operation since. Operating margins are periodically checked, and in steady state operation the multiplier's error rate has been about one every two months, or one error per 5×10^{11} multiplications at 10^5 multiplications per second. Most of these errors appear to have been caused by defects in the wiring. Of the eight transistor failures, seven were caused by accidental application of overvoltage or overload, and one was a defective transistor. A ten-percent sample of the multiplier's transistors are now undergoing extensive tests.

The research reported herein was supported jointly by the Army, Navy and Air Force under contract with the Massachusetts Institute of Technology.

BIBLIOGRAPHY

W. A. Clark, The Lincoln TX-2 Computer Development, Proc Western Joint Computer Conference, Feb. 1957.

K. H. Olsen, TX-2 Circuitry, Proc Western Joint Computer Conference, Feb. 1957.

J. L. Mitchell and K. H. Olsen, TX-0, A Transistor Computer with A 256 x 256 Memory, Proc Eastern Joint Computer Conference, Dec. 1956.

R. L. Best, Memory Units In The Lincoln TX-2, Proc Western Joint Computer Conference, Feb. 1957.

J. W. Forgie, The Lincoln TX-2 Input-Output System, Proc Western Joint Computer Conference, Feb. 1957.

J. M. Frankovich, A Functional Description of the Lincoln TX-2 Computer, Proc Western Joint Computer Conference, Feb. 1957.

TWO-COLLECTOR TRANSISTOR FOR BINARY FULL ADDITION

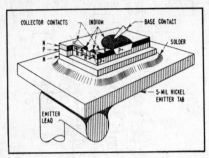

FIG. 1—Cross section of experimental two-collector full-adder transistor using point contacts

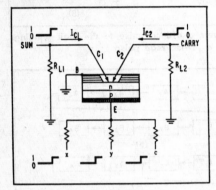

FIG. 2—Full adder circuit

MULTIELECTRODE transistors can be used to perform complex logical operations and provide amplification in computer circuits. The advantage of performing these operations in a single transistor is economy of component parts. For high-speed operation, where the capacitance and inductance of sockets and leads become speed limiting factors, there is an inherent advantage of having logical operations performed within the semiconductor body of the transistor itself.

Figure 1 shows a schematic cross section of an experimental two collector, full-adder transistor which utilizes point contacts as the collector electrodes.

▶ **Full Adder**—A full-adder circuit using this type of transistor is illustrated in Fig. 2. The input signals at terminals X, Y, and C and represent the binary numbers to be added. The currents I_x, I_y and I_c corresponding to those numbers are mixed at the emitter. The output signals are the two collector voltages representing the sum and carry. Load lines corresponding to the unequal load resistors R_{11} and R_{12} are shown superimposed upon the individual collector V-I characteristics in Fig. 3.

With zero emitter current, the operating points at each collector will be those labeled A_1 and A_2. As the emitter current is increased from a zero value collector 1, C_1 is favored for low emitter currents, because either it is physically closer to the emitter, or its effective intrinsic alpha initially exceeds that of C_2.

When the emitter current is increased until the C_1 operating point is at B_1, C_1 is saturated and subsequent emitter hole current may now go to C_2. As the emitter current is increased over the value needed to saturate C_1, the current in C_2 increases. As the emitter current is further increased, the electric field near C_2 becomes strong enough to begin to divert some of the current originally going to C_1. With continued emitter current increases, this diverting action becomes stronger, and finally C_2 collects essentially all of the injected hole current and is driven to saturation, as indicated by operating point B_2 in Fig. 3. Now C_1 is collecting cur-

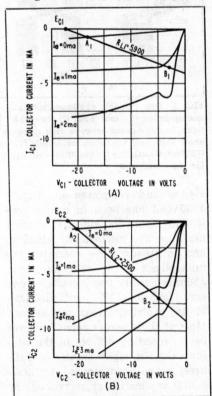

FIG. 3—Characteristic curves

rent and is in the off condition indicated by operating point A_1. The load resistor for C_2 must be roughly

half the value of that for C_1 plus that which was necessary to build up the strong electric fields in the base region to accomplish the robbing action.

After the second collector is in saturation, any additional emitter current will not be accepted at C_2, but instead will be collected at C_1 until it, too, is again saturated. If C_1 is considered as the sum collector and C_2 the carry collector, the operation of full addition is accomplished.

The transistor, in view of its multiple electrodes and properties of collector interaction, can be used in other circuits. For example, if the full-adder circuit of Fig. 2 has a lower value of load resistor for C_2 the binary logical connective neither-nor is obtained. The full added circuit with different value of input current can perform the logical operation not-both.

▶ **Junction Type**—Recent advances in the techniques of transistor fabrication have made it practicable to construct all-junction designs of the full adder transistor. In these designs, the formed point-contact collectors, which are difficult to produce to preassigned specifications are replaced by p-n hook collectors which are made by alloying and diffusion processes. Figure 4 shows a cutaway view of an experimental structure of this type which is currently being investigated. The details of the various processes involved in the assemblies of such a unit are indicated in the illustration.

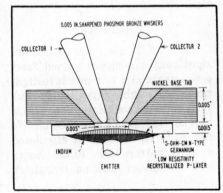

FIG. 4—Cutaway of full adder

Chapter 17

COMPUTER AUXILIARY EQUIPMENT

DIGITAL-ANALOG CONVERTER PROVIDES STORAGE

By H. N. PUTSCHI, J. A. RAPER and J. J. SURAN
Electronics Laboratory, General Electric Co., Syracuse, N. Y.

When high-speed or environment rule out mechanical converters, purely electronic units must be employed. This transistorized converter changes eight binary bits, received in parallel from a shift register, to 128 steps in amplitude of a 400-cps sine wave. One binary bit is used to obtain phase information. Operation is performed within a 4-millisecond sampling period occurring at an average rate of 20 cps

ELECTRONIC DIGITAL computers use digital systems throughout their arithmetic units, but the outputs required are often of an analog form. The digital-analog conversion is sometimes done in an electromechanical form as in shaft position servos.

Transistors were used in preference to vacuum tubes in the equipment to be described because of their reliability, low power consumption and small size. The converter is required to convert seven bits of binary information to 128 analog steps in amplitude of a 400-cps sinewave and convert an additional binary bit to requisite phase information.

These functions must be performed within a 4 millisecond sampling period, occurring at an average repetition rate of 20 cps. Input information is received in parallel form from a shift register and must be stored in the digital-analog converter between sampling period. A binary ZERO is represented by a −30 volt signal at the respective input terminal, a ONE by zero volts. The converter output works into a high-impedance load and a full-scale output of 20 volts peak to peak must be supplied. The equipment must operate over a temperature range from −50 C to +85 C.

Gates

The circuit diagram of the complete digital-analog coverter is shown in Fig. 1. To keep the current drain on the gate pulse signal within the permissible limits, a gate pulse current amplifier, common to all eight bit-converters, was used. Its current gain is in excess of that required to keep the drain on the gate-pulse source within the permitted limits. Diodes D_5 and D_6 provide correct d-c levels for the a-c coupled gate pulse.

Seven bit-converters have identical circuitry. Transistor Q_7 plus associated circuitry constitute a sample gate circuit. It is connected as a common-emitter amplifier and is normally cut off by its −16 v bias battery. If a ZERO (−30 v) is read into the gate from register connection X_1 the gate transistor is kept cut off even if the positive gate pulse arrives.

The gate pulse is also connected to the collector side of the gate transistor through a 33,000-ohm resistor and will appear at the gate circuit output with an amplitude of about 8 volts if transistor Q_7 is not conducting. Thus a ZERO read into the register results in a positive output pulse from the gate circuit. If a ONE (0 v) is read into the gate circuit from the register,

Table I—Resistance Matrix Outputs

Digital Input			Analog Output
E_1	E_2	E_3	E_0
0	0	0	0
1	0	0	1/8
0	1	0	1/4
1	1	0	3/8
0	0	1	1/2
1	0	1	5/8
0	1	1	3/4
1	1	1	7/8

Q_7 still remains cut off in the absence of the gate signal. In the presence of the gate signal the transistor is driven into saturation.

The collector circuit thus becomes a low-resistance path to ground. Since the collector voltage between gate pulses is about +8 volts, there will be an 8-volt negative pulse at the output of the gate circuit for the duration of the gate signal. To avoid accumulative charging of the 1-μf coupling capacitor and thereby losing the correct 8-volt collector potential between pulses, it is necessary to have a time separation between subsequent pulses that is long compared to the pulse duration.

An alternative gate circuit, which can operate at higher word rates, is illustrated in the phase reversal unit. The gate circuit output for a ZERO input is a positive, differentiated pulse fed to the base of Q_{11} of the storage flip-flop, driving it into conduction. The base circuit of gate transistor Q_{10} is the same as in the previous gate circuits and Q_{10} will therefore remain cut off with a ZERO at the register input.

The collector side of Q_{10} is connected in parallel with the second flip-flop transistor Q_{12}. If a ONE is read from the register, gate transistor Q_{10} is driven into conduction, bringing the collector voltage of Q_{12} down thus setting the flip-flop so Q_{11} is cut off and Q_{12} conducts. The differentiated pulse, which tends to set the flip-flop in the opposite direction, is swamped out because of its shorter duration.

Storage

To obtain reliable operation of the diode switches, several milliamperes must be drawn from the storage flip-flops. Because of the 20-volt peak to peak maximum analog output signal required, the tran-

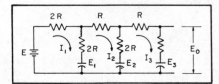

FIG. 2—Resistance matrix used to solve mesh equations to determine parameters of output circuit

sistors must be operated at collector voltages of up to 40 volts. The flip-flops will operate reliably using transistors with a current gain of 15. Test conditions for this gain were $V_c = 50$ volts, $I_c = 2$ ma and an ambient temperature of −55 C.

In the simple Eccles-Jordan cir-

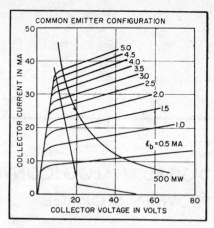

FIG. 3—Characteristic curves of transistors used as phase-reversal switches show broken load lines over which transistors are operated

cuit of Fig. 1, $(Q_8\ Q_9)$ the collector voltage swing is approximately 4 to 45 volts. To maintain the lower of these limits a low collector saturation resistance is required.

As has been mentioned previously, a ZERO read from the register will result in a positive output pulse from the gate circuit. The positive pulse from the gate circuit drives Q_8 into saturation, while Q_9 is cut off. Thus a ZERO from the register results in a voltage of +45 volts on the collector of Q_9 and a voltage of +4 volts on the

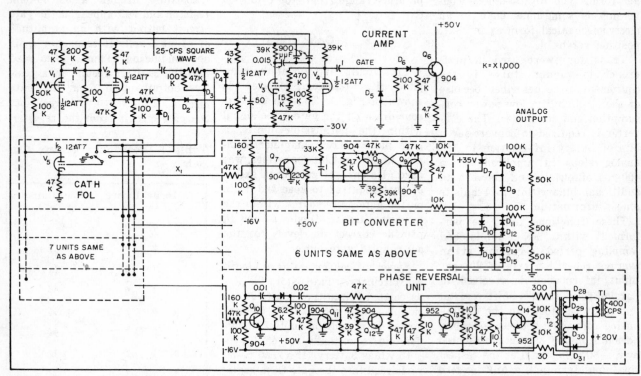

FIG. 1—Complete digital-analog converter. Electron-tube circuits check operation of transistorized converter and supply gate signal

collector of Q_8. A ONE read from the register reverses the voltages.

Solution of the mesh equations for the resistance matrix of Fig. 2 provides the following equation for its output voltage:

$$E_0 = E_3 - (E + E_1)/8 - (E_2 - E_1)/4 - (E_3 - E_2)/2$$

If $E_0 = 0$ and E_1, E_2 and E_3 are assigned the values of either 0 or 1, inserting all possible combinations of these values results in the outputs shown in Table I.

Thus, if a correct resistance ratio is maintained throughout the matrix, the output will be a true analog of the digital reference voltages. A zero for E_1, E_2 or E_3 means a short circuit to ground, not an open circuit!

The resistance matrix described above is expanded to seven digital inputs and provided with the necessary diode switches as shown in Fig. 1. The switching diodes providing single pole-double throw action for the first binary bit are D_7 and D_8. If the voltage on the collector of Q_9 is $+45$ volts, current will flow from this point through the 10,000-ohm resistor and D_7 to the $+35$-volt line, providing a low-impedance path between the resistance matrix tie-in resistor (100,000 ohm) and the $+35$-volt line, which is heavily bypassed and thus on ground potential for the 400-cps reference voltage.

The voltage at the tie point of D_8 and D_9 is $+20$ volts with 20-volt peak to peak of the 400-cps voltage superimposed. Thus the highest instantaneous voltage at this point will be $+30$ volts and the lowest instantaneous voltage will be $+10$ volts. The d-c voltage at the tie point of D_7 and D_8 is slightly over 35 volts and D_8 is therefore cutoff during the entire reference voltage swing.

If a ONE is read from the register, the voltage at the collector of Q_9 is about $+4$ volts. Thus current is flowing from the tie point of D_8 and D_9, through D_8 and the 10,000-ohm resistor to the collector of Q_9. Diode D_8 will never cut off during the entire period of the reference voltage. Since the impedance of the conducting diode D_8 is small compared to the 10,000-ohm resistor, practically all of the reference voltage will

appear at the tie-in point of the resistance matrix. The voltage at the tie point of D_7 and D_8 never goes above $+30$ volts. Thus D_7 remains cut off during the entire reference-voltage swing.

The two load extremes on the reference source will be no-load (all converters ZERO) and a load of about 1,250 ohms (all converters ONE).

To keep reference-voltage fluctuations as a result of these varying loads to a minimum, a load compensation circuit is included in each bit converter. In the first bit converter it consists of diode D_9 and the 10,000-ohm resistor connecting it to the collector of Q_8. This collector is at a low potential whenever the collector of Q_9 is at a high potential and vice versa. Thus either D_8 or D_9 will be conducting at all times, and the variation in load of the reference signal source will be greatly reduced.

Phase-Reversal Unit

In the phase-reversal operation either one of the diode pairs D_{28}, D_{29} or D_{30}, D_{31} is biased in the conducting direction, while the other pair is cut off. The plate side of all four diodes is maintained at a d-c potential of $+20$ volts from the $+20$-volt battery through the center tap on the secondary winding of T_1. The center tap on one of the primary windings of T_2 must be kept at a more positive potential to keep the corresponding diodes cut off. The center tap on the other primary winding of T_2 is at a lower d-c potential, maintaining conduction through the corresponding diodes.

Since the primary windings on T_2 are connected to oppose each other, transferring the reference signal input from one of these windings to the other will shift the phase of the reference signal output voltage by 180 degrees. The d-c drawn through the forward-biased diodes must be equal to at least the peak a-c necessary to produce the required output voltage.

After quite a bit of experimentation, a diode bias current of 15 ma and a transformer T_2 with an open-circuit shunt inductance of four henrys were used. The transistors driving the phase-reversal

switches Q_{13} and Q_{14} are operated along a broken load line as shown in Fig. 3. As long as the collector voltage exceeds $+20$ volts, the transistors operate with a 10,000-ohm load resistor connected between collector and the $+50$ volt supply. As the collector voltage becomes less than $+20$ volts, the diode pair starts to conduct and the 300-ohm resistor connecting the collector to the center tap on T_2 becomes the decisive part of the load. This keeps the transistor from exceeding its dissipation rating during any part of the operating cycle.

The electron-tube circuits of Fig. 1 constitute the test equipment used to check the operation of the digital to analog converter. All necessary signals are obtained from a 25-cps square wave input to provide the digital-analog converter with a repetitious series of two 8-bit binary numbers. The two numbers can be preselected by appropriate settings of the bit selector switches in the test equipment. The test equipment also provides the gate pulse.

The outputs available from the register connections for corresponding settings of the bit selector switches are: bit selector switch in ZERO poition, 0 volts; bit selector switch in ONE position, $+30$ volts; bit selector switch is ZERO — ONE position, 0 volts for positive half cycle of 25-cps input pulse and $+30$ volts for negative half cycle of input pulse; bit selector in ONE — ZERO position, $+30$ volts for positive half cycle of 25-cps input pulse and 0 volts for negative half cycle of input pulse. Voltages are with reference to the simulator chassis.

Input Signals

The input simulator must be kept at a -30 volt potential with reference to the digital-analog converter chassis. The 25-cps input pulses of negative polarity are amplified and limited in amplitude by V_1. Tube V_2 is a phase splitter and the outputs available at plate and cathode are 180 degrees out of phase. Phase splitter outputs are limited to 0 volts and $+35$ volts by the diodes D_1, D_2 and D_3, D_4 respectively. These voltages are fed to the appropriate bit selector switch positions, in each block.

Tubes V_3 and V_4 constitute a monostable multivibrator with a 4-millisecond pulse time. In its stable state, V_4 is conducting; V_3 cut off. A positive trigger pulse at the grid of V_4 will turn this tube off and consequently turn on V_3 maintaining this state until the

0.015 μf coupling capacitor is charged. The pulse at the plate of V_4 is roughly rectangular and has a 4-millisecond duration.

There must be a gate pulse for every half cycle of the 25 cps input voltage to obtain a reading of subsequent words and thus, a positive

trigger pulse must be derived from positive and negative slopes of the input pulse. This is done by obtaining a differentiated pulse before and after the first amplifier stage.

PUNCHER TRANSCRIBES COMPUTER OUTPUT

By JAMES E. PALMER, JAMES J. O'DONNELL
and CHARLES H. PROPSTER, JR.

Commercial Electronic Product Division, Radio Corporation of America, Camden, New Jersey

Transistor circuits consisting of two-input resistor gates, flip-flops and other delay circuits are combined in plug-in assemblies to provide logical and driving operations for card puncher that produces business machine cards at the rate of 150 a minute. Each card is checked by reading completed card and comparing its output with original input

LARGE DIGITAL DATA processing systems are often required to produce accounting machine cards punched with information in a business machine code. These cards may be used as permanent records, to communicate with other data processing facilities or as output documents in the form of bills or checks.

The transistorized transcribing card punch to be described is used with the Bizmac computing system. It converts large volumes of data stored on magnetic tape into punched accounting machine cards at the rate of 150 cards per minute and provides many accuracy control features to assure correct data punching. Transistor circuits are used throughout.

Functional Description

The input to the transcribing card punch is received from a magnetic tape reader in seven-bit binary-coded decimal alpha-numeric code. Character rates of 10 to 30 kc are acceptable but the unit re-

quires that the magnetic tape messages be of fixed field format. All characters appearing on tape which do not have business-machine code equivalents are interpreted as blank columns on the cards, except for certain control symbols such as start message, end message and item separator which are deleted completely.

A plugboard permits rearrangement of data and insertion of special characters. Generally an entire message is punched into one card and the characters of the message may be rearranged in any sequence.

In all modes of operation the system checks itself by reading the cards after they have been punched and comparing them with the tape message from which they were punched. Parity check on input information provides further accuracy control. Errors stop card punching. The erroneous card is separated from the correct ones and the tape is backed up, coming to rest just ahead of the message which was punched in error.

Logical Operation

A block diagram of the transcribing card punch is shown in Fig. 1. The temporary storage contains three sectors that are switched cyclically around the read-in, punching and checking circuits. Three messages are in process at any one time. Simultaneously one is read in, a second is punched and a third is checked. Operation will be described by following one message through a complete machine cycle.

During the read-in cycle the temporary storage is cleared and connected to the tape reader. A full card cycle of 40 millisec is allotted for this purpose although read-in needs only part of that time. At the end of this cycle, the temporary storage sector containing the message just read in is switched to the punching circuits.

During the punch cycle, information is read from temporary storage into the coder to convert to business machine code. The output of the coder feeds a 12-chan-

* Formerly with RCA

Electronic circuits for transcribing card punch are packaged on plug-in assemblies and housed in racks shown

Cards in punch mechansim feed from left to right. Punches and solenoids are housed in hinged portion shown in open position

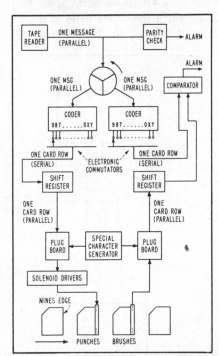

FIG. 1—Block diagram of transcribing card punch shows method whereby stored information is first used to produce punched cards and then used to check them

nel electronic commutator synchronized with the card advance.

The cards enter the system nines edge first and travel in a direction parallel to their shorter axes. They are moved intermittently past a punching station and punched one row at a time.

The commutator selects the output channel from the coder corresponding to the row to be punched. This card-row is sent serially to the shift register which converts the information from serial to parallel form. Eighty bits are read out simultaneously through the plugboard to a register which

drives the 80 punch magnets that punch one card row. The card is advanced to the next row and the process repeated. After 11 reptitions, one card is completely punched and is then transported to the checking station located exactly one card space away from the punching station.

During the check cycle the card is read, one row at a time, by 80 brushes and the information sent through the plugboard into a second shift register. This shift register converts parallel information to serial forms. Its output is sent into a single-channel comparator.

Meanwhile, the temporary storage sector containing the original information has been switched from the punching circuits to the checking circuits. The information in the temporary storage is trans-

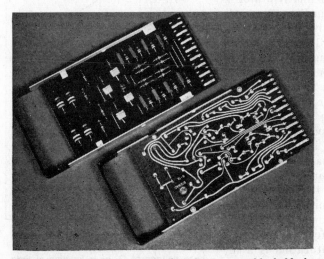

Universal gate, inverter and flip-flop plug-in assembly holds four of the basic elements shown in Fig. 3

Wiring in prototype model of transcribing card punch shows simplicity and neatness resulting from use of cabled wiring

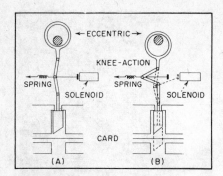

FIG. 2—Knee-action punch mechanism is shown at top of punch stroke (A). Solid lines in (B) show condition when no punching is done while dotted lines indicate condition for punching

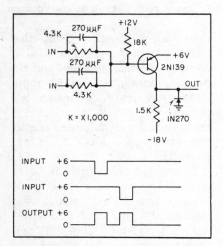

FIG. 3—Universal gate, inverter and flip-flop circuit. Response to various inputs is shown below

lated by a second coder and another 12-channel commutator selects and routes the appropriate channel to the comparator where a check is made against the information read back from the brushes.

Punch Mechanism

Figure 2 illustrates the knee-action punch mechanism. Eighty such mechanisms are arranged side by side permitting an entire row to be punched.

At the beginning of each punch cycle, the solenoid armature is driven into the solenoid by the mechanical motion of the knee. In this position the solenoid, if energized, moves no mechanical mass. To punch it prevents the knee from bending during the downward stroke of the eccentric. This re-

quires relatively little force at a time when the solenoid is capable of exerting its maximum force; that is, when the armature air gap is smallest.

Transistorized Circuits

The basic circuit chosen to meet the logic requirements of the trancribing card punch is shown in simplified form in Fig. 3. This two-input resistor gate provides not only the gating function, but also signal amplification and standardization. If either input is lowered to ground potential, the transistor conducts raising the output to +6 volts. For negative going signals, defined as logical ONES, the element acts like an OR gate followed by an inverter. If a single input is used, the element functions as a simple power amplifier-inverter.

When both inputs are at +6 volts, the output will be at ground potential. Thus, the AND function preceded by inversion is obtained.

Typical delays of 0.2 μsec per stage are obtained when these elements are cascaded. The output resistor of each element can absorb 10 ma from the stages it drives and 1.6 ma are required by each gate input. Hence, up to six gate inputs may be driven by each element.

The 18,000-ohm resistor supplies leakage currents when the transistor is not conducting. The necessary flip-flop or storage function is provided by cross-connecting two of the basic elements.

An accurate time delay or pulse-forming one-shot multivibrator is shown simplified in Fig. 4A. Application of a negative-going pulse to the input terminal triggers Q_1 into conduction causing its collector to swing from ground level to +6 volts. This 6-volt swing is coupled through capacitor C which is initially charged to +6 volts. This raises the base potential of Q_2 to about +12 volts and cuts off the transistor. The base potential then begins an exponential decay toward ground. When the base of Q_2 reaches the vicinity of +6 volts, Q_2 begins to conduct. The voltage swing required for the transition from cutoff to saturation is a few tenths of a volt and this transition is assisted by the gain of Q_2 and

the feedback through Q_1. Thus, the time required to turn on Q_2 is a small fraction of the total one-shot duration. Therefore, wide variation in transistor characteristics can be tolerated without compromising circuit performance. Only in the case of short duration one-shots, where transistor storage effects become important, does variation of transistor parameters become appreciable. The standard pulse width of this system is 3 μsec. Similarly, transistor saturation voltages and diode-voltage drops have little effect as they are small compared to the 6 volts through which the time-determining capacitor must discharge.

Figure 4B shows that supply voltage variations have little effect on circuit operation. If the +6 volt

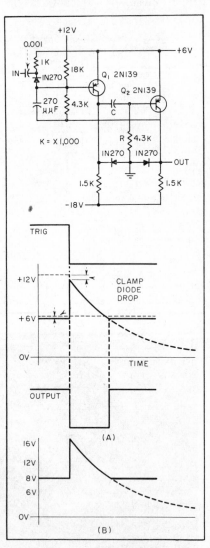

FIG. 4—Time delay or pulse-forming one-shot multivibrator (A). Waveforms show base voltage of Q_2 as a function of time with correct supply voltage. Effect of increasing supply voltage is shown in (B)

supply should rise to +8V then C would be initially charged to +8V also. When Q_1 turns on in response to a trigger, its collector swings through 8 volts and the base of Q_2 moves to +16V. The decay takes place as before but starts from a much higher voltage. However, the base of Q_2 must only drop to +8 volts before Q_2 begins to conduct. This is half way between its starting point and ground just as in the case of a +6 volt supply. Hence, the duration of the output pulse remains the same.

Shift Register

The two-cell shift-register in Fig. 5 has provision for parallel input, gated-parallel output, serial input, serial output and reset; thereby making a single circuit universally applicable to all shift register requirements.

Each flip-flop cell is set by the application of a ONE (ground level) to its parallel input terminal, and reset at once by the application of a ONE to the reset terminal. Shifting is accomplished by the application of a ZERO (+6 volts) to the advance terminal. This advance signal must endure for at least 5 μ sec and must be separated from the previous advance signal by at least 25 μ sec.

The gated-parallel output is derived from a diode gate consisting of D_5 and D_6 or D_7 and D_8 which allows a one to be presented at the output terminals of all cells which are set whenever a ONE is applied to the gate terminal. Conversely, ZEROS will be presented at the output terminals of all cells which are reset at the time of the gate signal.

Shift-register operation may be explained by assuming Q_1 and Q_3 are conducting and both transistor pairs contain ZEROS. The first cell, consisting of Q_1, Q_2 and associated circuitry, is set by the application of a ONE to the first parallel input terminal labeled P_{1-A}. This causes Q_2 to conduct turning Q_1 off. The first cell now contains a ONE and the collector of Q_1 is at ground level while the collector of Q_2 is at +6 volts.

The advance terminal is held at ground level externally and 6 volts appears across the series combination of R_3 and C_3 charging C_3 to

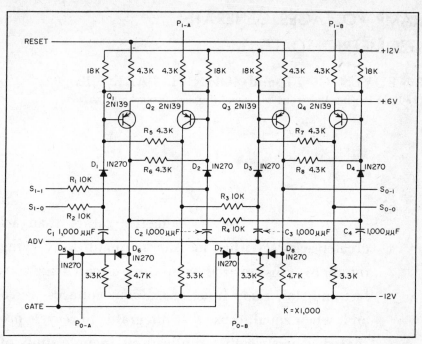

FIG. 5—Universal shift register is made up of a number of cells in cascade. Circuits are packaged four cells per plug-in unit

+6 volts. No voltage is applied across R_4 and C_4 as the collector of Q_1 is at ground level.

After C_3 has been charged the shift register is ready to receive an advance signal. When the advance terminal is raised to +6 volts, the anode of diode D_3 is raised to about +12 volts by the voltage doubling effect of the charge on C_3. This biases D_3 in the forward direction and allows the charge stored in C_3 to flow into the base of Q_3 turning it off. Flip-flop action turns on Q_4 and the second cell of the shift register now contains a ONE.

The levels impressed on the serial input terminals S_{1-1} and S_{1-0} determine the charge on C_1 and C_2 and therefore determine the effect of the advance signal on the first

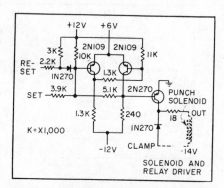

FIG. 6—Punch-solenoid driver uses medium-power transistors to supply required driving currents

cell. If S_{1-0} is at ground level and S_{1-1} at +6 volts prior to the application of the advance signal, then the first cell is switched to the reset or ZERO state through D_2. If S_{1-1} is at ground level and S_{1-0} at +6 volts, the first cell does not change state in response to the advance signal since it already contains a ONE. The shift-register output is fed through the plugboard to a punch-solenoid driver. Four-hundred ma at 14 volts are required by the punch solenoids at a 20 percent duty cycle. As shown in Fig. 6 this power can be supplied by medium power transistors.

A flip-flop is included as part of the package to provide storage and to serve as a preamplifier for the output transistor. The flip-flop is set by a ONE and reset by a ZERO. In addition to reducing the required driving current for reset this feature allows the circuit to be used as a simple current amplifier, without storage action, by connecting the set and reset terminals together and driving them both with a one. Although the output transistor is required to supply large currents, it is still operated safely as the transistor is saturated to minimize internal dissipation.

Credit is due K. L. Chien, C. T. Cole, H. H. Cremer, N. C. Florio and R. F. Bov for their aid.

RAMP VOLTAGES GENERATE
GEOMETRIC SCALE

By EUYEN GOTT and JOHN H. PARK, JR.
The Johns Hopkins University, Radiation Laboratory, Baltimore, Maryland

Use of geometric rather than arithmetic scales provides greater accuracy when measuring quantities varying over wide ranges. Step inputs to transistorized computing circuit trigger two integrators to furnish fast and slow ramp functions. Ramp voltages are compared at discriminator and, when equal, cause fast integrator to recycle producing a chain of pulses spaced geometrically. Applications include study of pulse amplitude, pulse-width and pulse-position-modulation systems as well as logarithmic operations in electronic computing

SUCCESSIVE READINGS of a uniformly divided scale form an arithmetic progression called an arithmetic scale. This kind of scale is satisfactory as long as the variation of the measured quantities is small. In practice however, quantities varying over a ratio of 10:1 or more are not uncommon. In such cases, the relative accuracy of the measurement varies considerably.

To obtain a constant minimum relative accuracy a scale whose successive readings form a geometric progression is proposed and transistor circuits for producing such a scale are described.

Constant Accuracy

The word accuracy is used to denote relative accuracy hereafter for the sake of simplicity. It is defined as

$$A(x) = 1 - |m - x|/m \qquad (1)$$

where $A(x)$ is the accuracy, x is the actual magnitude of the quantity and m is its measured value. Although x is a continuous variable, m is restricted to a particular set of values $m_0, m_1, m_2, m_3, \ldots$ rendered by the scale used.

For an arithmetic scale, the m's are usually integers. A geometric scale requires $m_n = km_{n-1}$ with $k > 1$.

In either scale, the particular points where readings are taken can be so arranged that one of the three possible cases [$m \leq x$, $m \geq x$ and $m >$ or $< x$] is true. This is shown in Fig. 1.

For simplicity, only $m \geq x$ will be discussed although the same reasoning can be applied to the other cases. In practice, the case of $m >$ or $< x$ with m_i occurring at the midpoint of the ith interval

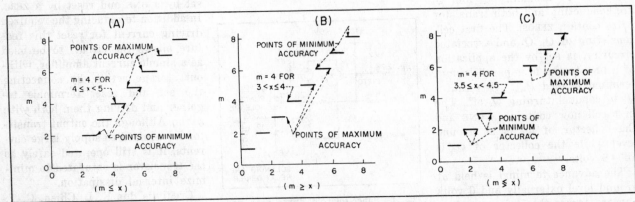

FIG. 1—Graphical representation of ways of setting the reading points of a scale. Method works equally well for both arithmetic and geometric scales

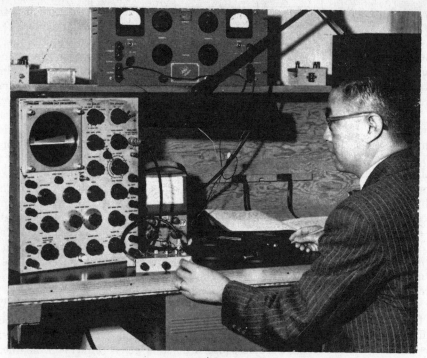

Compact transistorized unit produces geometric progression display on cro

reduces the possible difference between m and x to a minimum.

Applying the condition of $m \geqq x$, Eq. 1 becomes

$$A(x) = x/m \qquad (2)$$

From Eq. 2, $A(x)$ can be represented by a broken curve consisting of an infinite number of straight lines as shown in Fig. 2A. These straight lines start and end at the points where readings are taken—points of minimum and maximum accuracy. A closer look at the points of minimum accuracy for the two kinds of scales reveals that for the arithmetic scale, minimum ac-

curacy occurs when $m - x$ is almost equal to unity. Substituting $m = x + 1$ into Eq. 2 gives

$$\min A_a(x) = x/(x + 1) \qquad (3)$$

This is shown in Fig. 2A by the dotted hyperbola which indicates that the minimum accuracy is poor when x is small. It gradually improves when x increases. The actual accuracy also ranges between zero and unity when x varies.

For a geometric scale the measurement is least accurate when x differs from m by almost one division. For example, if $m = m_8$, then $x = m_7$. Hence Eq. 2 becomes

$$\min A_g(x) = m_7/m_8 = m_n/m_{n+1} = 1/k \quad (4)$$

Thus a constant minimum accuracy can be obtained with the geometric scale. By choosing k nearly equal to unity, $\min A_g(x)$ will also be close to unity and the range of variation of the actual accuracy will be small.

Double-Sweep Method

To produce a geometric scale, two linear ramps are employed, as shown in Fig. 2B. The slow ramp starts at $x = 0$ and the fast ramp starts at $x = m_0$. The slopes of the two ramps are functions of k. For

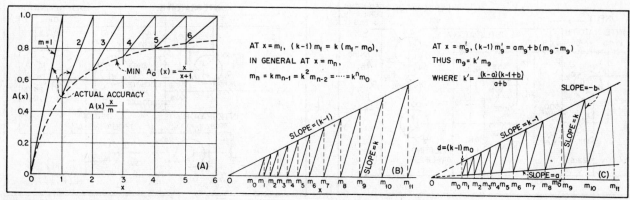

AT $x = m_1$, $(k-1) m_1 = k (m_1 - m_0)$,

IN GENERAL AT $x = m_n$,

$m_n = k m_{n-1} = k^2 m_{n-2} = \cdots = k^n m_0$

SLOPE = $(k-1)$

SLOPE = k

(B)

AT $x = m'_9$, $(k-1) m'_9 = a m_9 + b(m_9 - m_9)$

THUS $m_9 = k' m_9$

WHERE $k' = \dfrac{(k-a)(k-1+b)}{a+b}$

$d = (k-1) m_0$

SLOPE = $k-1$

SLOPE = k

SLOPE = $-b$

SLOPE = a

(C)

FIG. 2—Dotted hyperbola in (A) indicates minimum accuracy is poor when x is small. Ramps required to produce geometric scale (B) are modified so low ramp starts at same time as fast ramp (C)

257

example if $k - 1$ is the slope of the slow ramp, then k will be that of the fast ramp.

The two ramps meet when $x = m_1 = km_0$. At this time the fast ramp is cut off and restarted from the initial zero level. It meets the slow ramp again at $x = m_2 = k^2m_0$. By starting and stopping the fast ramp repeatedly a geometric scale is obtained.

This is an ideal situation. In practice, since ramp functions are usually produced by charging capacitors, it is not possible to restart the fast ramp as soon as it meets the slow damp. An increasingly longer waiting period is necessary each time. It is also found that under certain circumstances the starting level increases gradually. Taking these two factors into account, a slightly different value for the ratio m_n/m_{n+1} is obtained. The details are shown in Fig. 2C.

A further modification in Fig. 2C allows the slow ramp to start at $x = m_0$ but with an initial level of $d = (k - 1)\ m_0$. This means that the two ramps can be started simultaneously after m_0 seconds from the beginning of the measurement. At the end of the measurement the same control can be used to stop the two ramps together.

Block Diagram

Figure 3 is a functional block diagram of the transistor circuit for producing a geometric scale.

The input consists of controlling step functions. A positive step comes at m_0 seconds after the beginning of the measuring period and puts the circuit into operation. At

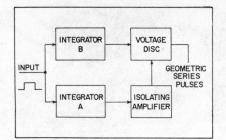

FIG. 3—Block diagram of transistor geometric-series pulse generator

the end of the period a negative step rests the circuit.

Ramps

The slow ramp is produced by integrator A and the fast ramp by integrator B. The voltage discriminator constantly compares the outputs of these two integrators. As soon as their instantaneous magnitudes become equal, a pulse is generated by the discriminator which discharges integrator B. At the end of the pulse, integrator B restarts the fast ramp again and the whole process is repeated. Thus a series of pulses whose consecutive spacings increase at a constant rate is produced continually until the two integrators are stopped by the controlling input signal.

The isolating amplifier is inserted between integrator A and the voltage discriminator so the discriminator will not disturb the slow ramp.

Circuit Description

A circuit employing both point-contact and junction transistors is shown in Fig. 4. Type npn transistors in the grounded-collector con-

nection are used for the integrators and amplifier. For good results, high-α and high-collector-resistance junction transistors must be specified. The discriminator uses one point-contact transistor.

A transistor version of the compensated bootstrap circuit is employed for the integrators. Compensating resistors R_4 and R_7 should be so chosen that linear outputs are obtained at the base of Q_1 and at the emitter of Q_2. The use of the emitter of Q_2 as the output terminal for driving the isolating amplifier, Q_3, makes it easier for Q_3 to isolate the slow ramp from the sudden disturbances at the voltage discriminator.

Linearity

When other circuits are connected to the integrators their effects on the linearity of the ramps should be as small as possible, thus high back-resistance silicon junction diodes are used for D_1, D_2 and D_3.

The designed initial amplitude of the slow ramp can be obtained by choosing proper values for the resistors R_1 and R_2.

The grounded-collector isolating amplifier has as its load the base-collector loop of the voltage discriminator. Thus, R_9 must be high enough to make the isolating amplifier see a positive load at all times.

The discriminator uses an emitter-input, negative-resistance point-contact transistor multivibrator. The slow ramp is fed by the isolating amplifier to its base through resistor R_9 and the fast ramp is fed to its emitter through diode D_3.

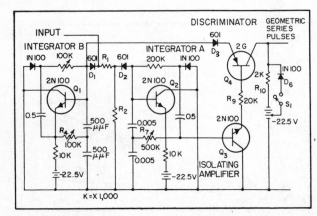

FIG. 4—Circuit uses npn junction transistors in integrators and amplifier, point-contact transistor in discriminator

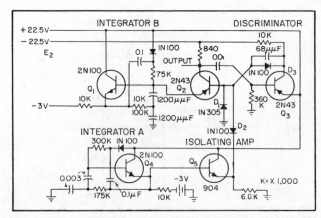

FIG. 5—Change in discriminator circuit allows use of junction transistors throughout

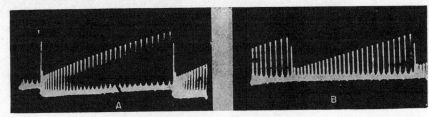

FIG. 6—Uneven base line caused by gradual rise of valley point and collector cut-off current (A) is eliminated by compensating R_{10} with diode D_6 (B)

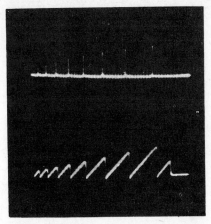

FIG. 7—Output pulses at collector of Q_2 in circuit of Fig. 6

A circuit using only junction transistors is shown in Fig. 5. It differs from Fig. 4 mainly in the voltage discriminator circuit (Q_2 and Q_3). Here the two ramp functions are compared between the emitter and the base of Q_2. The emitter is always at the potential of the fast ramp while the base is at the potential of the slow ramp as long as D_2 is conducting. Normally Q_3 is conducting in the transition region, D_2 is biased in the forward direction and D_1, D_3 are biased in the reverse direction.

When the magnitude of the fast ramp just surpasses that of the slow ramp, Q_2 starts to conduct, cutting off Q_3. The base of Q_2 is thus forced toward E_2, and D_2 and D_3 become reverse and forward biased respectively. Q_2 continues to conduct until the potential at the base reaches ground (integrator B is therefore returned to ground) and D_1 conducts cutting off Q_2. The circuit then returns to normal.

Even Base Line

The output pulses of the circuit of Fig. 4 are shown in the two photographs of Fig. 6. Delay time m_0 was omitted in taking the two pictures.

The rising base line of Fig. 6A is caused by the fact that the slow ramp acts as a part of the supply voltage in the base-collector loop of the voltage discriminator. This causes both the collector cut-off current I_{co} and the valley point of the negative-resistance curve to increase linearly with time. The increasing valley point makes each pulse stop at a higher level than the preceding one and the increasing I_{co} flows through load R_{10} when there is no pulse.

If an even base line as shown in Fig. 6B is desired, the effect of both I_{co} and the valley point can be reduced by making R_{10} small. However, this has the disadvantage of reducing the amplitude of the output pulses as well.

A better method to obtain an even base line without sacrificing pulse amplitude is to put a biased diode, D_6 in parallel with R_{10}. This reduces the effect of I_{co} to a negligible degree. Diode D_6 conducts in the absence of a pulse and stops conducting when a pulse comes. The design criterion is that the original current flowing through D_6 before the start of the first pulse must be at least equal to the highest I_{co} anticipated at the peak of the slow ramp.

Figure 7 shows the waveforms of the fast ramp and the output pulses at the emitter and the collector respectively of Q_2 in the circuit of Fig. 5.

Further Advantage

A constant minimum accuracy is not the only advantage of a geometric scale. Compared with an arithmetic scale of the same minimum accuracy when measuring a certain range of values, the geometric scale gives a much smaller digital number at the output. When the variation of the quantity to be measured is large, the simplification in equipment necessary to handle the data is tremendous.

For example, if a number varies from 20 to 20,000, an arithmetic scale will have a minimum accuracy of 95 percent and a fifteen-stage binary counter or a five-stage decimal counter will be necessary to handle the data. Using a geometric scale with a constant minimum accuracy of 95 percent, the highest output number is only 142 instead of 20,000 and an eight-stage binary counter or a three-stage decimal counter is enough.

Owing to the linear relationship between voltage and time of a ramp function the geometric scale can be used to measure either time or voltage. In either case, the circuit operation starts m_0 seconds after the reception of the quantity to be measured. If the quantity is time, the end of the measurement coincides with that of the quantity and no delay is involved.

If the quantity is voltage, the measurement ends when the instantaneous magnitude of the slow ramp equals that of the quantity. The time delay in this case depends on the slope of the slow ramp and the voltage to be measured. Another voltage discriminator should compare the slow ramp with the received voltage, which can be stored in a holding capacitor. This voltage discriminator signals the equality of the two voltages and discharges the holding capacitor simultaneously.

Applications

The geometric scale can be employed for the analysis and coding of pulse amplitude, duration or position modulation. In addition, the geometric scale can also be used in computers to reduce multiplication and division to addition and subtraction respectively, to find the logarithm of a number, or to obtain the nth root of a given quantity.

The authors acknowledge the help of M. I. Aissen and H. Blasbalg in ideas and S. Edelson and F. W. Schaar in bench work.

This research was supported by the U. S. Air Force through the Office of Scientific Research of the ARDC.

GENERATING CHARACTERS FOR
CRT READOUT

By KENNETH E. PERRY and EVERETT J. AHO

Lincoln Laboratory, Massachusetts Institute of Technology, Lexington, Massachusetts

Analog device displays alphabetic or numeric characters on face of cathode-ray tube by deflecting spot to trace out each desired character smoothly and continuously. Necessary X and Y deflection voltages for scope are obtained by Fourier synthesis technique that involves combining sine and cosine terms of first five harmonics of 30-kc fundamental frequency. Each character is traced in about 30 microseconds. Transistorized gated oscillators, flip-flop serial counters and emitter-followers feed ten toroidal transformers having one set of secondary windings for each character desired

ALTHOUGH MANY PLANS have been devised in the past for scribing numeric and alphabetic characters on a scope face by spot deflection, a new analog circuit recently developed for this purpose has some advantages in both simplicity and versatility.

The Arabic octal numerals zero through seven each may be represented as a segment of a continuous closed curve given in cartesian coordinates by the equation $y = f(x)$. In general, y is a multivalued function of x, but the curve can be represented by two parametric equations: $y = f_1(t)$, $x = f_2(t)$ where $t_0 < t < t_1$ and where f_1 and f_2 are single-valued functions of t. If t is the time, then these functions define the continuous motion of a point along the curve. They must be single-valued functions, since the spot cannot be in two different positions at the same time.

If the tangential speed of the point is known at all times (specifically, if it is constant), then the parametric equations are defined by $y = f(x)$. Thus, if $f_1(t)$ and $f_2(t)$ represent the voltage waveforms that are applied to the y and x deflection amplifiers, the desired curve will be traced on the scope face. Since most of the symbols are not closed curves, an unblanking function must be pro-

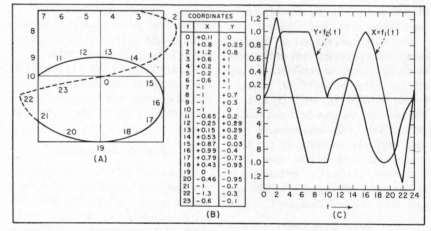

FIG. 1—Waveforms at right, obtained by measuring coordinates of numeral five as at left, will generate this numeral when applied to X and Y inputs of oscillocsope

vided to intensify the desired segment.

Equations for Numerals

A function of the type just described can be expanded into a Fourier series of sine and cosine terms:

$$f_1(t) = A_0 + A_1 \sin \omega t + B_1 \cos \omega t + A_2 \sin 2\omega t + \ldots.$$

where $\omega = 2\pi (t_1 - t_0)$ and $t_0 = 0$. The expression $(t_1 - t_0)$ is the time required for the spot to trace the entire closed curve.

The procedure for finding the coefficients A_n, B_n is as follows: The desired character is drawn on graph paper as in Fig. 1A, including a retrace segment which closes

the curve. To ensure that all characters can use the same unblanking function, closed figures like zero and eight have redundant retrace segments tacked on as an appendix. Twenty-four points are laid off along the curve at roughly equal intervals the actual number being arbitrary. These points divide the time $(t_1 - t_0)$ into 24 equal intervals. The x and y coordinates of each point are tabulated as in Fig. 1B, with t_0 taken as the center of the retrace segment. These tabulated values represent the two functions $f_1(t)$ and $f_2(t)$, as plotted in Fig. 1C. These functions may be analyzed by any one of several graphical and numerical integration methods.

The method now used[1] is a purely graphical one where each x or y value is laid off as a vector at an angle equal to $(n\omega t)$. When these vectors are added head to tail, the projections of the resultant vector give the coefficients A_n, B_n. When the coefficients have been determined it is possible to synthesize desired waveforms by electrically adding sine and cosine waves of correct frequency and amplitude.

Synthesizing System

The circuit for synthesizing the desired voltage waveforms from artificially generated sine and cosine waveforms uses five harmonics with a fundamental frequency of 30 kc. Ten tuned circuits (five sine and five cosine) are simultaneously shock-excited into oscillation by a gate 33 microseconds wide to give one cycle of 30 kc, two cycles of 60 kc, three of 90 kc, four of 120 kc and five of 150 kc.

These ten signals are fed through emitter-follower buffers to the primaries of ten toroidal transformers. Secondaries are wound on these toroids, with direction of winding and number of turns determined by the sign and magnitude of the Fourier coefficients. When these secondaries are connected in series and one end of the series circuit is grounded, the desired voltage waveform appears at the other end.

Figure 2 is a complete block diagram of the prototype system. The circuit as depicted here will display the numerals 0 through 7, four rows deep (32 characters). This can be displayed on any oscilloscope having an external unblanking connection.

A 120-kc sine wave is fed into a clock generator which shapes the signal into a square wave. The prime side of the clock generator output is commutatively coupled to flip-flop F_1, the first of a chain of eight serial counters[2]. The logic levels used are +5 volts and −5 volts. The unblanking function is

Harmonic generator, with ten character-forming toroidal transformers in vertical row at left. Transistorized shock-excited oscillators are at right, buffer-emitters at center, and control and cycling circuitry is on plug-in cards sliding into grooves of lower compartment

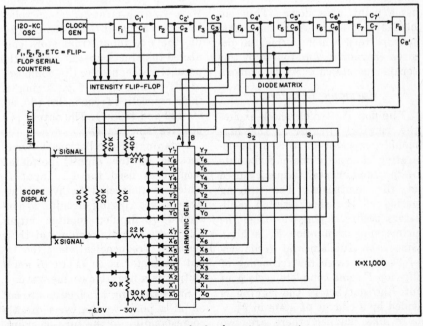

FIG. 2—Block diagram of Fourier-synthesis character generator

261

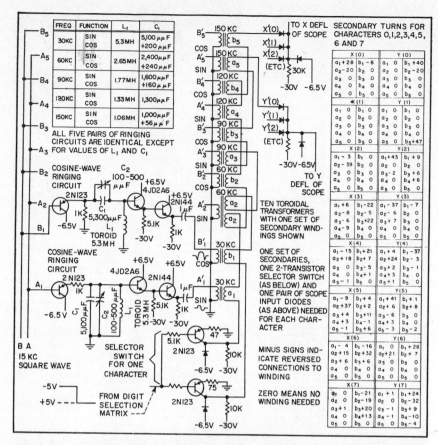

FIG. 3—Circuit of harmonic generator for producing one character. Each additional character requires additional selector switch and additional set of toroidal transformer secondaries feeding scope input terminals as at top of diagram

generated in the intensity flip-flop, controlled by F_1, F_2, F_3 and the clock generator. The intensity pulse starts one-half clock cycle or about 4 microseconds after the prime side of flip-flop F_3 goes up and ends 4 microseconds before the same point goes down. This unblanks that segment of the Lissajous pattern which forms the desired character. One-fourth of this continuous closed curve is blanked.

Harmonic Generator

Flip-flop F_3, which shock-excites the ringing circuits in the harmonic generator of Fig. 3, is operating at exactly one-half the rate of the fundamental frequency used in the synthesis. The ringing period of the shock-excited oscillators occurs during the time the prime side of flip-flop F_3 is high. Since the fundamental frequency of 30 kc is twice the frequency of flip-flop F_3, one complete cycle goes into the slot before the ringing is ended by a change of state in F_3.

In like manner, there are two cycles of the second harmonic, three of the third, etc., all initiated and terminated at the same time. The sine waves and the cosine waves are generated in parallel-resonant and series-resonant circuits respectively. Input A in Fig. 3, which is connected to five sine-wave ringing circuits, is controlled by counter output C_3'. When C_3' goes up, the five input transistors connected to point A are cut off and the parallel resonant circuits composed of L_1, C_1 and C_2 in Fig. 3 ring at their respective frequencies (30, 60, 90, 120 and 150 kc). The output is a positive sine wave.

Damping of oscillations is small because of the high-Q powdered iron cores used for L_1. Input B, which is connected to five cosine-wave series ringing circuits, is controlled by flip-flop counter output C_3. These circuits oscillate at their resonant frequencies when the input transistor is on (point B low). Output is a negative cosine wave.

Since the ringing circuits are cut off at a point in the cycle exactly corresponding to the turn-on point, there is no damping transient and

the operation is not duty-cycle sensitive. In other words, at the instant of turn-off the voltage on the capacitor and the current through the inductor are very near to the quiescent values. This would be exactly true except for the losses during ringing. It is only necessary to leave the circuit off long enough for this small amount of lost energy to be replaced.

The values of L and C in Fig. 3 are determined by setting $\sqrt{L/C} = R$ where R is the critical damping resistor, arbitrarily chosen as 1 k, L and C are unknown.

Solving first for L in terms of C and substituting this result in the equation $\sqrt{LC} = 2\pi f$, then solving for C, L can then be found from either equation. Trimmer C_2 has a range of from 100 $\mu\mu f$ to 500 $\mu\mu f$ and is adequate for adjusting the ringing circuit for any L and C inaccuracies.

Each ringing circuit is followed by an emitter-follower buffer amplifier which also drives the base of a power transistor in an emitter-follower amplifier configura-

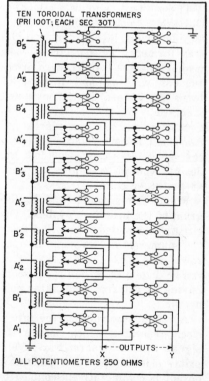

FIG. 4—Simulator circuit in which potentiometers duplicate changing of turns on toroidal transformer secondaries, for trying out effects of various combinations of coefficients before putting windings on transformers permanently

tion. The output of the power transistor is coupled through a 1-μf capacitor to the primary of a toroidal transformer.

Referring to Figs. 2 and 3, X_0 and X_0', X_1 and X_1', etc, or Y_0 and Y_0' or Y_1 and Y_1', etc, on the harmonic generator block are the terminals to the series secondary windings on the toroidal transformers. Every time flip-flop F_3 cycles, these circuits have $f_1(t)$ and $f_2(t)$ waveforms on them. These secondary waveforms will not, however, be passed through the OR diodes to the scope unless the X and Y inputs are high.

The d-c levels of the unprimed ends of the secondary windings (X and Y in Fig. 2) are controlled by the state of their associated switches. When a switch output is high, the corresponding OR diode (Fig. 2) is forward-biased and the signal on that particular secondary is transferred to the scope.

Transistorized Switch Circuit

The switches are *pnp* transistors in the grounded-emitter configuration shown in Fig. 3. The collector controls the d-c level of the associated secondary winding in the harmonic generator. The base inputs have two states. When the base is high the collector is at -6.5 v and its associated secondary winding sees an open diode in the OR circuit preceding the scope (Fig. 2). When the base is low, the collector will be at ground or some small negative voltage, determined by the fixed resistor at the emitter. The purpose of this resistor is to adjust the level of the synthesized waveform $f_1(t)$ and $f_2(t)$.

In the original graphical analysis for $f_1(t)$ and $f_2(t)$, no attempt was made to compute the d-c Fourier coefficient A_0 since the zero frequency cannot be accommodated in the transformers. Therefore, some of the numerals would be displaced from their proper relative positions on the scope face. It is this discrepancy in the d-c level that is adjusted by the resistors.

The diode matrix selects the number to be displayed under control of flip-flops F_4, F_5 and F_6. A different number will be displayed during each unblanking pulse. Only

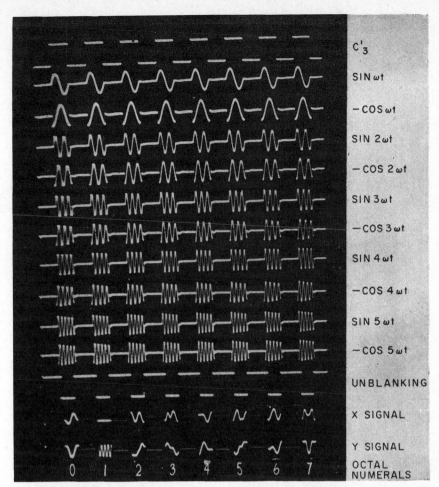

Waveforms involved in generation of eight Arabic numerals by synthesis

one output is low at any time. This voltage turns on a pair of switching transistors in the selection-switch package.

The four resistors on the X input of the scope (Fig. 2) are used to generate an eight-step ladder of voltages at the same rate as the unblanking function, thus displacing each numeral consecutively.

The three resistors on the Y input, in conjunction with the slower-running flip-flops F_7 and F_8, displace the whole row of eight numbers vertically four times.

Toroid Construction

The ten toroidal transformers in the harmonic generator each consist of a General Ceramics F-108 ferrite core with 100 turns machine-wound evenly around the entire toroid, then covered with insulating tape. With the ten cores mounted at right angles to the panel, the secondaries can be placed on by hand as they are needed. A set of series secondaries consists of

a single length of No. 24 Formvar wire wound through and around the ten toroids. Ample space is available to accommodate additional windings on the toroids for generating other characters.

A simulation device was built to try the effect of various combinations of coefficients in generating various characters. The circuit is shown in Fig. 4. The toroid primaries are substituted for the toroids in the harmonic generator, and the 250-ohm potentiometers are adjusted to the proper coefficient values. The resulting character can then be observed.

The research work on this project was supported jointly by the Army, Navy and Air Force under contract with Massachusetts Institute of Technology.

REFERENCES

(1) T. C. Blow, Graphical Fourier Analysis, ELECTRONICS, p 194, Dec. 1947.

(2) R. H. Baker, Boosting Transistor Switching Speed, ELECTRONICS, p 190, March 1, 1957.

CRYSTAL CLOCK FOR AIRBORNE COMPUTER

By C. W. PEDERSON
Bell Telephone Laboratories, Whippany, N. J.

TRADIC, the transistorized airborne digtal computer developed at Bell Telephone Laboratories, is a system designed to take full advantage of the desirable features offered by transistors. One device required in this system is a timing clock operating at a frequency of 1 mc ±0.05 percent. This clock must provide sine-wave output of 25 mw ±10 percent into a 500-ohm load

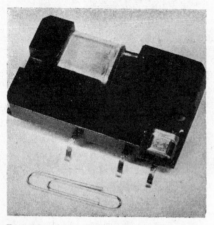

Transistor and crystal unit lead are soldered into the package after the rest of the oscillator is encapsulated

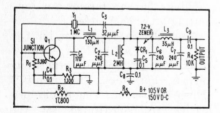

Schematic circuit diagram of the oscillator

over a temperature range of 0 to 50 C and must fit into a space no larger than $1\frac{7}{8} \times 2\frac{7}{8} \times \frac{7}{8}$ in. The oscillator described below was designed to meet these rigid requirements.

A single-transistor, grounded-emitter circuit was selected. The frequency of oscillation is determined primarily by the quartz crystal Y_1. Capacitors C_1, C_2 and inductor L_1 form a low-pass pi network that provides the required phase shift to sustain oscillations at the operating frequency.

To provide a nearly constant output over the temperature range stated above, output limiting obtained by diode CR_1 and capacitor C_5 was employed. The limiting introduced some distortion, so an output filter was added to provide a sine-wave output. This filter is

Internal construction of transistor crystal-controlled oscillator showing printed wiring and encapsulating techniques

composed of the low-pass pi network C_6 C_7 L_3.

Resistors R_1, R_2 and R_3 are necessary for current stabilization. Shunt feed is employed in the transistor collector circuit, with isolation from B+ being provided by L_2. Capacitors C_4 and C_8 serve as r-f bypasses and C_9 couples the 1-mc output to the load.

Capacitor C_3 permits fine frequency control and is adjusted so the output frequency has a nominal value of 1 mc. Resistor R_4 provides circuit protection by preventing a high current surge in the event

that a loose connection should occur at the ground terminal when the supply voltage is on.

Power for the oscillator is taken from a d-c supply of either 105 or 150 volts. Since only 40 to 55 volts is needed for the oscillator circuit, a series dropping resistor R_5 has been added. This dropping

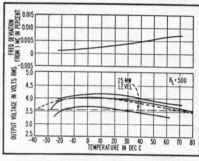

Frequency-temperature curve (top) and output voltage vs temperature (below) for several laboratory models of the oscillator

resistor equalizes to a certain extent the output of transistors with different gains. The d-c input power to the oscillator is about $\frac{1}{3}$ watt.

The curves show typical oscillator output voltage versus temperature for several laboratory models and also the typical frequency-temperature relation. While output voltage depends upon the characteristics of the particular transistor in the circuit, in general, this voltage was found to fall between 3.8 and 4.05 v at room temperature.

Performance of the laboratory models indicates that the circuit will provide the desired sine-wave output over a temperature range from −20 to 85 C. While the frequency deviation is well within the specified ±0.05 percent, even better frequency stability may be obtained by using a thermistor network in place of C_3 to compensate for temperature effects.

INDEX